TRACE ANALYSIS:
A structured approach to obtaining reliable results

Trace Analysis:

A structured approach to obtaining reliable results

Edited by

E. Prichard

Laboratory of the Government Chemist, Teddington, UK
(On secondment from the University of Warwick, UK)

with

G. M. MacKay and J. Points

Laboratory of the Government Chemist, Teddington, UK

THE ROYAL
SOCIETY OF
CHEMISTRY
Information
Services

LGC
Laboratory
of the Government Chemist

VAM
VALID ANALYTICAL MEASUREMENT

The front cover shows Hexagram 62, Hsiao Kuo, one of the 64 hexagrams of the I Ching, 'The book of change' of Chinese philosophy. The name signifies 'Preponderance of the small' or 'Predominance of the small'.

A catalogue record of this book is available from the British Library

ISBN 0-85404-417-5

Published for the Laboratory of the Government Chemist
by the Royal Society of Chemistry,
Thomas Graham House, The Science Park, Cambridge CB4 4WF

Typeset by
Land & Unwin (Data Sciences) Ltd, Bugbrooke, Northants NN7 3PA
Printed in Great Britain by Hartnolls Ltd., Bodmin, UK

Foreword

At the most general level there are two ways of improving the reliability of trace analysis measurements: raise the level of understanding of those carrying out the work and introduce quality assurance principles to the laboratory. Both of these approaches have, in fact, been advocated over many years for all analytical measurements by the UK's Valid Analytical Measurement (VAM) Initiative. VAM is a programme of work sponsored by the Department of Trade and Industry as part of the UK's National Measurement System. This book was prepared as part of a 3 year VAM project on quality assurance of trace analysis. It aims to address the first approach to improving reliability, *i.e.*, raising the level of understanding of those who undertake trace analysis.

In recent years the demands on analytical chemists have increased and working at the trace or ultra-trace level poses new problems. New analytical techniques and automated methods appear to enable the analyst to measure huge numbers of samples at these levels in a short period of time. With the present economic pressures, many laboratories do not employ sufficient numbers of experienced and knowledgeable staff and allow insufficient time to keep up with new developments. Standard operating procedures prevent a number of errors occurring but are often applied as a recipe. If they are followed blindly with no effort to understand the technique or methods the analyst is unlikely to recognise the significance of changes to the methodology or its application. All these factors contribute to some wrong results being produced. This book aims to alleviate the present situation by providing a compilation of relevant 'expert' knowledge in a format which will be readily accessible to many analysts.

The VAM project also developed a *Protocol for Quality Assurance of Trace Analysis*. The protocol provides a checklist for those aiming to introduce quality assurance principles to the trace analysis laboratory and gives particular emphasis to the integration of a quality system with the practice of sound science. The protocol is complemented by a more extensive document, *Guidelines for Achieving Quality in Trace Analysis*, which amplifies and explains it. The protocol and guidelines provide a digest of 'best practice' related to both the science of trace analysis and its quality assurance. Readers seeking technical advice on trace analysis from this book are strongly advised to apply it within a framework of the type set out in the protocol and guidelines.

Reliable trace analysis depends greatly on the way in which the initial sample is taken. It is important to take account of the boundary which is reached when the sampling error accounts for more than about 40% of the overall measurement error. Under these circumstances further effort to improve the accuracy of the laboratory's procedures will have little effect on the overall reliability of the results. Sampling is a topic outside the scope of this book but general advice may be obtained from several publications on sampling prepared under a separate VAM project.

Mike Sargent
VAM Trace Analysis Project Manager
June 1995

Contents

Contributors

The following members of staff from the Laboratory of the Government Chemist (LGC) have contributed to this book: Tim Catterick, Neil Crosby, Howard Hanley, Dave McDowell, Indu Patel, Graham Reed, Mike Sargent, Helen Watts, and Ken Webb. Other contributions are from: Jack Firth (Consultant), Andrew Fisher (University of Plymouth), Ernest Newman (Consultant), and Geoff Telling (Consultant).

The Editors gratefully acknowledge the following who gave valuable advice during the preparation of the book: from the LGC, Derek Craston, John Francis, Ian Lumley, Sheila Merson, and Selvarani Packirisamy. In addition: Dick Hoodless (Consultant), Mike Masters (Dionex), and Paul Worsfold (University of Plymouth) are also thanked.

Elaine Jaundrell of LGC gave valuable help with the word processing.

CHAPTER 1

Achieving Valid Trace Analysis

1.1 Introduction

1.1.1 What is Trace Analysis?

The term 'trace analysis' is widely used to describe the application of analytical chemistry (the measurement of amount of substance) under circumstances where the amount of analyte is very small. As such, its scope is as broad as that of analytical chemistry. The range of inorganic analytes is relatively small and comprises around 100 elements together with organometallic compounds and the common anions. However, in the field of organic chemistry several million compounds are known to exist and many are of interest at trace levels. Furthermore, materials presented for analysis of inorganic or organic species span an enormous range of composition and properties.

In the past, almost all analysis was undertaken using the so-called 'classical' techniques which involved dissolution of the sample, removal of any interfering species by precipitation and/or complexing agents, followed by determination using titrimetry, gravimetry or colorimetry. Such procedures required good manipulative skills and a deep understanding of the basic chemistry involved, even when used to determine relatively high concentrations or amounts of analytes. Many of these 'classical' techniques are capable of trace analysis but each analysis is generally time consuming and difficult. Hence in the heyday of 'classical' analysis the number of routine trace analysis measurements was quite small. However, from around the 1930s a range of physico-chemical tools was developed and applied to the solution of analytical problems. Techniques based on spectroscopy gained quite wide acceptance from the 1930s, particularly for trace element analysis in fields such as metallurgy, and were further developed in the 1950s and 1960s. Similarly, the chromatographic separation of organic compounds was developed in the 1950s and allowed sensitive determination of many important species using spectroscopic, electrochemical, or other detectors.

These developments in instrumental analysis enabled the analyst to routinely determine lower and lower concentrations of analytes and to resolve or separate very complex mixtures. This ability stimulated demands for trace analysis from industry and from those interested in applications such as environmental and consumer protection, forensic science, and clinical analysis. It is probably fair to say that the apparent ease with which these instrumental techniques could detect and measure analytes at ever lower concentrations led to an unwarranted confidence in the ability of the analyst to produce trace analysis data cheaply and reliably. Only as experience with each new or improved technique accumulated was it realized that many such trace analysis applications are reliable only if the instrumental determination is preceded by quite elaborate chemical manipulations to overcome interference effects. This need increases the cost, time, and expertise required and introduces the problems of contamination and loss.

In spite of the fact that the term 'trace analysis' is widely used by analytical chemists and their clients it does not actually have a clear or unambiguous definition. Many analysts would apply the term to determinations made at or below the part per million level, *i.e.* 1 ppm $\equiv$ 1 µg g^{-1} $\equiv$ 0.0001%, or 1 mg l^{-1} for liquids. Other analysts would define the term more generally as applying to an analysis where the concentration of the analyte is low enough to cause difficulty in obtaining reliable results. This may be caused solely by the low concentration of analyte in the matrix but other factors may also be important. Factors such as analyte losses, contamination, or interference may influence the perceived difficulty of analysis at lower concentrations and it may not be possible, or useful, to assign a numerical limit to trace analysis. Generally, the amount of sample available for analysis is plentiful but in applications such as clinical or forensic analysis there may be only small portions for use. This in turn will limit the mass of analyte presented to the detector, even though the initial concentration might be quite high. All such applications requiring special precautions to be taken are considered to fall within the scope of trace analysis and will be included in this book.

1.1.2 The Importance of Trace Analysis in Today's World

Measurements based on analytical chemistry are important to almost every aspect of daily life. They are critical to the success of many business sectors, the effectiveness of many public services depends on them, and everyone benefits from the use of such measurements to safeguard health, safety, and the quality of the products consumed or used. Some examples of the applications of analytical chemistry are given in Table 1.1.1. Such applications cover an enormous range of concentrations, from major constituents of materials down to contaminants present at parts per billion or below. Nevertheless, it is true to say that an ever increasing proportion of all analytical measurements can be described as trace analysis.

Trace analysis measurements play a key role in many areas of interest to industry and commerce, to governments, and to individuals. For example, the development and production of many new materials, of microelectronic devices, and of safe pesticides has been dependent on the availability of specific trace analysis techniques. Similarly, trace analysis is used in the first instance by governments to set many regulatory limits for purposes such as protection of the environment, or protection of the consumer, or to protect the health and safety of the workforce. Subsequently, trace analysis must be used by both industry and government to monitor or enforce these limits. Trace analysis is also essential to

Table 1.1.1 *Use of analytical measurements*

Business and commerce

Sectors:	agriculture		transport
	medical services		iron and steel
	building products		water supply
	non-ferrous metals		leather production
	chemical industry		waste treatment
	paper and board production		materials processing
	defence industries		
	petrochemicals and plastics	*Applications:*	quality control of bought-in materials
	energy production and distribution		quality control of finished products
	power generation		process control and optimization
	food and drink processing		environmental monitoring
	pharmaceuticals		compliance with health and safety regulations
	forensic science		
	refuse disposal		checking of contractual specifications
	healthcare		research and development
	textiles		
	horticulture		

Public services

	trading standards	public health
	administration of justice	forensic science
	environmental monitoring	vehicle emissions
	consumer protection	customs controls
	product licensing	

ensure the smooth flow of trade between companies or countries. For example, a manufacturing company purchasing materials or components will need to know that its suppliers are meeting an agreed specification. Obviously, checking such specifications may require a wide variety of physical or chemical measurements but trace analysis data are often vital, particularly with high technology products or materials intended for human consumption or application. Similarly, international trade is subject to extensive controls and regulations, many of which depend on trace analysis data.

1.1.3 Difficulty of Achieving Reliable Trace Analysis

In general, trace analysis is an extremely demanding activity requiring extensive knowledge, skill, and experience. The particular problems presented by trace analysis can be summarized as follows.

(i) The concentration of the analyte to be determined is much lower than that of the other constituents present in the matrix.

(ii) Contamination arising from reagents, apparatus, or the laboratory environment is more critical and may lead to false results.

(iii)   Losses of analyte by adsorption, degradation, or during analytical operations are more critical at very low concentrations and may even result in failure to detect substances actually present in the matrix at concentrations well above the anticipated detection limit.

(iv)   Constituents of the matrix may interfere with the detection system used, leading to falsely high values, resulting in the need for more extensive purification and/or more selective detectors.

(v)   The results obtained with the commonly used instrumental techniques are less precise than those obtained using classical procedures.

(vi)   Generally, it is difficult to check the reliability of methods because there are relatively few reference materials available for the enormous range of trace analysis applications.

It must be emphasised that the factors in (i) to (vi) cause problems for the majority of laboratories. Several authors have published papers reporting the results of collaborative studies of trace analysis methods. Frequently the values reported on a supposedly homogeneous and well-characterised material show an enormous range, despite the fact that analysts taking part in such studies are usually experts in the specific application or method and often treat the samples more carefully than most routine determinations. For example, Sherlock *et al.* published the results of a UK inter-comparison exercise for cadmium and lead in selected foods.[1] The spread of results, from 30 laboratories, covered one and a half orders of magnitude and this paper did much to highlight the lack of agreement between laboratories and the need for quality assurance in analytical laboratories. The poor agreement arose in part because the levels chosen were close to the limits of detection of the routine methods used by many of the laboratories. Nevertheless, this is a common state of affairs for many 'real' samples. Even the results achieved by a sub-group of large, expert laboratories were spread over a much wider range than many analysts would have expected (Table 1.1.2). For lead in liver and cadmium in fish food, for example, the range was greater than 100% of the mean value. At the time the paper was one of the first widely read reports to raise doubts on the reliability of analytical data at low concentrations and to point out that users of analytical data must be aware that 'the customer gets what he pays for'.

Another source of examples highlighting the difficulty of trace analysis is the European Union's 'Community Bureau of Reference' (BCR) which has carried out an extensive programme to prepare certified reference materials. These products are carefully prepared and checked for homogeneity and stability. The analyte concentration is then usually determined by a number of specialist laboratories using several different techniques. In this way a consensus value for the concentration of the analyte is obtained with a statistical estimate of the measurement uncertainty. The results reported by the BCR programme show that initially there is a large spread of results, even from expert laboratories, the situation improves as the study proceeds and analysts have the opportunity to meet to discuss their problems. One such study is shown in Figure 1.1.1.[2] The first collaborative study for the determination of a chlorinated biphenyl in fish oil

Table 1.1.2 *Overall mean concentration and range of means*
reported by the specialist laboratories[1]

Sample	Concentration (mg kg⁻¹ dry weight)		
	Overall	Range	
Lead			
Cabbage	0.31 (5)	0.23 – 0.41	
Kale	4.8 (7)	4.1 – 6.2	
Fish flour	0.27 (5)	0.09 – 0.33	
Fish meal	2.5 (7)	2.27 – 2.95	
Liver A	0.21 (5)	0.04 – 0.37	
Liver B	8.9 (7)	7.8 – 10.3	
Cadmium			
Cabbage	0.41 (5)	0.32 – 0.52	
Kale	0.21 (5)	0.14 – 0.25	
Fish flour	0.04 (5)	0.007 – 0.08	
Fish meal	0.96 (6)	0.74 – 1.1	
Liver A	0.66 (6)	0.52 – 0.98	
Liver B	7.7 (7)	6.8 – 8.7	

( ) No. of specialist laboratories submitting data in parentheses.

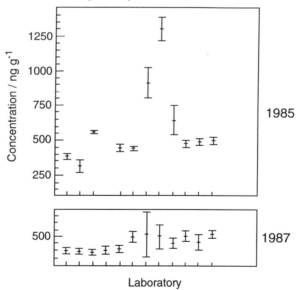

Figure 1.1.1 *Improvement in inter-laboratory agreement for the analysis of a pentachlorobiphenyl in two similar fish oils in 1985 (herring) and, in 1987 (mackerel). The data are given as laboratory means with a ± 2 s confidence bar where s is the standard deviation*
(Reproduced by permission from B. Griepink, *Fresenius' J. Anal. Chem.*, 1990, **337**, 812. © Springer-Verlag 1990)

showed a variation in results from 300 to 1400 ng g^{-1}. In the second study carried out 2 years later, after the group of laboratories had studied the reasons for the discrepancy and consequently removed sources of bias, the variation had been reduced to a range of 300 to 700 ng g^{-1}, with the majority of participants in a range spanning only 100 ng g^{-1}. Both of these studies also show a wide variation of estimated uncertainty between different laboratories.

A graphic illustration of the decreasing reliability of chemical analyses at lower levels was given by Horwitz *et al.* in 1980.[3] Evaluation of the results from over 200 collaborative studies conducted by the United States Association of Official Analytical Chemists (AOAC), showed empirically that the coefficient of variation between laboratories was a function of the concentration. These studies covered a wide range of analytes and matrices with analyte concentrations ranging from 10% down to 1 ppb (1 ng g^{-1}). The relationship is shown diagrammatically in Figure 1.1.2. Clearly, reproducibility is a major problem in trace analysis. Individual laboratories will be able to achieve better within-laboratory precision than that shown in Figure 1.1.2 but in today's world comparability between laboratories is

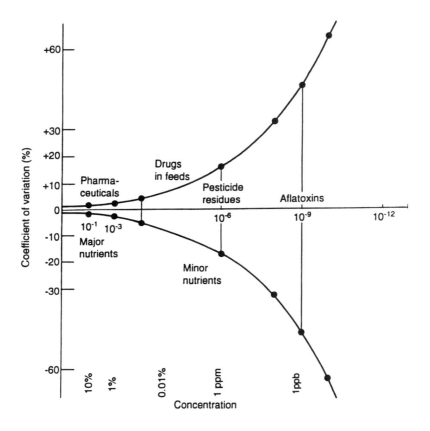

Figure 1.1.2 *Interlaboratory coefficient of variation as a function of concentration* (Reproduced from *The Journal of AOAC International* 1980, **63**, 1344. © 1980 by AOAC International)

all important. The data reported in Figure 1.1.2 demonstrate only the achievable precision and not the bias of a method. This latter factor will usually introduce further uncertainty into analytical results.

These examples clearly demonstrate the need for all laboratories to evaluate and improve the quality of analytical data, particularly at trace levels. They also show the benefits that can be achieved when analysts work together to improve methodology and techniques.

1.1.4 How Can the Reliability of Trace Analysis be Improved?

This book highlights areas where problems might occur and possible ways of dealing with them. It is, however, worthwhile to reflect briefly on why, apparently, so many trace analysis measurements go wrong. There may be laboratories that knowingly and unscrupulously produce poor quality data but, in general, this seems an unlikely source of such widespread problems. The laboratories participating in the collaborative studies mentioned earlier knew that their results would be evaluated at length and closely compared with those of other organizations. If anything, they would have carried out their analyses more carefully.

There do, however, seem to be two other much more likely reasons for incorrect results. Firstly, and as mentioned earlier, trace analysis is often extremely difficult. It presents problems which push the abilities of analytical chemists to their limits. An analysis referred to as 'routine' may well be founded on complex chemical principles, requiring difficult practical manipulations and involving measurements using very expensive, state-of-the-art scientific instruments. Yet all of this will frequently be entrusted to a relatively inexperienced operator with little opportunity to turn to more experienced or knowledgeable colleagues for help. Thus it is important to raise the level of understanding of those carrying out trace analysis. In particular, the easy availability of extremely sensitive instruments means that many such measurements are made by personnel who do not consider themselves as 'analysts'. In some cases they may not be chemists or have no relevant scientific training and may regard the problems of trace analysis as solved once the instruction manual of the instrument has been mastered. Hence the reliability of trace analysis measurements should improve if those using these instruments have ready access to clear, understandable advice on the pitfalls of the common techniques and ways to overcome them. This advice needs to be complemented by information on handling and preparing samples for trace analysis prior to the instrumental measurement itself.

The other reasons for poor trace analysis data seem to stem from the economic and other pressures now facing most laboratories. There is the temptation to neglect well-known precautions designed to reduce contamination, or to apply a method to determine analytes at levels below those for which the method has been validated. Carrying out measurements in an area saturated with analyte will inevitably cause errors in trace analysis. The need to deal with large numbers of samples in a short period may lead to mistakes caused by human error or by inexperienced staff. Sample pre-treatment may be undertaken by different analysts from those making the end determination leading to potential problems unless there is close collaboration. These problems are, of course, quite common and not unique to trace analysis. As might be expected, there is also a common approach to avoiding them. This is based on the introduction of *quality assurance principles* into the workplace. This approach requires agreement on what needs to be done to ensure that the analytical service provided is fit for its intended purpose. It is then possible to set up a *quality system* which sets out clearly what must be done and makes provision to monitor day-to-day operations to ensure that the specified work has been carried out.

1.1.5 The Purpose and Structure of This Book

The book covers all aspects of trace analysis from receiving the sample at the laboratory to the end determination and suggests ways of dealing with both organic and inorganic analytes in a variety of matrices. The aim in preparing this publication has been to bring together many of the practical considerations which have been found to be important over a wide range of trace analysis applications. Thus it should be useful both to analysts entering the area for the first time and to those undertaking a new type of determination. Equally, analysts who already regularly undertake trace analysis in a specific application area may find the book a useful source of new ideas or alternative approaches. The ultimate aim has been to provide all analysts with at least some of the practical advice which may be available to those fortunate enough to work alongside an experienced and knowledgeable trace analyst to whom they can turn when problems arise. The book does not attempt to describe detailed methods or protocols for particular analytes in specified matrices, although useful references can be found in several of the chapters. Similarly, individual analytical techniques have not been discussed in detail as there are many publications in which this has been done already. Considerable attention is given, however, to guidance on factors that should be taken into account when selecting appropriate methodologies and techniques for a particular application. We have attempted to show the relative merits of many methodologies and techniques; although practical details are constantly changing, the relative positions often remain the same. No doubt areas are omitted and many hints and tips are left out. These can always be added in the next edition, if you let us know.

In order to complement the protocol and the guidelines mentioned in the Foreword, the structure of this book is broadly similar to these documents and follows the various stages that comprise a trace analysis method.[4,5] Included in this introductory chapter is information on the approach to quality assurance and trace analysis and the selection, development and validation of trace analysis methods. An entire chapter is devoted to sample pre-treatment, (*i.e.* safeguarding the sample on arrival at the laboratory and preparation of a test portion of the sample for analysis). Methodologies for inorganic and organic analytes are then described separately since, in general, they represent very different analytical problems. In each case a chapter is devoted to the problems of sample preparation with a second chapter covering the most widely used instrumental measurement techniques. Additional chapters cover speciation and techniques which may be used for both inorganic and organic analytes. A separate chapter deals with signal processing and some aspects of preparing and reporting results.

1.1.6 References

1. J.C. Sherlock *et al.*, 'Analysis – Accuracy and Precision?', *Chem. Br.*, 1985, **21**, 1019.
2. B. Griepink, 'Certified Reference Materials (CRM's) for the Quality of Measurement', *Fresenius' J. Anal. Chem.*, 1990, **337**, 812.
3. W. Horwitz, L. Kamps, and K.W. Boyer, 'Quality Assurance in the Analysis of Foods for Trace Consituents', *J. Assoc. Off. Anal. Chem.*, 1980, **63**, 1344.
4. M. Sargent, 'Development and Application of a Protocol for Quality Assurance of Trace Analysis', *Anal. Proc.*, 1995, **32**, 71.
5. 'Guidelines for Achieving Quality in Trace Analysis', ed. M. Sargent and G. A. MacKay, The Royal Society of Chemistry, Cambridge, UK, 1995, ISBN 0-85404-402-7.

1.2 The Approach to Quality

1.2.1 Overview

As mentioned in Section 1.1.4, reliable trace analysis data are achievable through a combination of sound analytical science and the application of quality assurance principles. It cannot be emphasised too strongly that, no matter how good the science, reliable measurements require the implementation of a system for assuring quality in the laboratory. Furthermore, such a system can offer considerable advantage in ensuring that the reliability of the measurements is readily and widely accepted by those who use the data. The nature and complexity of the quality system will depend on the requirements of the laboratory's clients, the type of work it undertakes and the policies of its management or parent organisation. Nevertheless, all trace analysis laboratories, regardless of size, will benefit from appropriate application of quality assurance principles.

The everyday use of the word 'quality' is frequently synonymous with excellence. It is true that scientific excellence is a vital aspect of achieving reliable trace analysis but most analysts are also well aware of the need to achieve a balance between excellence and the time and cost of the analysis. Under these circumstances it is generally more helpful to use a more recent interpretation of quality which seems to have developed with the application of quality systems in manufacturing industry. A product or service is deemed to be 'of good quality' provided it meets a specification previously agreed between the supplier and the customer. This approach is very helpful in analytical chemistry provided all those concerned with a particular analysis understand exactly what is intended. In order to achieve this an analytical specification meeting the customer's requirements is defined using the concept of 'fitness for purpose'. This is discussed further in Section 1.2.2. Having defined the specification it is essential to ensure that subsequent measurements meet it and also that those who use the measurement data, either immediately or in the future, are aware of any consequent limitations. These two goals are achieved through a series of activities which comprise the quality system of the laboratory.

Quality systems are a relatively new feature of the analytical laboratory. They represent a substantial additional burden for most organisations, both to implement and maintain. Their introduction requires a management commitment to develop and resource a quality assurance programme covering either the entire laboratory or, at least, major areas of work. This programme will need to encompass all the resources and methods used in producing the data, specify quality control measures to monitor data quality on a routine basis, and introduce procedures to ensure that the quality assurance programme is being properly implemented. An additional, major feature of all quality systems is their documentation. In analytical laboratories this usually comprises a quality manual together with, for example, written instructions for methods and techniques. Good documentation is a key feature of an effective quality system and in many laboratories its preparation is seen as the major task of implementing a quality assurance programme. The documentation also needs to be fit for purpose and not over elaborate! It is also essential to remember that the quality assurance programme does not 'finish' on completion of the documentation. Reliable trace analysis measurements depend on establishing an effective programme and applying it on a daily basis.

The following areas are addressed by the quality assurance programme and its documentation:

(i)   selection and validation of the analytical methodology;
(ii)   the resources used for trace analysis;
(iii)   laboratory operations for sample handling and analysis;
(iv)   quality control, monitoring, and auditing.

Selection and validation of the analytical methodology must take place before actually undertaking any analyses. It begins with a planning process which should aim to establish what is required to achieve 'fitness for purpose' (Section 1.2.2). Analytical methodology is discussed in Section 1.4. The other three areas are reviewed briefly in the sections following the discussion of fitness for purpose.

1.2.2 Fitness for purpose

In order to achieve reliable trace analysis it is essential to agree an analytical specification and undertake thorough planning of the work. This will involve discussions between the analyst and the initiator of the work to establish the exact nature of the problem, why the work is being done, and what will happen to the results. It must be made very clear what *decisions* will be taken following completion of the analytical work. Only then can the analyst decide what is needed to obtain results which will be 'fit for the purpose intended' and select an appropriate analytical method for the determination. Aspects which need to be considered include:

(i) which analytes are to be determined;

(ii) what are the likely concentrations and what detection limit(s) will be required;

(iii) what other constituents are present in the matrix and whether they would be likely to cause interference;

(iv) the accuracy and precision demanded by the particular problem and the use to which the analytical data will be put;

(v) the degree of uncertainty at a specified confidence level which is acceptable to the initiator.

These considerations are reviewed in more detail in Section 1.4.2. It will also be necessary to discuss some of the parameters involved in validating the chosen method (Section 1.4.3). The possible need for confirmation of results will also need to be discussed. Finally, aspects such as data processing and any requirements for interpretation and reporting of results should be considered.

In many cases, reliable trace analysis depends critically on the way in which samples are taken and transported to the laboratory. It is essential that the analyst plays the fullest possible part during initial discussions with the customer/sponsor of the work to ensure that a proper plan for sampling is drawn up and agreed. Such a plan must include the number and type of samples to be taken, the procedure to be followed, and the place and time of sampling. All of these factors can have an important effect on the results and the conclusions drawn from them. In all areas of sampling it is most important to remember that unless the portion taken for analysis is truly representative of the bulk material, the analysis could be in error and hence wrong, and possibly expensive decisions may subsequently be taken. It has been shown that once the analytical error has been reduced to one-third of the sampling error, there is no further point in attempting to improve the analytical process. The problem of sampling is, however, outside the scope of this book.

As well as the 'technical aspects' a number of other factors must be discussed before the analytical specification can be agreed. Most customers are concerned not only with the reliability of the analysis but also with its cost and the time taken to produce a result. Hence the correct balance must be established so that, in the eyes of the customer, the entire analysis is 'fit for purpose'. The decision on cost and time will, of course, influence not only the specification but also the subsequent choice of method. In many cases the initial planning will need to take account of other constraints which may be important to the customer or the analyst. For example, legal requirements, confidentiality and safety will often require consideration. For some purposes (*e.g.* data produced for regulatory authorities) there may be special requirements on the extent of method validation and the amount or type of quality assurance (*e.g.* compliance with a prescribed, external quality assurance scheme).

1.2.3 The Resources for Trace Analysis

The resources needed for trace analysis include the laboratory or other workplace, appropriately qualified and experienced staff, the instrumentation or other equipment, and consumable supplies. Quality assurance requirements for the laboratory's resources are discussed in the *Guidelines for Achieving Quality in Trace Analysis*[1] and in textbooks on laboratory quality assurance.[1-5] Some technical aspects particularly relevant to reliable trace analysis are discussed briefly here.

Trace analysis differs significantly from macro-analysis in that small amounts of *contamination* prior to the final determination stage of the analysis can give rise to significant errors. Contamination may arise from construction materials, from the environment, or from equipment and chemicals used in the analytical procedure. The most common contamination problem is the spurious addition of the analyte itself. The need to avoid contamination is a critical consideration when evaluating the suitability of resources for trace analysis. Thus it is best if trace analysis and bulk analysis are kept in separate locations, *e.g.* different laboratories. Where this is not possible a clean cabinet could be used or a special part of the laboratory reserved for trace analysis. Other operations in the area must be taken into account including the use of fume cupboards. Samples containing high concentrations of any of the analytes that may be determined at trace levels should not be allowed into the trace analysis area.

The laboratory and its facilities can often be arranged to allow tasks to be carried out in well-defined areas. Fittings should be made of materials compatible with the analysis being performed and resistant to attack by the chemicals likely to be used in the area. The requirements for the determination of inorganic and organic analytes will be different and these will need to be taken into account. For example, contamination from metal fittings may occur where trace metals are determined but such fittings are widely specified in laboratories undertaking analysis of trace organics. Polishes and other commercial cleaning agents should not be used on work surfaces. Where old wooden benches are polished precautions need to be taken such as the use of sheets of plate glass on the bench top. Trace analysis at very low levels may require the use of a specially built clean room. Where this is not possible through financial or other constraints the cleanest available room and/or a clean cabinet should be employed. Fumes or particles in the atmosphere can be controlled to some extent in ordinary laboratories by proper air circulation.

It is advisable to keep containers and apparatus covered to reduce the risk of contamination of samples or reagents. No eating, drinking, smoking, or application of cosmetics should be permitted in the trace analysis working area. Some cosmetics or jewellery may also cause contamination and should be avoided. The use of gloves and, in some circumstances, hair coverings will also reduce the risk of contamination. It is important that the precautions taken to avoid one contaminant do not introduce a different contaminant, *e.g.* dithiocarbamates from rubber or plasticisers from some plastics.

Laboratory equipment that is used for trace analysis should be kept exclusively for that purpose wherever feasible. Consideration can usefully be given to maintaining sets of laboratory apparatus for specific analyses or specific applications to avoid cross-contamination. Glassware can be borosilicate for most purposes, but for low level analyses of boron, sodium, aluminium, or silicon it will be necessary to use silica (except for Si), platinum, or poly(tetrafluoroethylene) (PTFE) apparatus. The cleaning process should be specified in the method for a particular analysis. Where detergents are used thorough rinsing is essential and may require a final rinse with organic solvents, aqueous acids, or alkali. If plastic-ware, particularly disposable ware, is used it should be carefully checked for loss or contamination problems.

The central laboratory supplies of deionised water or gases, may require further purification for certain analyses. A system for checking the purity of these supplies is also important. Careful, regular checks of gas supplies used with instruments should be made to ensure they are of adequate purity. Gases may be purified by the use of appropriate filters such as molecular sieves, anhydrous salts, activated charcoal, and oxygen traps. Specialised equipment may be necessary to achieve water of high purity. Reagent and solvent blanks should be measured to check for potential problems. If a high blank analysis response is obtained then checks of each reagent will be necessary to identify the source of contamination. Any reagent found to cause problems should be purified or replaced. Reagent contamination is relatively easy to measure by the same procedures used to analyse the sample. When checking reagent specification the nature of the impurity is as important as the percent purity.

As with other types of analysis, the performance of trace measurements within a quality assurance regime will require the use of *calibration standards* and *certified reference materials*.[6,7] It is sometimes suggested that, in view of the relatively large errors that may occur in sampling or analysis at trace levels, the requirements for these materials may be relaxed. This is not the case. Commercially available standards are often used, ideally supplied with adequate documentation and evidence of proper quality control procedures. The best available standards should be purchased, preferably certified reference material designated for calibration purposes. Where it is not possible to buy standards of known purity and composition, analytical standards will need to be prepared and their purity established according to a written protocol. It is not acceptable to use standards of unknown purity or composition such as, for example, a bottle labelled simply 'lead nitrate'. It is also important that the values obtained from standards and reference materials are traceable to a recognised authority, ideally a national or international body. Where 'ideal' materials are not available it is important to carefully consider the best alternative. This may entail, for example, choosing a certified reference material with a closely related sample matrix which gives similar analytical problems. It may be necessary to purchase materials which do not clearly state their traceability status; often, however, the supplier will provide additional information if requested.

Analytical standards should not be stored with samples or in areas subject to contamination. It is important to remember that the quality of analytical standards and stock solutions will deteriorate with time. Many changes that may occur during storage can be ascertained by visual inspection (*e.g.* solid deposits, turbidity, change of colour). For solutions in organic solvents, recording the weight before and after the removal of each aliquot provides a means of checking that there is no loss by evaporation during storage. Where it is intended to keep stock solutions of analytical standards for extended periods consideration must be given to possible reaction with the solvent or losses due to adsorption or decomposition on the surface of the container. It may be possible to add preservatives such as complexing agents, oxidising agents, or acids to extend the shelf life. It is, of course, essential to ensure that these materials do not contain significant concentrations of the analyte.

1.2.4 Laboratory Operations

Laboratory operations range from the receipt and recording of samples, through sub-sampling, analysis, and data handling, to the reporting and archiving of results. Only aspects which apply specifically to trace analysis are described here. Note that the precautions recommended apply both during the everyday work of the laboratory and when developing and validating the analytical method (Sections 1.4.2 and 1.4.3).

All operations undertaken on the sample will need to be checked to ensure that the composition of the sample is not altered in a way that would affect the concentration or identification of the analyte being determined. Procedures and practices used in the laboratory should be chosen to minimise the risk of contamination or other errors during the analysis. As mentioned in Section 1.2.3, contamination can be introduced from the laboratory and its staff and from sampling containers, glassware, equipment, reagents, solvents, and internal standards.

Precautions to avoid contamination begin with receipt of samples by the laboratory. Samples should be unpacked in a suitable area free from possible contaminants and taking into account any safety requirements. Samples that are known to be unstable can either be analysed immediately or else deterioration minimised by appropriate processing and storage. It is important to inspect the physical appearance and condition of the sample when it arrives and to note any unusual features. This will help to confirm the identity of the sample and may provide warning of any deterioration during transit. Where a sample is received in poor condition the analyst should ask the sender or other responsible individual whether they wish to proceed with the analysis.

It is important that samples are stored in suitable containers under the appropriate conditions and in a manner that prevents the possibility of degradation and contamination, including cross-contamination to and from other samples. Samples and standards should not be kept in the same storage area, so avoiding contamination of the samples. Several factors need to be taken into consideration during sample storage and preservation. The analyte itself may be unstable due to the effects of environmental factors such as temperature, moisture, and light and it may also interact with the container. The choice of container will depend on both the sample matrix and the analyte(s). Air-tight and light-tight containers are normally employed. Stability is usually better at low temperatures and special precautions may also need to be taken to avoid the loss of volatile constituents. If the sample is frozen it should be allowed to reach room temperature before sub-sampling or analysis. When a homogeneous sample is frozen it may need to be re-homogenized prior to sub-sampling or analysis due to the separation of ice crystals.

It is sometimes possible to apply a trace analysis method directly to the sample received from the customer but frequently a *laboratory sample* will need to be prepared. Samples should be visually checked for any separation of components before analysis is commenced and blended if necessary. If the entire sample received by the laboratory is not to be used in the analytical method a representative sub-sample will need to be taken. Sub-sampling for trace analysis requires precautions to avoid contamination, loss of analyte or change to the chemical form of the analyte or the matrix of the sample. In most cases it is best to prepare sub-samples for analysis immediately on receipt of the sample and to store these sub-samples under suitable conditions. The preparation of the representative sub-sample is an important step in achieving the correct result and the nature of the sample will dictate the steps which need to be taken. Sometimes blending is disadvantageous, *e.g.* where the process may result in an increase in activity of the enzymes in the matrix. In some cases separated components are analysed, *e.g.* sediments in water samples, herbage in soil samples. For some samples it may be necessary to remove irrelevant material from the sample before analysis, *e.g.* with some soil samples stones should be removed. Samples that are readily soluble in a solvent can be sub-sampled after dissolution and mixing. Any vessel used to hold or process the sample is a potential source of contamination, and its composition should be considered with care. This includes blenders, grinders, sieves, *etc.* used during the homogenisation step. Once homogenised, samples should again be stored under conditions that will prevent deterioration of the analyte and matrix.

In some cases the laboratory sample may be used directly for measurement by an instrumental technique. More commonly, trace analysis requires an *analytical method* of which the instrumental measurement is just one stage. For example, it is almost always necessary to dissolve or extract the sample or to break down the sample matrix. It is also frequently necessary to add additional stages such as dilution, concentration, or removal of interfering components of the sample matrix. It is important to remember that the stability of a sample, standard, or solution will affect not only the initial storage requirements but also the time and conditions of storage between the steps of an analysis. The extraction or dissolution conditions (temperature, *etc.*) should avoid breakdown or loss of the analyte(s) and minimise interference or contamination. It is important that the documented procedure is followed and that sensible precautions are taken. For example, care must be taken to avoid overheating during wet oxidations or during extraction by mechanical maceration with organic solvents. Similarly, care is needed to avoid analyte losses through spillages and absorption or adsorption processes. Chemical operations need regular checks during the routine use of a method.

The final *measurement step* of a trace analysis entails many of the same precautions and problems as analysis at higher levels. Additional problems may arise because many modern instruments exhibit such high specificity that they encourage measurements of low analyte concentrations in the presence of high levels of other species. This can cause problems in a busy laboratory analysing a wide variety of samples, particularly when using automated instruments with a high throughput of samples. It is particularly important to keep a close watch on the various batches of samples that may pass through such an instrument during the day. Remember, for example, that a species that is regarded as a 'trace analyte' in one batch may be a relatively large 'matrix component' in another, leading to carry-over problems. More insidiously, a matrix component which is innocuous for one determination may degrade the instrument response to another analyte.

These problems are, of course, precisely the ones which effective quality assurance precautions seek to avoid, or at least to identify. Thus in undertaking trace analysis it is essential to be meticulous with regard to frequent measurements of blanks, standards and *check samples*. The samples may be reference materials, previously analysed samples or, in some cases, a solution which contains a known amount of analyte and other species at concentrations corres- ponding to those arising from the sample. The elaborate analytical procedure which precedes many trace level measurements can also introduce unexpected problems. For example, traces of solvents used in the earlier stages of the method may introduce serious measurement errors at low analyte concentrations. Care should be taken in selecting solvents for, say, extraction or clean-up, to ensure that they are also compatible with the measurement technique. Components that may be affected include the sample introduction system, chromatography columns, atomisation systems, and the detector. In order to detect such problems it is essential that blanks and check samples are put through the entire procedure, including any homogenisation. It is rarely sufficient to introduce them at the measurement stage.

The *signal processing step* can be a major source of error in trace analysis, particularly where computerised techniques are used. This is not, in general, due to any fault with such techniques but rather because the analyst may not understand their operation and limitations. In some cases the manufacturer may not have disclosed sufficient information to allow potential problems to be identified. Even the precautions laid down by a good quality assurance programme may fail if the instrument operator does not realise that conditions are outside those specified for a given method. For example, signal-to-noise ratio is critical for trace determinations made at the limits of an instrument's capability but this ratio may not be readily apparent to the operator of a modern instrument who relies on a computer print-out of results. The precision of a measurement is dependent on the signal-to-noise ratio at the measuring step. Where a minimum ratio is specified in the methodology, steps must be taken during the analysis of a batch of samples to ensure that the minimum is being achieved. If it is not, any data obtained will not lie within the limits specified in the method. Similarly, many instruments automatically correct for signal or other blanks. This can lead to serious errors if very large corrections are made to small signals. Thus the analyst or instrument operator needs to be alert to the *size* of blank signals even where correction is made by the computer. If a blank is unacceptable the entire batch of analyses is invalid. A blank is unacceptable where the value would adversely affect the results of the analysis. It is important to trace the source of any high value for the blank as it will add to the uncertainty of the measurement.

An additional requirement in many trace analysis applications is the need for *confirmation of results*. This is different from investigation of unusually high or low values or other unexpected results, which is usually done by a repeat analysis of a duplicate sub-sample and a third sub-sample. Additional confirmation of all results is necessary where the initiator requires greater confidence than is provided by the routine analysis. Confirmation of both null and positive results can be equally important, and confirmation of both identity and concentration may be necessary. It is important to remember that confirmation is different from assessing repeatability. Confirmation requires the analysis to be performed by more than one technique, whereas repeatability is assessed when the analysis is performed several times with one technique. The two determination techniques should rely on different physical principles. Where the final measurements are made on a solution it is also necessary to consider the relevance of using different dissolution, extraction, or digestion procedures. The extent of additional confirmation will be determined by practicality and the importance of the particular result. Particular problems occur when, for example, statutory limits are set at or close to the limit of determination. The need for additional confirmation should be part of the discussion between the analyst and initiator before the analysis proceeds.

The approach used by the trace analysis laboratory for *reporting results* will, in principle, not differ from that used for analysis at macro-levels. Experience shows, however, that care needs to be taken by the laboratory to avoid misunderstandings by those who use its data. For example, many laboratories still provide just a single number without any indication of the uncertainty inherent in that result. A value such as 0.15 ppm may assume a different significance if the user is aware that it is actually a statement that the laboratory is 95% confident of the 'true' value lying between 0 and 0.3 ppm. Similarly, the reporting of results as 'not detected' or 'not determined' can cause confusion. The laboratory should evaluate its own limit of detection and limit of determination using a sample matrix, not merely pure solutions of standards. The limit of determination can be assessed by using samples containing known amounts of the analyte. In some cases a sample found not to contain the analyte can be spiked with the analyte to produce a concentration close to the anticipated limit of determination. These limits should be clearly stated with the analytical results, together with the criteria by which the terms have been applied. In some trace analysis applications an arbitrary 'reporting limit' is set by the initiator or a regulatory authority. This is a limit below which the analyte concentration is deemed to be 'zero'. Again, it is essential that the laboratory indicate clearly if this has been done together with the numerical value of the limit. There is also confusion with regard to the practice of 'correcting for recovery'. This refers to the systematic error which can arise when an elaborate sample preparation method is consistent but fails to give '100% recovery' of the analyte. Good measurement practice stipulates that a correction should be made for such errors but there are some trace analysis applications where initiators or regulatory authorities require otherwise. The laboratory should always indicate the magnitude of such errors and state whether or not its results incorporate a correction.

1.2.5 Quality Control, Monitoring, and Auditing

Quality assurance goes over and above the keeping of quality control charts. A quality control system is useful in the detection of errors, but can only result in action being taken to rectify any faults. It does not prevent errors being made in the first place. A quality assurance system is designed to prevent errors occurring and can improve both quality and efficiency.

This section describes measures to confirm data quality on a routine basis (quality control) and checks to ensure that the quality assurance system is operating satisfactorily (monitoring and auditing). Both quality control and monitoring entail experimental checks of data quality. The distinction arises in the way these checks are made. Quality control is generally a routine activity undertaken by the laboratory staff and documented with the analytical methods or other laboratory procedures. It is sometimes referred to as internal quality control. Monitoring usually provides a more independent check involving laboratory management or an external organisation. Auditing entails inspection and review of the quality system as a whole to see whether it is operating according to plan and achieving the anticipated effects. It may be operated internally or externally. Monitoring and auditing are sometimes referred to as external quality control.

Establishing a well-defined *quality control* (QC) regime is part of setting up a routine analytical method. Its aim is to ensure that calibration is maintained, to monitor instrumental/analytical drift, and to check the consistency of results. This is best achieved by periodic analyses of QC samples which resemble as closely as possible the sample matrix and offer a range of analyte concentrations. The QC procedures should be clearly defined in a quality manual or other documentation. The level and type of QC will depend on the nature of the analysis, frequency of analysis, batch size, degree of automation, and method difficulty/reliability. Quality control data should be recorded over an extended period of time. Over this period the variation in method performance can be monitored by recording the analysed value of the QC sample, usually by plotting it on a chart. Some guidance on the purpose and implementation of internal quality control procedures has been produced by the Analytical Methods Committee of The Royal Society of Chemistry.[8] Details of control charting and other quality control procedures are covered in special publications.[7,9-13] Another useful source of material is the Codex guidelines.[14]

The level of quality control adopted should be sufficient to ensure the required reliability of the results. As a guide, for routine analysis the level of internal QC would typically be not less than 5% of the sample throughput, *i.e.*, 1 in every 20 samples analysed should be a QC sample. For more complex procedures, 20% is not unusual and on occasions even 50% may be required. For analyses performed infrequently, tests to confirm the performance of the method should be carried out on each occasion. This may typically involve the use of a matrix reference material containing a certified or known concentration of analyte, followed by replicate analyses of the sample and a spiked sample (a sample to which a known amount of the analyte has been deliberately added). Similar precautions should also be taken when it is necessary to undertake 'ad hoc' analysis at short notice.

It is important that regular checks be carried out to confirm that contamination is not occurring. This can be done by retaining as blanks those samples which are known to contain less than a detectable amount of the analyte. Where this is not possible consideration should be given to producing artificial blanks from materials known to be free from the analyte of interest. At least one blank and preferably two should be analysed with every batch of samples, no matter how small the batch and how recently a blank has been run. In QC of trace analysis the blank result is particularly important since it will affect all other results. The repeatability between blank duplicates will also provide information on the variability of contamination.

Achieving low blanks is essential for trace analysis but regular assessment of the overall operation of the method should also be made. This can be done by carrying out recoveries of analytes close to the levels expected in samples. These recoveries should, where possible, be determined in the same matrix as the sample. Recoveries may be ascertained either by spiking a portion of a current sample or by the analysis of materials of known composition. The latter may be certified reference materials, similar to the sample, or previously analysed (check) samples. In the case of spiking it is important that the entire methodology is checked by making an addition of the analyte to the sample before any treatment is carried out. For some analyses the use of check samples may be more convenient or appropriate than carrying out spiking experiments. Regular analysis of a check sample is useful to highlight any change in the performance of a method or an analyst. With very large batches a reasonable number of check samples will need to be included. Where appropriate, previously analysed samples may be retained for future use as check samples. When a check sample result is shown to be outside its permitted range the results for all samples analysed since the last correct check sample should be rejected. Appropriate corrective action should, of course, be taken before re-running the samples.

Internal *monitoring* is usually based on the inclusion of check samples of known value with sample batches. Often these are similar to QC samples and the analyst is aware of their significance but 'blind samples' (*i.e.* samples which are included with other work without prior warning) are also widely used in some areas. External monitoring may simply involve the inclusion of check samples by the originator of a batch of samples. There are also co-operative schemes known as *proficiency testing* (PT) that arrange for distribution of a test sample and subsequent collation and analysis of the data provided by participants. These schemes allow laboratories or their customers to compare results from individual laboratories against the 'true value', which may be obtained either by the organisers or as a consensus value. Laboratories should, where possible, take part in appropriate PT schemes, which are a valuable way of checking that results are comparable to those obtained in other laboratories. It must be realised that interlaboratory variability is greater than intralaboratory variability. Nevertheless, formal comparisons between laboratories, as in PT schemes, are a valuable means of identifying errors and improving comparability of results between laboratories.

The term '*quality audit*' is used to describe an assessment of the quality system and its operation. Internal audits are required under all formal quality systems and may be conducted as a senior management function or by a designated quality manager or quality audit unit. External auditing is similar but provides a completely independent assessment which can be made known or made available to interested parties such as senior management, customers, or regulatory authorities. Most external audits form part of widely recognised *accreditation* or *certification* schemes. These are generally based on quality systems that comply with an accepted national or international standard, and auditing is carried out by officially approved organizations. The three most widely used international standards are the ISO 9000 series (certification), ISO Guide 25 (accreditation), and Good Laboratory Practice.[15-17] Relevant descriptions of these schemes and advice on undertaking quality audits can be found in books on laboratory quality assurance.[2,3,9] An audit checklist is given in the *Guidelines for Achieving Quality in Trace Analysis*.[1]

1.2.6 References

1. 'Guidelines for Achieving Quality in Trace Analysis', ed. M. Sargent and G. A. MacKay, The Royal Society of Chemistry, Cambridge, UK, 1995, ISBN 0-85404-402-7.
2. F. M. Garfield, 'Quality Assurance Principles for Analytical Laboratories', Association of Official Analytical Chemists, 2nd edn, Arlington, VA, 1991.
3. J. K. Taylor, 'Quality Assurance of Chemical Measurements', Lewis Publishers, Chelsea, MI, 1987.
4. 'Quality Assurance for Analytical Laboratories', ed. M. Parkany, Special Publication no. 130, The Royal Society of Chemistry, Cambridge, UK, 1993, ISBN 0- 85186-705-7.
5. G. V. Roberts, 'Quality Assurance in Research and Development', Marcel Dekker, New York, 1983.
6. J. K. Taylor and N. M. Trahey, 'Handbook for SRM Users', NIST Special Publication 260–100, National Institute of Standard & Technology, Washington, DC, 1993.
7. 'Uses of Certified Reference Materials', ISO/IEC Guide 33, Geneva, 1989.
8. Analytical Methods Committee, 'Internal Quality Control of Analytical Data', *Analyst (London)*, 1995, **120**, 29.
9. E. Prichard, 'Quality in the Analytical Chemistry Laboratory', J. Wiley, Chichester, UK, 1995, ISBN 0471-95541-8 (5).
10. M. Thompson, 'Variation of Precision and Concentration in an Analytical System', *Analyst (London)*, 1988, **113**, 1579.
11. M. Thompson and R. Wood, 'Harmonised Guidelines for Internal Quality Control in Analytical Chemistry Laboratories', *Pure Appl. Chem.*, 1995, **67**, 649.
12. A. F. Bissell, 'CUSUM techniques for Quality Control', *Appl. Stat. (UK)*, 1969, **18**, 1.
13. J. M. Lucas and R. B. Crozier, 'Fast Initial Response for CUSUM Quality Control Schemes: Give your CUSUM a Head Start', *Technometrics*, 1982, **24**, 199.
14. 'Supplement One to Volume Two, Pesticide Residues in Food', FAO/WHO Codex Alimentarius Commission, Rome, 1993, Section 4.2, 'Guidelines on Good Practice in Pesticide Residue Analysis'.
15. BS EN ISO 9001:1994, 'Quality Systems – Model for Quality Assurance in Design/ Development, Production, Installation and Servicing', BSI, Chiswick, UK, 1994.
16. 'General Requirements for the Competence of Calibrating and Testing Laboratories', ISO Guide 25, Geneva, 3rd edn, 1990.
17. 'The OECD Principles of Good Laboratory Practice', Environment Monograph no. 45, OECD, Paris, 1992.

1.3 The Approach to Trace Analysis

1.3.1 The Special Needs of Trace Analysis

Many sample matrices are a complex mixture of large numbers of ingredients. This is particularly true of biological systems. For example, foods consist of fats, proteins, carbohydrates, water, minerals, vitamins, and other additives. A potato contains around 90% of its weight as water and the rest is mainly starch, but 150 other compounds have also been identified in potatoes. Orange oil is known to contain at least 12 different alcohols, nine aldehydes, two esters, four ketones, and 14 hydrocarbons. Nearly 100 volatile compounds have been identified in blackcurrants and it is certain that there are other compounds in all these products which have not yet been identified. Processed foods may contain a number of added chemicals such as colours (natural or synthetic), flavours, preservatives, antioxidants, emulsifiers, sweeteners, *etc.* which manufacturers use to improve the shelf life, acceptability, or palatability of the product. Only a few such compounds will be present in any one food but several thousand chemicals have been approved worldwide for use in this way. There are, for example, many ingredients in a typical infant milk food (see Table 1.3.1) even though the proximate analysis might appear on the packet as: fat 28%, carbohydrate 56%, ash 2%, and protein 12%, belying the complex nature of the product.[1] Hence, the analytical problem is not simply the detection, identification and quantification of extremely small amounts of analyte but achieving this in an excess of other complex compounds.

Table 1.3.1. *Ingredients of an infant milk food*[1]

Skimmed milk powder	KOH
Electrodialysed whey	$KHCO_3$
Lactose	$ZnSO_4$
Oleo oil	Vitamin E
Coconut oil	Nicotinamide
Soya bean oil	Vitamin A
Oleic (safflower) oil	$CuSO_4$
Soya bean lecithin	Calcium pantothenate
$CaCl_2$	Vitamins B_1, B_2, B_6
$NaHCO_3$	β-Carotene
Vitamin C	KI
Calcium citrate	Folic acid
$FeSO_4$	Vitamins D_3, B_{12}

1.3.2 The Structure of Trace Analysis Methods

As described in Section 1.1.1, classical analytical methods were once the main tool for trace analysis and were relatively straightforward. They involved getting the sample into solution, removal of interferences, followed by a determination step usually based on titrimetry or gravimetric methods. Nowadays, chemical analysis is a multi-stage operation as follows:

(i)   definition of the problem;
(ii)   sampling;
(iii) isolation (or separation) of the analyte from the matrix;
(iv)   removal of interfering substances (clean-up);
(v)   concentration of the analyte;
(vi)   determination step;
(vii) data processing and interpretation of the results;
(viii) solving the problem.

The analytical approach used in trace analysis is essentially to remove the analyte of interest from its environment of other compounds (the sample matrix) in a series of steps which progressively enrich the analyte by comparison with the other substances. At the final stage the analyte is presented to a detection system which further discriminates between the analyte and other remaining compounds. As the analyte concentration is normally very low, it may be necessary to incorporate concentration steps into the procedure. Of course, this will normally concentrate both analyte and any other substances present in the same ratio. Hence, the need for discrimination at the detection stage.

1.3.3 Classification by Analyte and Matrix

Trace analysis methods can be conveniently classified on the basis of the key aspect noted above, the need to separate the analyte from the sample matrix prior to measurement, *i.e.*:

(i)   determination of an inorganic analyte in an inorganic matrix;
(ii)   determination of an inorganic analyte in an organic matrix;
(iii) determination of an organic analyte in an organic matrix;
(iv)   determination of an organic analyte in an inorganic matrix.

The separate procedures to be followed are illustrated in Figure 1.3.1 and summarized below. Detailed information on the procedures is given in Chapters 3 and 5.

Inorganic Analyte/Inorganic Matrix

A typical example would be the determination of trace elements in rocks. The principal difficulty with this type of sample is often dissolution as much silicaceous material is insoluble. It is, however, essential to achieve total solution since analytes at trace levels are readily adsorbed by such constituents giving rise to low results on analysis. Once the sample has been totally solubilised, the concentration of the analyte can then be enriched by the removal of other potentially interfering ions. This can be achieved by precipitation (either of analyte

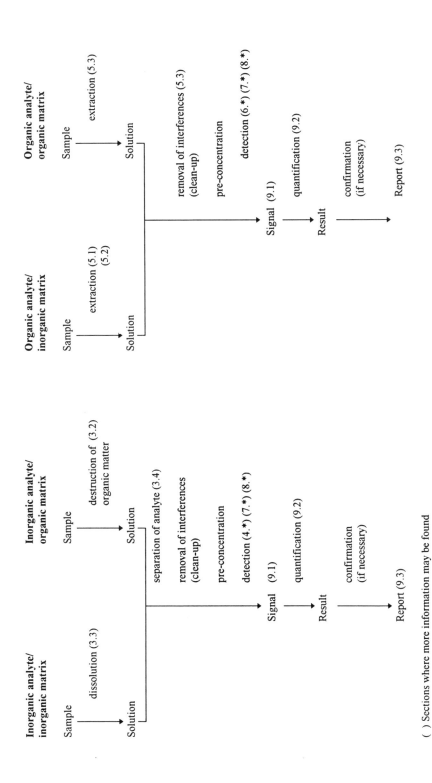

Figure 1.3.1 *Flow chart for trace analysis*

or other ions), by complexation, or by ion-exchange chromatography. The remaining solution can then be progressed to the detection and measurement stage.

Inorganic Analyte/Organic Matrix

In some cases a direct determination without clean-up or concentration may be possible where the organic matter concentration is fairly low. For example, direct aspiration of soft drinks into an atomic absorption spectrophotometer may give satisfactory results for quality control where the limit of detection required is not too low. In this case the flame is used to destroy the organic matter simultaneously with the detection stage. In most cases it is necessary to remove the organic matter totally prior to determination of inorganic analytes. This not only prevents any interference from organic compounds at later stages of the method but also enriches the analyte *vis-à-vis* other constituents. Organic matter is commonly removed either by combustion or by the use of oxidising acids. Both systems have their advantages and disadvantages depending on the analyte to be determined. This is discussed further in Chapter 3. Once the organic matter has been removed, further purification and concentration can be achieved where necessary as noted above.

Organic Analyte/Organic Matrix

This is, perhaps, the most difficult situation encountered in trace analysis. Since, for obvious reasons, the organic matrix cannot be destroyed it is necessary to find some means of separating the analyte from the matrix. Solvent extraction is employed most frequently for this purpose. Whilst the extraction system employed must be sufficiently strong to remove the *maximum* quantity of analyte from the matrix, at the same time it should remove the *minimum* quantity of co-extractants. These are often conflicting requirements. As mentioned earlier, additional purification and concentration may also be included in the method prior to the detection and measurement stage.

Organic Analyte/Inorganic Matrix

This type of trace analysis method is less common than the other three categories. Separation of the analyte from the matrix is often straightforward, relying on the volatility of the analyte or extraction using organic solvents which leaves the bulk of the matrix as a residue to be discarded.

1.3.4 Reference

1. N. T. Crosby, 'Determination of Veterinary Residues in Food', Ellis Horwood, London, 1991, ISBN 0-7476-0065-1.

1.4 Analytical Methodology

1.4.1 Terminology

The wide scope of trace analysis has already been referred to in Section 1.1. Not surprisingly, the analytical approach to solving trace analysis problems can be equally wide. The importance of understanding why and for what purpose an analysis is being carried out, leading to the concept of fitness for purpose (Section 1.2.2), cannot be over-stressed. This approach emphasises that it may not always be necessary, desirable, or cost effective to use the 'best' method available. This situation presents the trace analyst with a major problem. He or she can consult the scientific literature and find at their disposal a vast array of potential solutions to their specific task. The problem lies in choosing one that meets their needs, for which they have adequate experience and resources, and for which they can adequately demonstrate fitness for purpose. To make matters worse, especially for the beginner, a variety of sometimes conflicting terminology is used to describe aspects of the methodology such as its intended range of application or its 'quality'. In this section we attempt to give a brief guide to the terminology and a simple, systematic approach to achieving a methodology that is fit for a particular task. For this purpose, some of the terms commonly used to describe analytical methodology can be conveniently classified according to three distinct hierarchies:

(i)   the breadth of application of the methodology;
(ii)   the extent of knowledge of analyte concentration provided by the methodology;
(iii)   the degree of confidence required from the methodology.

Classification by Breadth of Application

The hierarchy shown in Table 1.4.1 reflects a move from general purpose methodology at the top of the list to a very specific application at the bottom. For a given analytical determination, the literature will often provide a vast variety of methods, procedures, protocols, *etc.* based on the same technique. Understanding the significance of these terms is important but only expert judgment, and proper evaluation, will reveal whether any of the existing solutions are adequate for the current application.

Table 1.4.1 *Classification of methodology by breadth of application*

Technique	
Method	Increasing
Procedure	specificity
Standard operating procedure	
Protocol	↓

A *technique* is a scientific principle that can be used to provide data on the composition of a material. For example, atomic absorption spectrometry is widely used to provide data on elemental composition. It is relatively rare that a technique can be applied simply and directly to analysis of a 'real sample'; it is more likely to form the last stage of more complex methodology and hence its use is often referred to as the *end-determination*.

A *method* is based on a particular usage of a technique and incorporates all the methodology required to use the technique for a specific application. For example, determination of copper in an alloy by atomic absorption spectrometry requires dissolution of the alloy and, usually, dilution of the solution to bring the copper concentration within the calibration range of the spectrometer.

A *procedure* comprises a set of written instructions for the use of a particular method. The level of detail of the instructions varies widely between procedures but it is rare to find every possible aspect of the method specified in precise terms. Many procedures published in the open literature, or even in compilations of 'standard methods', provide relatively little detail on well-known aspects of the method in order to save space. In some cases the lack of detail is deliberate, allowing the experienced analyst flexibility in applying the method. Whatever the reason, the scope for varying interpretation of many widely circulated procedures is a frequent cause of variability in data between laboratories.

A *standard operating procedure (SOP)* is mostly commonly used within an organisation. It should provide sufficient detail and explanation to ensure that the method is applied consistently by different analysts and at different times. Most quality systems require the implementation of SOPs and stipulate that they must be followed without exception unless a change is specified, and recorded, by an authorised person.

A *protocol* is similar to a SOP and is the most specific implementation of a method. Every aspect of the method will be precisely defined and must be followed without exception. Protocols tend to be used for critical data, such as those required to enforce a regulation, and are often prescribed by official bodies.

Classification by Extent of Knowledge of Analyte Concentration

Many trace analysis applications do not require a full determination of the analyte and this is an important consideration for the analyst needing to save time or money. A common hierarchy of terms reflecting this aspect of analytical methodology is shown in Table 1.4.2. This classification is often thought of as

Table 1.4.2 *Classification of methodology by extent of knowledge of analyte concentration*

Qualitative	Increasing
Semi-quantitative	knowledge
Quantitative	about analyte

reflecting the relative 'accuracy' of methods but that interpretation has been avoided here because the term 'accuracy' carries with it additional implications.

Qualitative methods provide an indication as to whether or not a particular analyte is present but offer little if any information on the amount. Qualitative methods can be divided into two distinct groups: confirmation of the presence of a particular analyte and identification of unknown analytes detected in a sample. Examples of the first group include colorimetric reactions, *e.g.* pregnancy tests, card tests to detect the presence of heavy metals in foods and pharmaceutical products. Examples of the second group include GC–MS and the most up-to-date, sensitive FTIR and NMR instruments. If it is necessary to confirm the absence of an analyte then the sensitivity (or limit of detection) of the test used needs to be considered, to avoid false negatives. Many tests to confirm the absence of impurities in pharmaceutical products fall into this category. Where positive results are obtained it is usually necessary to use a quantitative method to provide further information.

A *semi-quantitative* method provides an estimate of analyte concentration, perhaps to within an order or magnitude or sometimes much better. Many of the classical 'spot test' methods fall into this category and can often be used at higher trace levels. A frequent use of such methods is in the form of 'test kits' comprising appropriate reagents and instructions. A number of rapid test kits for various trace analyses are currently available or under development. These range from cholesterol in blood to veterinary residues in meat, milk, and offal. The kits have a defined limit of detection and a limited response range. However, this range can be extended by suitable dilution of the test solution. For example, by comparison against a colour chart, an approximate or semi-quantitative value can then be obtained. Examples of where semi-quantitative analysis can be used in field situations to give a 'ball park' figure for the concentration include toxic gases in the atmosphere, water test kits for NO_{2-}, dissolved oxygen, chlorine, *etc.*, and soil test kits for anions. More sophisticated semi-quantitative techniques are based on the use of automated instruments such as inductively coupled plasma mass spectrometry (ICP–MS). In this case a result will be produced using a rapid method with a larger uncertainty than the full quantitative version.

Quantitative methods provide a well-defined value for the amount of analyte. It is important to remember that the uncertainty (error range) of the value will vary widely between methods and even between different written procedures for the same method; there is no definition of a 'quantitative method' in terms of measurement uncertainty. Each specific application of a quantitative method requires the analyst to ascertain what will be fit for that purpose and to develop and validate the method to meet that need. In many cases the analyst will require a quantitative method with quite tightly prescribed tolerances. For example, the customer may need to enforce a contamination limit of 2 ppm. Under circumstances where most sample results are expected to fall only slightly below this limit, a typical measurement error of ±30% would not be acceptable. However, errors of this magnitude would be quite acceptable where the limit is the same but most samples are expected to lie around the 0.5 ppm level.

Classification by Degree of Confidence

The degree of confidence required by the customer or other users of the data refers to the assurance that can be offered of both the qualitative identification of the analyte and the quantitative determination. This aspect is related to the consequences of making a wrong decision due to the errors of individual measurements. In some cases the consequences may be modest and the degree of confidence needed for each measurement can be correspondingly low. For example, in a large geological survey it may be important to gather many results quickly and at low cost. Unexpected large errors for some of the measurements will have little consequence given the original sampling errors and the eventual statistical treatment of the data. The situation will be totally different for, say, clinical samples taken from individual patients. Here the consequences of a wrong decision could be life-threatening and the analyst may be required to offer a 95% confidence level or better that all results lie within a narrow, specified concentration range. Some terms used to indicate the degree of confidence offered by analytical methods are listed in Table 1.4.3.

Table 1.4.3 *Classification of methodology by degree of confidence*

Screening	
Routine	
Accepted (recognised)	
Standard	Increasing
Regulatory	degree of
Reference	confidence
Primary	

Screening methods are used where large numbers of samples have to be examined as quickly as possible at minimum cost. Very rapid methods are preferred so that the majority of the samples can be eliminated in an initial screen allowing more time-consuming methods to be applied to just a few samples for which a presumptive positive value has been obtained. The rapid methods used must be sufficiently sensitive to ensure that false negatives (*i.e.* samples that do contain low levels of analyte but give a negative response in the test) are not obtained. A few false positive results (*i.e.* samples that appear to contain the analyte when in fact they do not) should not be a major cause for concern since they will be eliminated in the subsequent analysis. As screening methods must permit a high throughput of samples at low cost, they may only be qualitative or at best semi-quantitative.

Routine methods represent the bulk of the day-to-day work of many laboratories. In general, they are developed in-house although they will frequently have been adapted from published methodology or procedures. Routine methods require validation to ensure they are fit for their intended purpose but their suitability is a matter of judgment by the individual analyst or organisation.

Accepted or *recognised* methods generally represent a consensus view from a body of analysts working in a particular application area. They are frequently developed through a committee approach, often under the auspices of professional or official bodies or trade organisations. In some cases their standing is based solely on the expertise available to the various contributors; in other cases validation or other data may be obtained on a collaborative basis.

Standard methods are similar to accepted methods but tend to be developed on a national or international basis by an organisation with some official status. They frequently comprise quite detailed written procedures. The implication is one of greater confidence but this is not necessarily the case. Many 'international standard methods' are developed by committees that undertake little if any experimental evaluation of their recommendations.

Regulatory methods are typically prescribed by an official body for use in enforcement of a specific regulation. As such they should have been subject to extensive testing, either collaboratively or by an agency with high standing in the field. They should, therefore, command a substantial degree of confidence. It is essential, however, that any laboratory using the method obtains adequate validation data to confirm that the method is working as prescribed in the hands of its own analysts.

A *reference* method is one which has been thoroughly investigated to ascertain the bias and precision of the values that it provides. The investigation should also allow description of an exact procedure for its use to produce data within the claimed limits. Reference methods may be used to assess the accuracy of other methods for the same measurement. They are frequently used for the characterisation of reference materials.

A *primary* method is one having the highest metrological qualities, whose operation can be completely described and understood, and for which a complete uncertainty statement can be written down in terms of SI units. Such methods are generally of interest to national laboratories contributing to the development of a national or international chemical measurement system.

Tailoring the Methodology to the Problem

Most analytical chemists recognise that problem solving forms a major and very important part of their job. Typically, a client, who may be a paying customer or a colleague within the organisation, poses the problem and expects the analyst to provide appropriate measurement data to solve it. In many cases, the client will have reached an independent view as to the appropriate data and will approach the analyst with a list of requirements. The key issue for the analyst is choosing the right tool for the job: a method which will provide data that are fit for purpose within constraints such as time, cost, safety, and the resources actually available. A simple, systematic approach to developing a strategy to solve the problem is shown in Figure 1.4.1. This approach can be applied to *any* problem which may be put to the analyst; ranging from a 'one-off' sample where the result is required within the hour to the complete certification of a reference material requiring many months work.

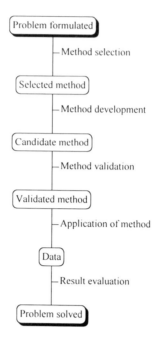

Figure 1.4.1 *Systematic approach to solving a problem*

Clearly, there must be differences in applying the approach set out in Figure 1.4.1 to problems of vastly differing complexity but, just as the methodology should be 'fit for purpose', so should the task of tailoring it to the problem. Thus, where time, cost, or other factors preclude a thorough examination of the method, the analyst may have little option other than to rely on prior experience in quickly working through each stage of the process set out in Figure 1.4.1. At best it may be possible to run some sort of check sample alongside the real one to see if the anticipated result is obtained from the former. At the other extreme, certification of a new reference material may require an exhaustive literature search to identify potentially useful methodology, a substantial laboratory project to develop it into a candidate method, and extensive validation trials involving collaboration with other laboratories. Regardless of the scale of the problem or the tailoring process, it is essential that the analytical chemist approaches every trace analysis in the methodical way suggested in Figure 1.4.1. The key issue is the extent of knowledge of the methodology when it is applied to the sample: does the analytical chemist have adequate grounds for asserting that the data will be fit for the agreed purpose and does the confidence with which that assertion is made match up to the client's expectations?

The two remaining sections of this chapter offer respectively advice on method selection and development and method validation. It cannot be over-emphasised, however, that reliable trace analysis will not be achieved unless appropriate samples are correctly taken and unless client and analyst are able to agree an appropriate analytical specification. Good planning is an essential aspect of trace analysis and the analytical chemist should make every effort to involve the client. The reasons for undertaking the analysis must be clearly understood and every effort should be made to discuss this aspect with the client. Similarly, the analyst is responsible for explaining to the client the implications and limitations of any analysis to be carried out. In many cases it will be necessary to resolve a conflict between the client's expectations for data quality and the time and cost of the analysis. Other constraints such as confidentiality, safety, and possible contamination from other work in the laboratory will also need to be resolved before method selection and development can begin. Various aspects of planning and the analytical specification were discussed earlier in Section 1.2.2.

1.4.2 Method Selection and Development

Earlier sections have emphasised the need to be absolutely clear as to why an analysis is to be undertaken and the decisions that will be based on the result. Once this has been done a suitable method must be selected and, if necessary, further developed to meet the objectives of the analysis. It is relatively uncommon in trace analysis to start with a completely clean sheet; the analyst will usually have several available sources of analytical methods in addition to his or her previous work. Such methods range from in-house ones (*i.e.* as developed by an individual laboratory), through methods published in the scientific literature, to those developed and perhaps tested by qualified organisations (*e.g.* trade or professional bodies and standards organisations).[1] There are also many trace analysis methods published by government departments to facilitate the enforcement of regulations (*e.g.* the Ministry of Agriculture, Fisheries and Foods in the UK and the Environmental Protection Agency in the USA). The amount and quality of existing data available for such methods will determine the additional work required from the analyst. This applies to both setting out a complete methodology for the task in hand and subsequently validating it against the analytical specification (see Section 1.4.3.).

The goal of the analyst in seeking an appropriate existing method is, ideally, to find one that closely matches the present requirement, has a comprehensive written procedure, and for which extensive and relevant validation data are already available. Needless to say, this ideal is seldom achievable. Few trace analysis problems are unique, however, and it is often possible to find an existing method which can be extended, adapted, or improved. Failing this, it may be possible to use components of several existing methods as building blocks. As a last resort, published research papers will provide ideas and guidance on techniques or methodology which can be incorporated into a new method. Laboratories with full access to the scientific literature will find that a vast amount of material is available; the problem is to find what is relevant and useful. *Chemical Abstracts* and *Analytical Abstracts* are widely used for searching the chemical literature but often leave the analyst with the problems of sifting and of access to the material which has been identified. General analytical chemistry journals, particularly *Analytical Chemistry* and *The Analyst*, offer a more direct route to useful methodology and also publish review articles. Much more specific information is also available from a wide variety of organisations with more narrow interests, including professional bodies, trade associations, instrument manufacturers, and chemical manufacturers.

Analytical methods should be evaluated by consideration of their known and proven performance characteristics. The most important of these are discussed below but it is up to the analyst to assess the relevance of any one given parameter for the job in hand. It may also be necessary to consider the availability and suitability of equipment and reagents, the cost of the work, and the time required for a complete analytical run. Do these fit in with the customer's requirements agreed during the planning stage? It may be helpful to prepare a simple checklist showing a target for each characteristic and its importance for a particular task (*i.e.* is it critical, important, secondary, *etc?*). Before selecting an existing method for validation, the analyst will also need to decide whether any claims made for it can be accepted, whether checks of some parameters need to be made, and whether development is needed to improve a particular characteristic.

Performance Characteristics

A systematic approach to method development can best be achieved by individual evaluation of key aspects of the methodology. Those most commonly considered for this purpose, and usually referred to as 'performance characteristics', are listed in Table 1.4.4. In some cases, several different names are used for essentially the same characteristic and some alternatives are shown in the table. The word 'accuracy' is widely used but has been avoided because there is some confusion amongst practising analysts as to its meaning: it is often used to describe the combined precision and bias but is also used as an alternative to bias. Use of the concept of 'uncertainty' is discussed with regard to method validation in Section 1.4.3. It is important to appreciate that there is no real boundary between development and validation. The latter requires attention to the same performance characteristics as does method development. Thus the laboratory undertaking development of a method should do so bearing in mind the subsequent goal of

Table 1.4.4 *Performance characteristics of analytical methods*

Limit of detection	(Determination)
Precision ⎫ Bias ⎭	⎧ (Accuracy) ⎩ (Uncertainty)
Specificity	(Selectivity) (Interferences)
Recovery	
Range	(Calibration range)
Robustness	(Ruggedness)

validation, there is no point in obtaining the same data twice. The concepts of method validation and some special requirements are discussed in Section 1.4.3.

The *limit of detection* or *limit of determination* (see below) is the most critical factor in selecting a method for trace analysis. Obviously, if the method selected cannot detect the levels of analyte likely to be present in the sample, then all other factors are irrelevant and an alternative method must be found, or developed, or an existing method modified. The latter might be achieved by taking a larger sample for analysis, by taking greater aliquots at various stages in the procedure, by improving the clean-up and so reducing background 'noise', or by the use of a more sensitive detection system. The 'safety margin' provided by the limit of detection of a method is particularly important in trace analysis when one has to decide whether a contaminant is present or not and whether it is present above a given legal limit. Ideally, the limit of detection of the method selected should be one-tenth of the lowest concentration to be determined. The limit of detection is defined as the smallest concentration of analyte that can be measured with definable statistical certainty, *e.g.* it may be taken as the mean of a given number of blank samples plus three times the standard deviation of that mean. Some analysts define a limit of determination as being the concentration of analyte that can be measured with a higher specified degree of certainty. Sometimes a limit of decision is also specified. The limit of decision is then typically defined as the mean of the blanks plus six times the standard deviation of the mean. This treatment assumes that bias and precision are constant over the range of values studied. *It is essential that these limits are measured by the laboratory itself;* data obtained elsewhere, even for a fully detailed procedure, may show significant differences. The detection or other limits must also be determined under the actual conditions of the analysis, *e.g.* in the presence of the sample matrix, not on the basis of signal-to-noise ratios measured for pure standards of the analyte.

The *precision* of the method is evaluated by making repetitive measurements and this should be done for the entire method, not merely the end determination step. If the precision achieved is not good enough, the method will have to be investigated step by step. The effect of instrument conditions and aspects of the method such as variations of temperature, shaking times, extraction conditions, and flow rates, will all need to be examined as appropriate to determine which factors are critical to reduce the variations observed. For many trace analysis applications, particularly for regulatory purposes, it is important to evaluate the precision of the method within a single laboratory (*repeatability*) and between laboratories (*reproducibility*). The latter can only be achieved by collaboration between laboratories, usually through circulation of a 'round-robin' sample. Such collaborative studies provide data on the precision of the method where all participants use the same procedure; studies allowing a range of methods may also provide data on bias (see below). Normally the relative standard deviation from a reproducibility exercise will be at least double the value for repeatability in a single laboratory.

The *bias* of a method is defined as the difference between the experimentally determined value and the true value (where known). The bias is the systematic component of measurement error, whereas the precision (see above) is the random component. The preferred way to determine the bias is by the analysis of a certified reference material containing the analyte at about the same concentration expected to be present in the test sample. Ideally, the reference material should be identical in composition to the test sample matrix but only rarely will this be so. The choice of reference material is usually a compromise, the best match is chosen. It is better to use any reference material than none at all, providing that the analyte concentrations are of the same order of magnitude. If no suitable material is available, it may be acceptable to use a well-characterised sample previously analysed in-house or by another laboratory. It is also possible to evaluate bias using additional measurements by a different method. Care must be taken, however, that the method is really different. Use of two different GC columns or even two different instrumental techniques will achieve nothing if the main source of bias is an earlier extraction step. It is important for both analysts and their clients to recognise that in trace analysis it is not always necessary, or reasonable, to require demonstration of a very low bias. For example, the permitted maximum level for fluorine in a complete animal feeding stuff is 150 mg kg^{-1}. If a sample is analysed and found to contain 50 mg kg^{-1}, it does not matter if the analysis is in error by even 100% as the level of contamination is still well below the permitted maximum. Where the concentration of a contaminant, or a permitted additive in a food, is close to the maximum allowed, then the error range becomes more important. For certain veterinary residues in foods, the European Union uses criteria for accuracy shown in Table 1.4.5.[2] These limits recognise that for such compounds at the levels stated, the analysis is very difficult and so a reasonable limit of variation must be permitted.

Table 1.4.5 *Criteria for the accuracy of methods to determine veterinary residues in foods*[2]

True content (μm kg^{-1})	Range (%)
≤ 1	–50 to +20
> 1 to 10	–30 to +10
> 10	–20 to +10

Reproduced from *Official Journal of the European Communities*, L118, 14 May 1993.

The *selectivity* or *specificity* of a method is a measure of the extent to which other substances affect the determination of an analyte. Such effects are commonly referred to as *interferences*. Failure to correctly identify, and overcome, interferences for all the circumstances under which the method will be used is a major source of error in trace analysis. A common mistake is to check for interferences only at relatively high analyte concentrations and not near the detection limit where such effects are often magnified. Proper verification of selectivity is particularly important where the detection method used for the end determination is not specific, for instance many methods using colorimetric or chromatographic techniques. Where time and resources permit, method development should include use of another technique to confirm that the analyte in question is being measured. For example, GC–MS is often used to verify a GC measurement. The next stage is to obtain a 'blank sample', *i.e.* a sample that is known not to contain the analyte. This sample can then be put through the analytical procedure to check that interferences do not arise from the sample matrix. It is also desirable to check for possible interferences in the method by the addition of known compounds to the 'blank samples'. These could be isomers, metabolites, or compounds similar in molecular structure to the target analyte. Equally, substances known to be present in the matrix but totally unrelated to the analyte could also be used, as a check on the specificity of the method. This approach is particularly important in the evaluation of immunochemical procedures.

Where a significant interference is observed it will be necessary to develop improved or additional clean-up procedures or to adopt a more selective end-determination. Use of different detectors, derivatisation, or use of chromatographic separations dependent on different interactions are all widely used strategies. Similarly, in spectroscopic analysis, it may be possible to make measurements at another wavelength or at more than one wavelength. The latter approach is used, for example, in spectrophotometry, where ratios of measurements at two wavelengths are often more specific to a compound than a single measurement, particularly, in the ultraviolet region.

The *recovery* is the fraction or percentage of the actual amount of a substance obtained after one or more manipulative stages of a method. Thus recoveries are commonly quoted both for the entire method and for stages such as extraction or clean-up. It is important to appreciate that in trace analysis this performance aspect includes not only losses of analyte but also gains due to *contamination*. Both aspects must be checked during method development; care is needed in case a fortuitous combination of losses and contamination leads to recoveries satisfyingly close to 100% at the particular analyte concentrations used during method development. Ideally, one should aim for recoveries better than 95% but for some difficult analyses at very low levels it may not be possible, for example, to achieve better than 50% extraction. The analyst must then decide whether the method can still give sufficiently reliable data for the intended application. A decision may also be needed on whether to correct results for extraction efficiency (see Section 9.3.2).

The use of the entire method with a 'blank sample', as mentioned above, is an essential check for contamination arising from reagents or elsewhere. The next stage in the development process is to add known amounts of analyte to a blank or other sample matrix, undertake the complete analysis, and calculate the recovery obtained for the analyte. The concentration of added analyte should be close to the level expected in the sample. It may also be necessary to carry out similar recovery tests at one-half, and at double, the expected concentration. If the recovery values obtained for the complete method are unsatisfactory, it will be necessary to carry out recovery experiments at each single stage of the method to ascertain where losses are occurring.

The method of addition of analyte to the 'blank sample' (spiking) can be critical. Addition of analyte as a solution in the same solvent as is used for extraction followed by immediate extraction, is not a satisfactory test of analyte recovery. The analyte solution should be left in contact with the matrix for a few hours (preferably overnight) before extraction to allow any analyte–matrix interactions to occur. The extracting solvent should if possible be different from the one in which the analyte is added. In many biological systems analytes are known to bind to constituents of the matrix giving rise to incomplete recoveries following extraction and analysis. Hence, a simple spiking and immediate extraction procedure may give falsely high recovery values. Spiked samples are seldom homogeneous so it is necessary to take the whole sample for analysis and not an aliquot portion. Improved homogeneity can be achieved in some cases by spiking the analyte solution onto an inert carrier such as chalk or starch, allowing the solvent to evaporate, and then mixing the carrier and matrix phases together. In spite of these precautions, spiking is a potentially unreliable method of estimating the true recovery of an analytical method. It is widely used because it is often the only viable option available to the analyst when an appropriate reference material is unavailable. It is often possible, however, to use external data to add to the analysts' knowledge of recoveries. For example, agrochemical companies frequently obtain recovery data for pesticide residue methods using crops which have been treated with radio- or isotopically labelled formulations of their pesticides.

The *range* of a method is the interval between the upper and lower concentrations of analyte in the sample for which it has been demonstrated that the method is suitable. Reference is often made to a 'linear calibration range' but, particularly in trace analysis, all aspects should be considered, not just the linearity of calibration of the end-determination. However, errors can arise when a trace analysis method is developed using a two point calibration (blank and one standard) and the assumption is made that the calibration is linear up to and well beyond the standard.

The *robustness* (or *ruggedness*) of an analytical method is a measure of its capacity to remain unaffected by small but deliberate variations in the main parameters of the method. It is not an exact, numerical characteristic of a method but does provide an indication of its reliability during normal usage. In developing a method, and preparing a written procedure, the analyst should attempt to identify operations which are particularly sensitive to small changes in experimental conditions, techniques, *etc*. It may be possible to modify the method to make such operations more robust or to use alternatives. Failing this, the written procedure should highlight aspects of the method requiring particular care.

1.4.3 Method Validation

The goal of method validation, as discussed earlier, is to ensure that the method as applied in the analyst's laboratory will provide data that are fit for their intended purpose. It is important to appreciate that validation approached in this way is a matter of judgment, not a precise scientific evaluation. The client has a problem that can be resolved using appropriate data. The analyst has a method with performance characteristics that may be far from perfect. Both client and analyst must work within varying constraints such as cost, time, and available skills or techniques. Validation is thus a matter of weighing up specific requirements against method performance under one particular set of circumstances. Hence a method judged to be valid in one situation may not suffice in another.

This is an issue which seems to cause widespead confusion. The analytical literature provides detailed descriptions of a wide range of 'validated methods', often prepared by acknowledged experts in a field and subjected to extensive testing. It seems illogical to many analysts that this should not suffice; and even more illogical that, given limited time and resources, they should have something useful to add. This reasoning misses the two key points of the foregoing summary:

(i)   validation of a method will apply to a very particular set of circumstances;
(ii) even when these precisely match the users' circumstances, it is essential for the users to demonstrate their own capabilities to use the method in their own laboratories.

Further confusion arises with regard to the link between method validation and measurement uncertainty. Ideally, all analytical results should be accompanied by a realistic evaluation of the uncertainty associated with each specific measurement. The data obtained during method development and validation are very useful in undertaking such an evaluation. Nevertheless, the precision and bias observed during a validation study do not provide a quantitative uncertainty value for future results obtained with the method. Using them as such is liable to cause a significant underestimate of the uncertainty of routine results.

The Validation Process

A search of *Chemical Abstracts* or *Analytical Abstracts* under 'method validation' reveals an enormous range of literature references. Many of these describe the validation of a specific method or technique but others cover the general philosophy of validation, relevant statistical or chemometric techniques, and validation in important application areas such as pharmaceutical analysis. All of the approaches advocated are essentially summarised by the validation process shown schematically in Figure 1.4.2. Clearly method validation should be a straightforward process giving the analyst little cause for concern! Indeed, in many cases much of the required data will already be available having been obtained elsewhere, during earlier work, or during method development. Provided sufficient, suitable reference materials are available one need only analyse these using the candidate method and compare results with the certified or expected values. Figure 1.4.2 includes a requirement to demonstrate a state of statistical control for the measurement system. During the validation procedure it is important that the reproducibility of the equipment used in the method is checked. If this is not done one cannot rely on the results obtained from repeat measurements using the whole methodology. The validation data should be available with the written procedure so that future users of the method can make their own judgment of the adequacy of the validation and its range of applicability.

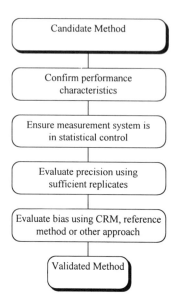

Figure 1.4.2 *The validation process*

The preceding description of method validation should, in an ideal world, apply to the entire range of problems facing the analyst provided the concept of 'fit for purpose' is applied to the validation process as well as the analytical specification. The method selection and/or development stage provides an evaluation of performance characteristics. Experience or experimental testing confirms that the method is being used properly and is under statistical control. Analysis of reference materials demonstrates whether or not the expected results are being achieved and, depending on the number analysed, may refine the preliminary evaluation of precision and bias. The major problem facing the analyst, in the real world, is that suitable reference materials are available only in very limited numbers or not at all for many important applications. The analyst must then seek alternatives. Determination of precision is relatively straightforward, provided an adequate amount of a stable and homogeneous sample is available for replicate analysis. Some alternative approaches for determination of bias are shown in Table 1.4.6, listed in order of increasing preference. The absence of *collaborative testing* from this list requires explanation. Collaborative studies of candidate methods are frequently undertaken, often under the auspices of a standards organisation or

Table 1.4.6 *Alternative approaches for the evaluation of method bias*

Analyse spiked samples	
Analyse samples of known composition	Increasing
Apply an independent method	preference
Apply a method of known accuracy	
Analyse certified reference materials	↓

other official body. They are widely regarded as the ultimate evaluation of a method, ensuring that it is subjected to detailed examination by a wide cross-section of users. However, many such collaborative tests provide an excellent evaluation of the interlaboratory precision of a method but no information on bias. As shown in Figure 1.4.3, a collaborative test can only determine bias if samples of known composition are available for circulation to participants or if the results from the candidate method can be compared with those obtained by a method of known accuracy. It is important to clarify the position with respect to bias when evaluating published, collaboratively tested methods. The absence of bias data does not negate the value of the other information available from the test but it does leave the user to resolve the most difficult aspect of method validation.

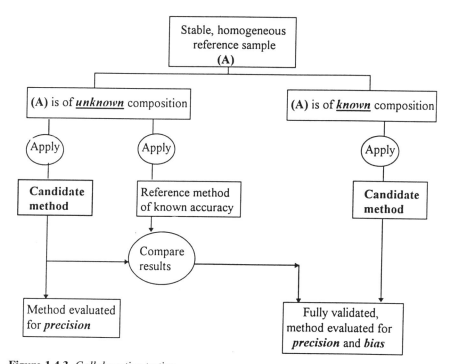

Figure 1.4.3 *Collaborative testing*

1.4.4 References

1. 'Official and Standardized Methods of Analysis', ed. C. A. Watson, The Royal Society of Chemistry, Cambridge, UK, 3rd edn, 1994, ISBN 0-85186-441-4.
2. 'Veterinary Drug Residues: Residues in Food-producing Animals and their Products: Reference Materials and Methods', ed. R. J. Heitzman, Commission of the European Communities Report, Luxembourg, 1992, ISBN 1018-5593.
3. J. K. Taylor, 'Validation of Analytical Methods', *Anal. Chem.*, 1983, **55**. 600A.

CHAPTER 2

Sample Handling

2.1 Sample Storage and Stability

2.1.1 Introduction and Overview

When asked to identify potential sources of error and uncertainty in an analysis, the majority of analysts will immediately think of high profile areas such as instrument calibration and contamination during the analysis. The possibility that the sample may have changed during storage, *i.e.* that the sample analysed is not the same as that received from the customer, is often added only as an afterthought. In fact, this is one of the key areas to consider when trying to ensure that results are valid. It is not that the risk of the sample altering during storage is necessarily high; the concern is that if there are any storage effects they are, at best, difficult to estimate and, at worst, likely to go completely unnoticed. No amount of repeat assays or method validation is worthwhile if the problem is with the sample itself.

The obvious method of avoiding any risks of storage changes is not to store samples. Unfortunately, few laboratories find it practical to analyse samples immediately upon receipt. In many cases samples have to be stored; for example if engaged in a survey where samples received at different times have to be analysed in the same batch. However, as a general rule samples should only be stored for the minimum possible time.

If samples have to be stored, the stability of the analyte in the sample matrix ideally should be fully investigated under identical conditions. In practice, this is seldom possible. The limited resources of most laboratories preclude such lengthy and expensive trials. However, information may be obtained from the customer or from published literature. These should both be investigated, particularly if the analyst is dealing with an unfamiliar analyte or sample type. Quick tests of the stability of standards should be performed as a matter of course.

The decision on how long and under what conditions a sample can be reasonably stored must eventually fall to the analyst (in consultation with the customer). This will very often be based on very little hard data, relying more upon experience and common sense. It must also take into account the purpose of the analysis: if the sample is a vintage bottle of wine, the effect of a few more weeks on the shelf will be negligible, whilst if the analyst is interested in the degree of metabolisation of a drug in a urine sample each hour could be vital. These examples would be fairly obvious. It is unfortunate that most samples fall into the grey area between.

2.1.2 Storage Conditions

Conditions of storage prior to analysis should be agreed in advance with the customer. This may involve storage in a cupboard, storeroom, refrigerator, freezer, or cold room as appropriate. The choice will depend on the properties of the sample and the need to protect the sample from light, elevated temperature, or humidity (see Table 2.1.1). It may be necessary to use a maximum/minimum thermometer to check for temperature fluctuations during storage. There are also a variety of physical and chemical methods of arresting or slowing sample degradation (see Table 2.1.2). The intent of all such methods is to alter the nature of the sample, and so it is vital to verify that the integrity of the analyte is not affected.

Table 2.1.1 *Storage conditions for analytical samples*

Storage condition	Appropriate sample types	Inappropriate sample types
Deep freeze (−18°C)	Samples with high enzymatic activity, *e.g.* liver Most sample types Less stable analytes	Fresh fruit and vegetables Samples which liquefy on thawing Aqueous samples
Refrigerator (4°C)	Soils, minerals Fresh fruit and vegetables Aqueous samples	Samples with possible enzymatic activity
Room temperature (dark)	Dry powders, and granules Minerals Stable analytes	Fresh foods
Desiccator	Hygroscopic samples	Samples which are more hygroscopic than the desiccant.

In all cases samples for trace level analysis must be stored in a separate physical location from analytical standards, and any other material which may contain a high concentration of the analyte. Storage is usually in a completely segregated and dedicated room. It may be necessary to take precautions to avoid cross-contamination between sample storage areas and other laboratory areas; this can include changing laboratory coats when entering and leaving the storage area, and disposable adhesive mats to prevent any material being walked between rooms.

Most analytes and sample matrices are more stable at low temperatures, and so freezing the sample is the usual first choice method of storage. At deep freeze

Table 2.1.2 *Physical and chemical methods of preserving samples*

Method	Examples of applications	Critical aspects
Freeze drying	Breads, biscuits, *etc.* Aqueous samples	Unsuitable for volatile analytes
Irradiation	Aqueous samples Biological samples	Stability of analyte must be established
Antioxidants	Liquids and solutions	Stability of analyte must be established. Check specific interference effects
Anticoagulants	Blood and clinical samples	Check specific interference effects
Autoclaving	Sterilising body fluids	Stability of analyte must be established.

temperatures ($-18°C$), most enzymatic and oxidative reactions are reduced to a minimum. However, changes in biological samples can occur during freezing and thawing as these processes disrupt the cellular structure. Some fruits and vegetables soften and blacken upon thawing. Fluid may be released from ruptured cells which can reduce sample homogeneity if ignored.

Samples which cannot be frozen or which do not need to be frozen, such as soils intended for elemental analysis, plastics, paints, or any other sample where both the matrix and the analyte are non-volatile and stable at ambient temperatures, are usually stored at 0 to $5°C$. It may not even be necessary to refrigerate the sample, many are stable at room temperature. For example, the benefits of refrigerating a packet of biscuits which has already spent a year on a supermarket shelf are fairly minimal.

For pesticide residues in food the Codex Alimentarius Commission guidelines on good practice in pesticide residue analysis recommend storage of samples at $-20°C$ where the analysis is not to be carried out immediately.[1] The Department of the Environment Standing Committee of Analysts has made recommendations for the storage conditions of waters and effluents.[2] Some samples have to be examined immediately on receipt, *e.g.* analysis for residues of certain volatile or labile pesticides (*e.g.* the dithiocarbamate fungicides) and unstable matrices.

There can be specific problems with blood samples. Anticoagulants or preservatives often have to be added if blood is to be stored, and some of these have been found to interfere with some analyses; particularly analyses for drugs and pharmaceuticals. For example, heparin was found to interfere with the HPLC determination of frusemide.[3] If blood is stored in plastic transfusion bags, phthalates can leach out at concentrations of 35 to 75 mg l^{-1}, a particular problem if the blood is to be used for the preparation of calibration standards.[4]

Other preservation methods are occasionally used for some sample types. Freeze drying can be a useful method of preserving friable samples with a moderate moisture content (such as breadcrumbs), and is also an effective way of preconcentrating aqueous samples. It is not appropriate for volatile analytes, *e.g.* mercury, as there are likely to be significant losses during the freeze drying process. Irradiation of samples for long-term storage is used particularly where it is desirable to minimise bacteriological activity in a sample; for example, to inhibit the growth of moulds in water samples. Antioxidants can be added to liquids and solutions to prolong the stability of unstable analytes such as vitamins or unstable matrices such as vegetable oils. The list of possible techniques is extensive, and can only be touched upon here.

Precautions may also have to be taken to prevent loss or gain of moisture, and to prevent photochemical degradation by storage in the dark or in glass containers protected by aluminium foil. Samples containing volatile constituents should be kept in well-sealed containers and preferably stored in the cold to reduce the vapour pressure of such compounds.

All samples should normally be allowed to reach ambient temperature before analysis. Care should be taken to avoid hygroscopic samples taking up water, both when stored in a deep freeze or a refrigerator and when warming back to room temperature.

In all types of trace analysis, the analyst needs to be satisfied that no changes have occurred, as a result of storage, that could vitiate the results of that analysis. Examination as soon as possible on receipt of the sample is strongly recommended, provided that the scope of the analysis and the methods to be used are clear and have been agreed. Storage conditions and the length of storage should be recorded.

2.1.3 Storage Containers

The choice of container will normally be made by the sampling officer, but the analytical chemist may also be consulted and asked for advice based on prior knowledge of the particular analyte(s) and the matrix involved. The prime functions of the container are to protect the sample from degradation by external agents and contamination from the local environment. The container must not contribute extraneous analytes to the sample and it must not adsorb analytes or other components from the sample. For some samples moisture changes must be avoided by the use of air-tight containers. In this way the original characteristics and composition of the sample will be preserved until the commencement of the analysis.

Glass containers are suitable for most samples, especially solids, since there is little chance of transfer of constituents to or from the glass and the sample. Most solids only interact with the container by abrasion. Where samples are light-sensitive, brown glass can be used or the container can be protected with aluminium foil or other suitable protective packaging. The container should be completely filled with the solid samples in order to leave the minimum of headspace so that any oxygen present does not cause oxidation at the sample surface. Containers are not normally filled to the brim with liquid samples, to allow room for mixing if the sample needs homogenising. When containers are not completely filled with sample, it may be necessary to add an inert filler. This is often an inert gas, such as nitrogen. For solids which may be subjected to physical tests, *e.g.* particle size distribution, the container can be padded with material such as paper to prevent movement and friction.

Polyethylene or poly(tetrafluoroethylene) (PTFE) containers can also be used for solids. PVC containers should not be used. However, compounds such as drugs which carry a charge on the molecule can 'plate' out onto the surface of the container by electrostatic effects. They are also unsuitable for liquid samples (or semi-liquids such as butters and other foods with a high fat content) that are to be analysed for trace organic compounds, because plasticisers can leach into the sample and cause problems at later stages in the analysis.

Glass containers can be unsuitable for storing liquid samples for elemental analysis. Liquids, especially if neutral or alkaline, may leach trace elements that may be adsorbed onto the surface of glass containers. Aqueous solutions containing trace elements should therefore be acidified if stored in glass. However, this raises the further problem of desorbing trace elements that were already present on the glass. To prevent this, rigorous cleaning procedures are required. Glass storage vessels should be percolated in acid to desorb any elements on the surface. Moody recommends pH 1 as being sufficient to achieve adequate desorption.[5] Inorganic acids have been the traditional cleaning agents.* Once cleaned and washed, glass containers should then be stored immersed in pure distilled water, which should be periodically changed to prevent the build-up of bacteria.

Other analytes, for example polynuclear aromatic hydrocarbons and pesticides, may be adsorbed onto the surface of glass containers. In these cases the container can be rinsed with the intended extraction solvent. If preparing a solution, the container should contain some solvent before the sample is added and mixed. Alternatively, polyethylene or PTFE containers may be preferred. Where polyethylene containers are used these should be of high density polyethylene to avoid the diffusion of volatile substances through the walls of the container. The Department of the Environment Standing Committee of Analysts have made recommendations for sample containers and, where appropriate, the use of preservatives for water and effluent samples.[2]

Any containers used must be clean and dry. Residual cleaning agents may contribute to contamination of the sample if not completely removed by washing. Detergents can contain borates and phosphates, and some impure solvents may contain organic residues. Chromic acid or dichromate solutions used for cleaning may lead to the adsorption of chromium on to active surfaces unless subsequent washing is thorough.

2.1.4 References

1. FAO/WHO Codex Alimentarius Commission, 'Guidelines on Good Practice in Pesticide Residue Analysis', Rome, 1993, Volume 2 and Supplement 1.

* There are safety precautions to observe when using inorganic acids to clean glassware. Special dispensations are now required to use chromic acid. Sulfuric

2. HMSO, 'General Principles of Sampling and Accuracy of Results, 1980. Methods for the Examination of Waters and Associated Materials', HMSO, London, 1980, Appendix to Chapter 4, p. 45, ISBN 011 751491 8.
3. R. S. Rapaka, J. Roth, T. J. Goehl and V. K. Prasad, 'Anticoagulants Interfere with Analysis for Furosemide (frusemide) in Plasma', *Clin.Chem.*, 1981, **27**, 1470.
4. N. P. H. Ching *et al.*, 'Gas Chromatographic Quantitation of two Plasticisers Containing Intravenous Fluids Stored in Plastic Containers', *J. Chromatogr. Biomed. Appl.*, 1981, **225**, 225.
5. J. R. Moody, 'The Sampling, Handling and Storage of Materials for Trace Analysis', *Philos. Trans. R. Soc. London*, 1982, **305**, 669.

General Reading

6. A. Mizuike and M. Pinta, 'Contamination in Trace Analysis', *Pure Appl. Chem.*, 1978, **50**, 1519.
7. K. Heydorn and E. Damsgaard, 'Gains or Losses of Ultratrace Elements in Polyethylene Containers', *Talanta*, 1982, **29**, 1019.

2.2 Sample Pretreatment, Homogenisation, Sub-sampling, and Potential Sources of Contaminants

2.2.1 Introduction and Overview

The first stage in any analysis must be to choose a suitable and representative test portion from the sample as received. This may involve a number of stages: pretreating the sample; mixing it; milling, grinding, chopping, or filtering it; rejecting a portion of it to leave a sub-sample of suitable size for analysis. The list of possible operations is vast. It is not surprising that the scope for introducing analytical uncertainty at this stage is also vast. Obtaining a representative portion of the sample is probably the most uncertain stage of most analyses, and it is certainly so for trace analysis.

The first problem the analyst faces is to decide what is meant by 'representative'. The purpose of the analysis must be carefully considered. 'Representative' is usually assumed to mean that the sub-sample is a homogeneous portion of the sample as a whole: *i.e.* if a 10 g portion of a 100 g sample is taken for analysis, then 10% of the total analyte should be contained within that sub-sample.

With a homogenised sample there is no indication as to how the analyte is distributed throughout the bulk. It may be that the sample is a piece of meat with tetraethyl lead concentrated predominantly in the strings of fat rather than the leaner flesh, or a sample of river water where polynuclear aromatic hydrocarbons predominantly lie in the sediment. The result of the assay is quoted as the mean weight of analyte per weight of sample (or per volume of sample) or the mean analyte concentration.

In some cases, however, the distribution of the analyte throughout the sample is critical. One extreme example is that of a mould containing mycotoxins in a consignment of peanuts. The sample could be as large as a ship's container full of nuts. Within this sample there could be one contaminated nut. This single nut, if eaten would cause severe illness. In this case, the analyst is not concerned with the average concentration of mycotoxins over the whole sample (which would be so low as to be unmeasurable) but with detecting if there are any individual nuts which are contaminated.

The case of the single mouldy peanut causes a sub-sampling problem for which there is no satisfactory solution. There are, however, less extreme examples of analyses where an analyst must exercise discretion before sub-sampling. A similar case would be a bunch of bananas sprayed with fungicides: the greatest concentration of analyte will obviously be on the skin. But who eats banana skins? Should the analyst sample the whole fruit or just the edible portion? Answers to this type of question should be agreed with the customer prior to the analysis.

Apart from the problem of representative or suitable sub-sampling, the other major source of uncertainty in the sample preparation stage is the possibility of contamination. Nearly all sample preparation techniques require such close physical contact between the sample and laboratory apparatus (and the analyst) that the risk of introducing extraneous substances into the sample is very high. There is also the possibility of analyte losses at this preparation stage. Section 2.2.6 lists a wide variety of potential contamination sources for many types of analyses.

2.2.2 Sample Pretreatment

Many samples require pretreatment to ensure that the sub-sample is suitably representative or, in some cases, to protect the analyst. This is a critical stage in the analytical process, particularly for foods. Such preparation procedures should have been discussed and agreed with the customer during the commissioning of the work prior to sampling. The type of pretreatment will be specific to the sample and is crucial to ensure that the sample is in the correct form to be analysed.

Solid Foods

Vegetables may need to be trimmed, peeled, and washed free of soil. The outer leaves are often removed. Fruit may or may not be peeled and/or de-stoned. Inedible parts such as bones and possibly skin may have to be removed from fish and meat. Some products may have to be cooked according to an agreed recipe before analysis. Nuts will probably be shelled and possibly skinned before analysis.

Defatting is an example of a commonly used sample pretreatment process applicable where the analyte is insoluble in non-polar solvents. Lipids generally interfere in many analytical processes and so are removed at the start of the analysis, if possible, by washing with a non-polar solvent such as hexane. Removal of fat may also assist in subsequent solvent penetration for extraction of the analyte.

The Codex Alimentarius Commission[1] and EC Directive 90/642/EEC[2] give guidance on the pretreatment of foodstuffs to be used for pesticides residue analysis. A report by the International Union of Pure and Applied Chemistry

Commission on Agrochemicals details some of the analyte losses possible when foods are washed, cooked, canned, or subjected to other preparation processes.[3] Some of these losses range up to 100%.

Liquid Samples

Liquids often contain a sediment or other solid matter in suspension. The presence of suspended material can affect the analyte. The suspended material may adsorb the analyte so it is important to check whether filtration, if used, has a significant effect on the analytical result. An example of this adverse effect of filtration can be seen in the polynuclear aromatic hydrocarbon levels found in river Thames water (Table 2.2.1). Hence, the effect of filtration, or centrifugation, on the analyte(s) present in liquid samples must always be investigated as they may influence the final result.

Table 2.2.1 *Effects of filtration on concentrations of polynuclear aromatic hydrocarbons (PAH) found in River Thames water* [a]

PAH	Concentration (ng l^{-1})	
	Filtered Sample	Unfiltered Sample
Fluoranthene	10.2	540
Benzo[k]fluoranthene	0.7	85
Benzo[b]fluoranthene	0.2	17
Benzo[a]pyrene	1.8	294
Indeno[1,2,3-cd]pyrene	Not detected	Not detected
Benzo[ghi]perylene	2.9	500

[a] D. C. Hunt, L. A. Philip, I. Patel, and N. T. Crosby, 'Determination of Polynuclear Aromatic Hydrocarbons in Food, Water and Smoke using High Performance Liquid Chromatography', *Analyst*, 1981, **106**, 135.

In some cases the analyte may be in suspension rather than in solution, *e.g.* metals in engine oils. Also liquids may settle into layers on standing. Therefore it is important that the sample is adequately mixed before taking the portion for analysis.

Moisture Content

Extraction of an analyte from a complex matrix such as foods is often dependent on the moisture and lipid content of the matrix. Hence, sample pre-treatment may involve drying (to remove excess moisture) or rehydration under controlled conditions of relative humidity to improve solvent penetration. An example of this effect is shown in Table 2.2.2; the addition of water improved the extraction of the pesticide dicrotophos from samples of tea.

Table 2.2.2 *Extraction of dicrotophos from tea, with and without dampening*[a]

Sample	Residues of dicrotophos (mg kg^{-1})	
	Dry Tea	Dampened Tea[b]
1	0.3	8
2	0.6	23
3	0.1	4
4	0.6	19

[a] A. P. Woodbridge and E. H. McKerrel, in 'Trace Organic Sample Handling', ed., E. Reid, Ellis Horwood, Chichester, UK, 1981.
[b] Dampened samples had an equal mass of water added 1–2 h before extraction. Both the dampened tea and the dry tea were then mixed with anhydrous sodium sulfate and tumbled with chloroform.

The addition of water or the use of an aqueous solvent mixture is important for the extraction of other organic analytes from dry foodstuffs or dehydrated foods. It is particularly necessary in aiding the permeation of solvent through freeze dried samples.

Autoclaving and Sterilisation

Many samples, such as body fluids, present a biological hazard to the analyst. These types of samples are normally analysed in toxicology or pathology laboratories, that have appropriate handling procedures and methods for rendering the samples safe. Laboratories that are unused to handling such samples may, however, still be occasionally required to analyse them, *e.g.* in surveys of polychlorinated biphenyl levels in human milk. In these cases, the samples are usually sterilised by a specially contracted agency before being received by the analyst. The usual technique is to autoclave the sample. It is therefore necessary to establish the stability of the analyte(s) to pressure and heat before proceeding with the analysis.

2.2.3 Ensuring Homogeneity of the Bulk Material and Taking a Sub-sample

As described in Section 2.2.1, in some cases it may not be desirable to homogenise the bulk material. However, these cases are the exception to the rule. The majority of sub-sampling procedures require the bulk sample to be homogeneous.

Most analysts are aware of the importance of homogenisation. Unless the portion taken for analysis is truly representative of the bulk material the result of the analysis will be erroneous, however diligently and accurately it is carried out. This applies equally to sampling in the field and sub-sampling within the laboratory, and is particularly important for trace analysis.

There should be no differences between sub-samples taken from the sample received in the laboratory. Any error introduced by sub-sampling becomes more important as the concentration of the analyte diminishes. There may be legislation or recommendations by standards organisations governing the sub-sampling protocol to be used. For example, the Food Chemicals Codex gives some guidelines for sub-sampling methods for certain analyses, and the Codex Alimentarius Commission makes recommendations for the reduction of samples for pesticide residue analysis.[1,4]

The sample submitted to a laboratory for analysis is unlikely to be much larger than 1–2 kg in the case of solids, or 2–5 l for liquids. However, many methods require the analyst to take a portion as small as 1 g, for example in trace element determination involving wet oxidation where higher sample weights would require larger volumes of acids and much longer digestion times. Classical methods for the determination of elements in organic matrices require only *milligrams* of sample for analysis.

Because the size of the sub-sample could be very small, it is usually necessary to assess the homogeneity of the sample prior to sub-sampling and analysis. Some samples, *e.g.* fertiliser blends, are visually heterogeneous, being made up of mixtures of particles of different size, density, and chemical composition. Many foodstuffs are inhomogeneous since they comprise proteins, fat, moisture, salts, fibre, *etc.* in a complex matrix. Muesli is a good example of a heterogeneous dry food, whilst meat and fish are typical heterogeneous moist foods. Solid samples containing fine dust particles must be treated with care since the fines may settle out during transport to the laboratory.

An example of the errors that can be caused by heterogeneous samples is shown in Table 2.2.3. Replicate analyses of tea bag contents for trace metals gave variable results by atomic absorption spectrophotometry. Subsequent work established that the tea bags contained a mixture of tea leaf and dust. The dust contains much higher concentrations of trace elements than the leaf (Table 2.2.4). Hence, the variation shown in Table 2.2.3 was caused by the variable ratio of leaf to dust in the 1 g sample taken for analysis.

Homogeneity cannot be assumed for any solid material. The analyst must always carefully consider the likely distribution of the analyte within the sample.

Solid Samples

If homogeneity is required, all solid samples should be thoroughly mixed before sub-sampling. For dry samples this can be done conveniently using an automatic mixer in which the container is rotated in an eccentric mode for a period of at least 1 h. Where the sample contains particles of different sizes it may help to reduce the inhomogeneity by grinding and sieving (see Section 2.2.4). It may be necessary to retain a portion of the original sample prior to grinding for identification purposes or to check whether the grinding process has brought about any change in moisture content.

Table 2.2.3 *Determination of the lead content of packet teas*[a]

Sub-sample	Pb (mg kg^{-1})	
	Assay 1	Assay 2
A	6.2	1.1
B	1.2	6.6
C	3.9	3.9
D	25	6.5
E	3.9	9.5
F	4.2	2.4
G	<0.2	0.6
H	0.6	0.4
I	<0.2	1.4
J	0.6	0.4

[a] N. D. Michie and E. J. Dixon, 'Distribution of Lead and other Metals in Tea Leaves, Dust and Liquors', *J. Sci. Food Agric.*, 1977, **28,** 215. Reproduced by permission of the Society of Chemical Industry.

Table 2.2.4 *Distribution of metals between leaf and dust fractions of tea*[a]

Sample	Leaves (%)	Dust (%)	Metal content (mg kg^{-1})							
			Lead		Iron		Zinc		Copper	
			Leaves	Dust	Leaves	Dust	Leaves	Dust	Leaves	Dust
A	44	18	0.2	17.2	150	1300	35	105	17	32
B	55	18	5.9	14.6	350	1500	40	140	20	33
C	33	19	0.4	8.5	150	1350	35	100	19	33
D	33	21	0.2	11.7	300	1400	35	125	20	33
E	34	17	8.6	10.0	200	1200	140	120	20	31
F	33	19	0.2	6.0	150	1400	30	100	20	34
H	78	6	0.2	3.7	200	950	35	70	26	33

Leaves: retained by 30 mesh sieve.
Dust: passed by 50 mesh sieve.
[a] N. D. Michie and E. J. Dixon, 'Distribution of Lead and other Metals in Tea Leaves, Dust and Liquors', *J. Sci. Food Agric.,* 1977, **28,** 215. Reproduced by permission of the Society of Chemical Industry.

Foodstuffs (with Appreciable Moisture Content)

Foodstuffs, *e.g.* meat products, pickles, fruit, and vegetables may need to be minced or chopped as a first stage in the homogenisation process since moist products cannot be easily sieved. Alternatively, high-speed food blenders can be used (see Table 2.2.5). Cooling may assist in this process where a high fat content is present or for fat products that are rubbery in texture, as is the case of cheese. Hard foods such as nuts or cheese can be grated or chopped up finely by hand. Further preparation using a pestle and mortar may also be required for solid samples where high-speed blenders cannot be used. The Colworth stomacher (Table 2.2.5) was designed for use in microbiological analysis to avoid contact with metallic parts and to prevent bacterial contamination. Precautions should be taken to avoid cross-contamination between samples. This usually entails dismantling mechanical blenders between samples and carefully washing and drying all parts that come into contact with the sample before reassembling. Depending upon the blender, this can be a laborious process.

Where comminution of the sample may affect the analyte (*e.g.* if livers are comminuted, enzymes are released which may affect analytes), segments of the product should be taken for analysis. The comminution or mixing processes must not be such that the sample separates into solid and liquid phases. Particular care should be taken to avoid separate phases when comminuted samples that have been frozen are subsequently thawed.

Comminution of foodstuffs with an appreciable water content (*e.g.* citrus fruits) can be aided by blending the sample with dry ice. This makes the blended sample much easier to handle, and avoids the problem of forming a separate liquid phase.

Liquids

Fats or viscous liquid samples such as syrups, molasses, or lubricating oils may benefit from warming before mixing and sub-sampling.

Liquids containing sediment, such as ground water samples, cause a particular problem. It is likely that the analyte will be unevenly distributed between the sediment and the liquid. The ultimate purpose of the analysis needs to be considered before deciding how to take a sub-sample. If a homogeneous sample is required, the bulk material should be thoroughly shaken or otherwise mixed before analysis. After shaking, they must be allowed to stand to allow any aeration to clear. Alternatively, if the analyst is interested in the liquid only, the sediment only, or feels that separating the two will not affect the result, the sample can be filtered or centrifuged before sub-sampling. The solid and the liquid can then be analysed separately.

Particulates, Smokes, and Vapours

Gases and dusts are sampled by pulling air through a simple filter or glass fibre disc. In these cases the whole filter or glass fibre disc should be taken for analysis.

Table 2.2.5 *Laboratory mixers and blenders*

Type	Function	Material	Capacity	Comments
Waring	Blender	Glass, stainless steel	1 to 4 l	Multispeed, multifunctional: cuts, shreds, grinds, chops, emulsifies, *etc.*
Silverson	Mixer/emulsifier	Stainless steel Heavy duty stainless steel	Up to 9 l Up to 12 l	For dispersal of solids in liquids, not for dry materials, high shear action
Ultra Turrax	Homogeniser	Stainless steel	0.3 ml to 20 l	High speed maintained irrespective of viscosity, dispersing action
Stomacher	Blender	Plastic (polythene)	5 to 3500 ml	Paddle action, no contact with sample. Little heat generated. Takes 30s
Tissue grinders	Homogeniser	Glass (ground surfaces), also PTFE/borosilicate	2 to 50 ml	Small amounts of tough tissue, can generate heat
Whirlimixer	Vortex mixer	–	Up to 25 mm diam. tubes	Sample container (tube) inserted for rapid, efficient, and controllable, mixing. Variable speed

2.2.4 Sample Particle Size Reduction

Where solid samples are thought to be heterogeneous it is recommended that the particle size is reduced as much as possible so that more representative portions can be taken for analysis. Errors introduced at the sub-sampling stage are inversely proportional to the number of particles in the sub-sample. Hence, the larger the sub-sample and the finer the particle size, the smaller the error introduced at this stage. Samples may also have to be ground and polished to a given specification for techniques such as X-ray fluorescence spectroscopy.

All techniques for reducing particle size by necessity involve close physical contact and abrasion between the sample and the apparatus. The risk of contaminating the sample at this stage is therefore high. Some specific potential contaminants are listed in Section 2.2.6, but the risk of contamination also depends upon the sample type. The harder the sample, the more abrasive the contact and the more material from the apparatus contaminates the sample. Samples such as rocks and ores are notorious for grinding grinders. The harder the sample, therefore, the higher the specifications of the apparatus construction materials.

Hard Materials

Several types of heavy duty laboratory mills and grinders are available for particle size reduction, variously described as jaw crushers, micro grinders, mortar grinders, or pulverisers. These are suitable for the harder samples as exemplified by the following materials: alloys, ores, slag, glass, cement, coal, clinker, ceramics.

Ball and hammer mills with powerful electric motors can generate a large amount of frictional heat (and noise). Hence, it may be necessary to check the moisture content of the sample before and after grinding and to make a suitable adjustment to the analytical result using Equation (1).

$$\text{Moisture factor} \; = \; \frac{100 - \text{moisture content of unground sample}}{100 - \text{moisture content of ground sample}} \qquad (1)$$

Normally, the moisture content of the unground sample is greater than the moisture content of the ground sample owing to loss of moisture caused by the frictional heat during grinding. Hence, the above factor will be less than 1.0. Therefore the result of any determination of an analyte on the ground sample should be multiplied by this factor to obtain the true concentration of the analyte in the original, unground sample. There are potential hazards in using this correction system. Sometimes a surface film can form on the unground material preventing drying and giving a falsely low value for its moisture content. Also, drying under heat can cause oxidation of sample material, increasing its weight and therefore decreasing its apparent moisture loss.

Soft Materials

Table 2.2.5 lists the different types of laboratory mixers and blenders most frequently used to comminute and blend soft, solid samples. More than one model of each type is generally available, the main difference between them being their capacity. Generally, such equipment is used at high speed for a short time so that little heat is generated.

In the extreme case where the heat produced is sufficient to cause degradation of the analyte, the sample should be ground with solid carbon dioxide, or frozen in liquid nitrogen before grinding. Grinding should be carried out in small bursts and not continuously to allow some cooling to occur.

Elastic materials such as rubber or plastic are best embrittled by plunging into liquid nitrogen before grinding. However, some plastics like polyurethane are still flexible at liquid nitrogen temperature. In this case manual cutting may be necessary. Foods are more easily ground if frozen. Leather and fibrous materials should be comminuted using a cutting mill.

Samples of Small Size

For smaller quantities of sample, a simple coffee bean mill may well be adequate. Only short pulses of grinding are normally required and little heat is produced. This is useful for dry foods, such as grain and nuts. Care should be taken to avoid the loss of any dust formed during the grinding. For very small quantities of sample, a pestle and mortar should be used. Some methods of analysis specify the use of a Waring blender to produce a slurry of the sample with the extraction solvent. This technique yields a material very finely comminuted and may allow smaller quantities of sample to be taken for analysis.

To avoid sub-sampling errors a large sample can be extracted and an aliquot portion of the extract taken for subsequent analysis.

Particle Size

The sample particle size may also affect the choice of which process to employ. Small particles would be lost in a large mill. Pestle and mortars are most suitable for matrices where the particles are already small.

2.2.5 Sample Reduction (Division) and Particle Size Separation (Sieving)

Even samples received in the laboratory may be too large and too heterogeneous simply to remove a small portion for analysis. The composite sample after grinding and sieving should then be systematically reduced in size preferably using an automatic sample divider.

Automatic Separators

Automatic separators give a greater consistency and reproducibility than reducing samples by hand. There are two major types which dominate the market: the riffler and the sample divider. Both are gravity fed and are designed primarily for powdered or granular materials. They are made of stainless steel and conform to BS 812 for aggregates, sands, and fillers or BS 1377 for soils.

A larger version made of tinned steel is available for sub-sampling coal. All can be used repeatedly until a suitably sized sub-sample is obtained. Rotary dividers are also available with a variable rotation speed and a vibratory feeder. Eight sub-samples are produced in either 250 or 500 ml containers.

Manual Separation

Alternatively, but less satisfactorily, the composite sample can be reduced manually by coning and quartering. The degree of homogeneity given by coning and quartering is compared with that by an automatic riffler in Table 2.2.6; it is apparent that manual separation gives a less homogeneous sample. To cone and quarter, the sample is arranged in a flat heap on a clean surface. Opposite quarters of the heap are then rejected, the remaining two are saved, mixed, and formed into a second heap. The operation is then repeated until a suitable amount remains.

Table 2.2.6 *Homogeneity of copper in fortified pig feed using different sample division methods*

Sampling method	Mean copper content ($\mu g\ g^{-1}$)	Standard deviation	Coefficient of variation (%)
Hand scoop	205.4	48.0	23.4
Riffle chutes	273.5	60.0	20.5
Coning and quartering	232.1	75.4	32.5
Rotating sample divider	244.8	39.6	16.2

The nominal copper concentration was 250 $\mu g\ g^{-1}$.
A total of eight sub-samples were taken from a separate 500 g portion of the fortified feed using each of the sampling methods.
The copper content was determined using ICP–optical emission spectroscopy (OES).

Where bulk samples of large units are received, the sample size may be reduced by combining opposite cut segments or suitable aliquots from each unit.

Particle Size Separation (Sieving)

It is often important that the particles of a sub-sample intended for analysis conform to certain size criteria. This is usually to maximise the surface area of the sub-sample and so increase the contact with the extraction solvent. Some statutory and recommended methods specify the particle size required, *e.g.* the British Standard method for metals in paints [EN 71, BS 5665 Pt 3 (Safety of Toys), 1989].

Sieves of various mesh sizes conforming to British Standards (also ASTM) are available covering the range 1–16 mm and 38–850 μm. They are usually 200 mm in diameter although other sizes can be obtained. They are made from stainless steel woven wire mesh fixed in a brass frame.

Such sieves can be used in a vertical stack and shaken or tapped manually. However, for more consistent results, power shakers are available which impart both a circular motion and a vertical vibration. The consistency of this motion is very important.

Solids are normally sieved in the dry state although wet sieving is practiced in some cases.

2.2.6 Sources of Contamination and Uncertainty

Processes such as grinding, milling, blending, and chopping necessitate a high degree of physical contact between the sample and the laboratory apparatus. Most sample preparation techniques also involve the analyst handling the sample. Therefore, there is a high risk of contamination of the sample at this stage. There is also the possibility of analyte losses as samples are subjected to harsh physical conditions.

It is impossible to give a comprehensive list of all potential interferences, but some of the more common are included in the following sections.

Contamination with the Same Substance as that being Determined (Enhancement)

This is primarily a problem in trace elemental analysis, where the level of analyte in the laboratory environment is significant compared with the level in the sample. The construction materials of any apparatus used must be carefully considered to ensure that they are not going to add to the concentration of the analyte in the sample.

Enhancement from contamination is rarely a problem in trace organic analysis, as it is unusual for most analytes to be present within the laboratory environment other than as carry-over from previous analyses.

Possible elemental interferents from commercial mills and grinders are listed in Table 2.2.7. This gives an indication of which construction materials to use for specific applications. The contamination from metals in grinders and mills is a major problem in trace element analysis of abrasive samples such as rocks and soils. The problem is sometimes underestimated, as it is not highlighted by performance assessment schemes and 'round-robin' trials where samples for analysis are always pre-ground, before distribution, to ensure homogeneity.

Examples of contaminants from other sources are given in Table 2.2.8.

Adsorption of the Analyte

Again, this problem is primarily of concern to the trace elemental analyst. The problem of adsorption onto glassware has already been discussed in Section 2.1, but substances can also adsorb onto drying agents, metal homogenisers and virtually anything else which makes contact with the sample. Polar organic

Table 2.2.7 *Materials and associated contaminants for a number of grinding methods*

Grinding material	Major elements	Minor elements	Measure of hardness (Moh)
Plastics[a]	C	–	1.5–2.0
Stainless steel	Fe,Cr	Ni,Mn[b],S[b]	5.0–5.5
Hardened steel	Fe	Cr,Si,Mn,C	5.5–6.0
Agate[c]	Si	Al,Na,Fe,K,Ca,Mg[d]	7.0–8.0
Fused zirconia	Zr	Si,Hf	7.0–8.0
Alumina ceramic	Al	Si,Ca,Mg	8.0–9.0
Silicon ceramic	Si,C	–	8.0–9.0
Tungsten carbide[e]	W,C,Co	Ta,Ti,Nb	8.5–9.0
Boron carbide	B,C	–	9.0–9.5

[a] Polystyrene and methacrylate.
[b] Possibly present.
[c] Rods or pestle and mortar.
[d] Usually less than 0.02% by weight.
[e] Tema mill.

Table 2.2.8 *Sources of elemental contaminants other than grinders*

Interference	Examples of sources
Aluminium	Dust, glass containers, low-pressure polyethylene, and PVC apparatus,
Antimony	Silica apparatus
Beryllium	Laboratory tissue paper
Bromide	Polycarbonate apparatus
Cadmium	Plastic stoppers
Calcium	Polyethylene, polypropylene, and PVC apparatus, dust
Chloride	Dust, polycarbonate apparatus
Chromium	Steel medical implements, *e.g.* needles, steel spatulas, and other apparatus, analyst's cosmetics
Cobalt	Steel medical implements, *e.g.* needles, steel spatulas, and other apparatus, anticoagulants for blood
Copper	Polyethylene containers, plastic stoppers, anticoagulants for blood, steel implements, *e.g.* needles
Iron	Polyethylene containers, any steel apparatus, dust
Lead	Airborne exhaust fumes, glassware
Magnesium	Glassware
Manganese	Polyethylene containers, steel medical implements, *e.g.* needles, steel spatulas, and other apparatus, glassware, plastic stoppers
Nickel	Steel medical implements, *e.g.* needles, steel spatulas, and other apparatus
Potassium	Teflon apparatus
Silicon	Dust
Sodium	Analyst's skin, dust, polyethylene, polypropylene, PVC, and Teflon apparatus
Strontium	Glassware
Titanium	Low-pressure polyethylene apparatus
Zinc	Dust, rubber apparatus, *e.g.* stoppers, anticoagulants used for blood, polyethylene and polypropylene apparatus, plastic stoppers

analytes, such as some drugs, can also be lost by adsorption. Some examples are listed in Table 2.2.9.

Table 2.2.9 *Examples of losses by analyte adsorption*

Adsorbed Analyte	Adsorbent
Any ionised element	Glass
(To minimise risk, acidify solutions and silanise glass: see Section 2.1)	
The drug propoxyphene	Glass
Phenoxyacetic acid herbicides	Glass
The drug oxprenolol	Calcium chloride (drying agent)
Chloroform	Silicone septa of GC headspace vials
Mercury	Polyethylene

Contamination with a Substance Other Than the Analyte which Interferes with the Analysis

Exogenous contaminants are the curse of trace organic analyses. The variety of compounds present within a laboratory which could potentially interfere with an analysis is virtually limitless. Whether a contaminant does interfere is dependent upon the analyte(s) sought and the determination method, as well as the degree of clean-up before determination. Interfering compounds can lead to chromatographic co-elution and even misidentification. Some common sources of such contaminants are listed in Table 2.2.10.

Contamination with a Substance Other Than the Analyte which Induces Chemical Change

There are many substances which, on contaminating a sample, alter the nature of the analyte. In the inorganic field, this includes complexing agents and competing ions. In organic analysis, it encompasses oxidising and hydrolysing agents, acids, and bases. For the analysis of metals, where the oxidation state is of interest, the presence of any oxidising or reducing agents must be avoided. Some examples are listed in Table 2.2.11.

Chemical Instability due to Elevated Temperatures

Many sample preparation techniques generate heat; usually by friction. This must be kept to a minimum if the analyte (usually organic compounds) is thermally labile.

Table 2.2.10 *Some sources of exogenous contaminants in trace analyses*

Sources of contaminants	Critical analyses
Sulfur compounds in gloves and rubber ware	GC / flame photometric detection
Phthalates in all plastic apparatus	GC / electron capture detection
Cosmetics, hand creams	All organic analyses
Polishes, cleaning aerosols	All organic analyses
Phosphate detergents	GC / phosphorus detection
Greases used on apparatus	All organic analyses
Tributyl aconitite in plastic gas lines	GC analyses
Impure solvents and reagents	All organic analyses
Exhalation from heavy smokers	Polycyclic aromatic hydrocarbons

Table 2.2.11 *Some sources of reactive exogenous contaminants in trace analyses*

Sources of contaminants	Critical analyses
Peroxide in diethyl ether	Oxidisable analytes, *e.g.* methadone
Acids in chlorinated solvents	Esters and amides hydrolysed
Aluminium in acidic solutions in metal or glass	Fluoride ion selective electrode

2.2.7 References

Applications References

1. FAO/WHO Codex Alimentarius Commission, 'Guide to Codex Recommendations Concerning Pesticide Residues. Volume 5 – Recommended Method of Sampling for the Determination of Pesticide Residues', Rome, 1989.
2. European Community (1990) Council Directive 90/642/EEC, *Off. J. Eur. Communities*, 1990, L350, 71–79.
3. P. T. Holland, D. Hamilton, B. Ohlin, and M. W. Skidmore, 'Effect of Storage and Processing on Pesticide Residues in Plant Products', *Pure Appl. Chem.*, 1994, **66**, 335.
4. Committee on Codex Specifications, 'Food Chemicals Codex', National Academy Press, Washington, DC, edn 3, 1981, ISBN 03 09 03 09 00.

General Reading

5. K. Heydorn and E. Damsgaard, 'Gains or Losses of Ultratrace Elements in Polyethylene Containers', *Talanta*, 1982, **29**, 1019.
6. J. Versieck, F. Barbier, R. Cornelis and J. Hoste, 'Sample Contamination as Source of Error in Trace Element Analysis of Biological Samples', *Talanta*, 1982, **29**, 973.
7. P. S. B. Minty, 'Drug Analysis in a Department of Forensic Medicine and Toxicology',

in 'Analytical Methods in Human Toxicology', ed. A. S. Curry, Pt II, Macmillan Press, Basingstoke, UK, 1986.

8. C. Veillon, 'Trace Element Analysis of Biological Samples – Problems and Precautions', *Anal. Chem.*, 1986, **58**, 851A.

General References

9. A. Mizuike and M. Pinta, 'Contamination in Trace Analysis', *Pure Appl. Chem.*, 1978, **50**, 1519.

10. H. T. Delves, 'Sample Preparation and Handling,' *Food Chem.*, 1992, **43**, 277.

11. J. R. Moody, 'The Sampling, Handling and Storage of Materials for Trace Analysis', *Philos. Trans. R. Soc. London*, 1982, **305**, 669.

12. W. E. Van der Linden, 'Definition and Classification of Interferences in Analytical Procedures', *Pure Appl. Chem.*, 1989, **61**, 91.

13. L. Kosta, 'Contamination as a Limiting Parameter in Trace Analysis', *Talanta*, 1982, **29**, 985.

14. J. Denton, 'The Increasing Importance of Proper Washing in the Laboratory', *Lab. Prod. Technol.*, October 1992, 14.

CHAPTER 3

Inorganic Analytes: Sample Preparation

3.1 Introduction

3.1.1 Overview

The role of sample decomposition and preconcentration in elemental trace analysis is crucial to obtaining a reliable result. The most common approach to elemental analysis is to achieve complete sample dissolution so that the analytes of interest are present in a clear solution. These solutions are measured by instrumental methods for which detection limits are becoming lower and lower. In order to make best use of the power of the analytical end-technique, the analyst should ensure that the decomposed and dissolved sample is free from contamination and that there are no losses of analyte.

It is possible, by using techniques such as X-ray diffraction and scanning electron microscopy, to determine the elemental composition of a sample by direct measurement. However, these methods are not usually suitable for the low levels of trace analysis. Trace analysis of solids can be performed with such techniques as X-ray fluorescence, laser ablation inductively coupled plasma mass spectrometry (LA ICP-MS), some electrothermal atomic absorption spectroscopy (ETAAS), and arc/spark mass spectrometry. The relative usefulness of such techniques is discussed elsewhere. This module deals with sample preparation procedures when they are required.

Time and labour considerations are important in any analysis. Throughout elemental analysis, the analyst will need to balance the time spent in sample clean-up against that taken by the end-determination. The view of most analysts is that unless there is a specific short-cut route available, for example hydride generation or the analysis of slurries (liquids containing solid particles, often silicate), it is best to ensure that solutions are clean (free of contamination and interferents) and within a suitable concentration range.

Hydride generation will involve the use of specific apparatus and be exclusive to a few elements and compounds. Any analyte can be tackled in a slurry and it can be regarded as a very convenient short cut. Slurries can be analysed by conventional spectroscopic techniques [ICP atomic emission spectroscopy

(ICP-AES), ICP-MS, ETAAS, and direct current plasma atomic emission spectroscopy (DCP-AES)] in much the same way as pure solutions. It is up to the analyst whether partial or total decomposition is necessary.

An area of metal analysis where the normal rules of decomposition and dissolution do not apply is speciation. The speciation of an inorganic element can be more important than its total content, for example the toxicities of inorganic and alkylated mercury or arsenic are quite different. Where such is the case, the integrity of the analyte should be preserved. This will preclude the use of the decomposition and dissolution steps outlined in this chapter. Instead, species such as methylmercury should be separated as organic compounds undergoing organic preparation, for example by HPLC using the final elemental analysis such as ICP-MS or ICP-AES as the detector.

This chapter concentrates on the use of acids and other destruction/decomposition media that are used to prepare the sample in the form of a solution or slurry such that it can be introduced into a spectrometer or other detection system. Where solids can be analysed the analyst can usually go straight to the determination stage.

The stages involved in preparing samples for inorganic analysis have been split into three sections. Section 3.2 deals with the options and problems associated with the destruction of organic matrices. The main variables are the technique and apparatus used, rather than the reagents (mainly acids). For this reason, the section concentrates on the techniques and equipment available, and how best to use them. Section 3.3 details the decomposition and dissolution of inorganic matrices. These could either be inorganic samples or the inorganic residue resulting from the decomposition of essentially organic matrices. The crucial aspect of the dissolution of inorganic matrices is the choice of reagent (again, usually acids), and so this section concentrates on the reagents available and compares their benefits and problems. Section 3.4 deals with the subject of preconcentration, *i.e.* obtaining the analyte extract in a form and concentration suitable for the final determination. The choice of techniques available and how they should be used is covered.

3.1.2 Where Errors Occur

Harsh conditions are often required during total decomposition to destroy matrices, and these increase the chances of analytical error caused by both losses and contamination. Contamination can occur from the environment, apparatus, and any chemicals used in the process. Losses can occur through volatilisation, adsorption, coprecipitation, and stable compound formation. For these reasons, it is important to consider ways of improving the analytical reliability of each stage of sample decomposition and dissolution.

Contamination

Contamination will always occur. The critical factor is whether or not a particular contaminant is significant. This is dependent upon the purpose of the analysis and the nature of the contaminant. Some examples of common sources of contaminants in sample preparation for elemental analysis can be found in Section 2.2.

The most common source of contamination is the vessels in which the sample preparation is carried out. This could be due to some form of residue from a previous sample, or because the container material itself is a contaminant. Various ways of overcoming such problems include rigorous cleaning procedures, use of alternative vessels or vessels of different composition, and the reservation of sets of apparatus for particular analyses. All these approaches are recommended.

Fall-out contamination from the laboratory environment can also take place during the course of analysis. This can be counteracted in various ways, the choice depending on the nature and extent of contamination and the quantity of analyte being determined. These include keeping the preparation covered, operating in a closed system, using laminar flow cabinets, wearing special clothing, and working in a high-class clean room provided with ultra-filtered air under positive pressure.

The reagents necessary for the dissolution process, usually acids, may be a source of contamination. It is necessary to select chemicals with appropriate specifications. For trace analysis the lowest level required will be Analytical Reagent Grade, and for many purposes a higher grade, such as HPLC or Spectroscopy Grades, must be used. In some instances, an even higher purity reagent may be required, and the analyst will need to consider if it should be purchased or specially prepared in the laboratory. It must also be remembered that laboratory water is an important reagent and the analyst must take steps to ensure that the water supply is of the desired quality.

For all contamination problems, the analysis of several blanks is an important precaution. There is always likely to be some low level contamination inherent in the laboratory/apparatus/reagent combination, and blank analyses will yield the information needed to correct for this. The analyses of blanks will not, of course, give any information on systematic errors. These are usually only detectable by analysing materials of known composition.

Adsorption

Adsorption of an analyte onto the surface of a container or filter is a problem commonly encountered in trace analysis. Sometimes the immediate pre-treatment of a glass surface can activate it to cause adsorption of analytes, but sometimes the phenomenon depends on the earlier history of the vessel. Certain elements, including arsenic, mercury and particularly silver, are prone to adsorption (or exchange) onto glass surfaces. Generally, silica surfaces are less likely to adsorb analytes. Losses of analyte can occur with other materials, for example iron may be lost from samples heated, and especially ignited, in platinum ware. The iron appears to enter the platinum which then needs careful but aggressive cleaning to ensure the iron is not released to contaminate subsequent samples.

Adsorption is an example of a systematic error and, as previously stated, can only be detected by analysing samples of known composition.

Volatilisation

Analyte volatility is another source of loss of analyte. This will mainly occur during dry-ashing.

3.1.3 Precautions Against Errors

In order to reduce errors, a few guidelines can be followed.

Keep Procedures as Simple as Possible

Minimise the number of handling operations between sample preparation and final determination. For example, in the cold vapour determination of mercury, or the determination of arsenic or selenium by hydride generation, the analytes can be volatilised directly into an interface with the spectrometer. (Where the sample contains stable complexes such as arsenobetaine or selenomethionine, an extra digestion step, such as photolysis, will be required to break up the complex).

Keep the Apparatus Used to a Minimum

Minimise the number of vessels used per determination. Try to work in one vessel if possible. Ensure that the vessel is no larger than is necessary, in order to minimise the surface in contact with the sample or its solution. Where possible, use high density polyethylene, polypropylene, poly(tetrafluoroethylene) (PTFE), or perfluoroalkoxy vessels. Low density (soft) polyethylene ware should be avoided as it causes problems in trace analysis.

Work in a Clean Environment

Always perform the work in as clean an environment as practicable. Work in fume cupboards, and preferably laminar flow cabinets, or even a 'clean room' if available. Exclude rubber and metal equipment, and use dedicated apparatus and closed systems whenever possible.

Minimise the Quantities of Reagents Used

Reducing the amount of reagents used might necessitate changing the technique employed. For example, in the wet digestion of organic matter, the use of a closed system will permit operation at an elevated temperature, this compensates for smaller volumes of the digesting acids.

Use Certified Reference Materials

Wherever possible, check the analytical procedure using certified reference materials and, if necessary, use them to evaluate the steps of the analysis.

Check the Robustness of the Procedure

The dependence of the results on temperature, duration of the process, and quantities of sample or reagents should be studied.

3.1.4 Choice of Acids

The acid must be compatible with the end determination. For example, oxidising acids cannot be used with a voltammetric determination. Sulfuric acid cannot be used with ICP-MS employing nickel cones.

The main acids used for elemental dissolution are hydrochloric and nitric, and their mixture known as 'aqua regia'. Other acids used less frequently and for specific purposes include sulfuric, perchloric, hydrofluoric, and orthophosphoric acids. Acids can be used either in their concentrated form or diluted. 'Concentrated' acids can range from about 30% solutions (hydrochloric acid) to almost 100% purity (sulfuric acid). For dissolution, dilute acids usually refer to molarities from 0.1 to 2 M.

Nitric Acid

This is normally obtained as the 70% azeotrope with water. It is commonly referred to as 'concentrated' but in fact nitric acid is available as 95% and higher concentrations (which must be used with great care). It is employed for its oxidising as well as its acidic properties, and because its salts are almost invariably soluble, is used for almost all matrices. It can form insoluble oxides of Al, Nb, Ta, Ti, Sn, Sb, and W.

Hydrochloric Acid

This is usually supplied in Britain as a 36% solution in water, known as the concentrated acid, but other concentrations (32% and 39%) are available. A reference to concentrated HCl solution in a publication from outside the UK could mean any one of these alternatives. It is non-oxidising; in fact it is reducing towards some higher oxidation states of metals, e.g., Ce(IV), Te(VI), Mn(IV), Mn(VII). It cannot be used for the destruction of organic matter but is widely employed for the attack of inorganic matrices. Chlorides of the metals are soluble except for Ag, Hg, Tl, and Pb. The main problem with hydrochloric acid is that it can form volatile chlorides such as those of Hg, Ge, As, Sb, and Se.

Aqua Regia

This is a 3:1 mixture by volume of concentrated hydrochloric and nitric acids. It has much more powerful oxidising and complexing properties than either constituent acid, due to the formation of chlorine and nitrosyl chloride (NOCl). It is generally used for the decomposition of difficult matrices, other than silicates, and for noble and other electronegative metals.

Sulfuric Acid

This acid is frequently used for the destruction of organic matrices, because it combines acidic, oxidative, and dehydrating properties. It is seldom used for inorganic matrices (though it is useful for some tasks, such as the dissolution of titanium dioxide) but is used for dissolving certain individual elements, as shown in Section 3.3, Table 3.3.3. The sulfates of Ca, Sr, Ba, and Pb are insoluble in water but soluble in sulfuric acid. It should be noted that the concentrated acid in the UK has traditionally referred to the acid of not less than 98% whereas in other countries the concentrated acid is often 95 or 96%.

Perchloric Acid

This is usually supplied either as the 60% solution or as the 72% azeotrope in water. It is a powerful acid and metal perchlorates (except that of potassium) are very soluble. It has little if any ability to form complexes. It is used in analysis mainly for the dissolution of steel samples and for the destruction of organic matrices. It has no oxidising properties in the cold but, when hot, becomes one of the most powerful laboratory oxidants. For this reason, if it is mixed with a significant amount of organic matter, especially easily oxidisable matter, in the cold it can appear safe, but on heating can cause explosive oxidations. It must therefore be used only for well-established digestions or to remove the last traces of organic matter. It is good practice to remove most of the organic matter with nitric acid prior to cooling down the solution ready for treatment with perchloric acid. Hydrogen peroxide can often be used as a substitute for perchloric acid.

Hydrofluoric Acid

This is supplied for analytical use as 40 or 48% aqueous solutions. It must be handled with care, and gloves, entirely free from any pinholes, must always be worn. It causes burns that are very painful and slow to heal and it passes through skin to attack bone. It is a weak, non-oxidising acid that forms volatile compounds with Si, Ge, Sn, Ti, Zr, and As. The fluorides of Mg, Ca, Sr, Ba, Al, Pb, and Cr(III) are insoluble. It is most generally used to destroy silicate matrices, from which SiF_4 and excess HF can be expelled by fuming with sulfuric or perchloric acids. Fuming with perchloric acid has the advantage of converting any insoluble fluorides into soluble perchlorates. After using HF it is good practice to add boric acid to combine with any residual traces of fluoride.

Orthophosphoric Acid

This is a weak, non-oxidising acid supplied in Britain as a syrupy solution containing 85 to 88% H_3PO_4. It forms many insoluble phosphates and so is not generally used for decomposition. However, due to its complexing properties, it is useful for dissolving chromites, ferrites, uranium oxides, and the phosphate ores of the lanthanide elements. Note that phosphoric acid damages nickel cones.

Use of Acid Mixtures

It is rare for only one acid to be used in the digestion of a matrix. Small amounts of other acids such as HF or $HClO_4$ can be employed to fulfil a particular function.

The use of HF in the dissolution of silicates has been mentioned, as has the associated addition of boric acid to complex residual fluoride. Niobium and tantalum ores may also be decomposed with HF, which forms complex fluoro ions of the metals. Certain niobium and tantalum ores can be decomposed with HF and HCl mixtures under slightly elevated pressure. A mixture of HF, HCl, and H_3PO_4 will also decompose niobium ores. The resulting solution is treated with tartaric acid to hold the niobium in solution.

3.1.5 Handling Acids Safely

The work covered in this chapter is potentially dangerous. The use of both highly corrosive and explosive acids means that every possible safety precaution should be taken. This will sometimes be in conflict with the need for a fast and efficient decomposition. For example, the rate of any dissolution or decomposition process is very dependent on the physical form of the sample. Before analysis, the sample should be ground to a small particle size to assist digestion. However, it must be borne in mind that a finely divided sample could react vigorously or even violently, so that it is prudent for the initial treatment of the sample to be with small volumes of the diluted acid. By the same token the analyst should only use a strong oxidising acid like perchloric acid at a stage when a small residual amount of organic matter remains that may be difficult to remove by other methods.

Acids must be used with caution. Hydrofluoric acid is particularly hazardous and should be used only when absolutely necessary, for example when silicates have to be removed from a sample residue. For obvious reasons do not use glass or other silica-based materials for carrying out HF digestions. When HF is used it is a legal requirement that a calcium gluconate gel is readily accessible. Particular care must also be exercised when using perchloric acid.[1] It is an extremely powerful oxidant when hot, and can cause explosions with large amounts of oxidisable matter, such as organic matrices, and in the presence of powerful dehydrating agents. Perchloric acid is used mixed with nitric acid, or with nitric and sulfuric acids, to decompose a range of biological matrices, especially plant materials. Except under established conditions, it is normally added towards the end of a digestion, to achieve final decomposition after the bulk of the organic matter has been destroyed. Perchloric acid is a very useful reagent but it must not be used without first being familiar with its hazards and methods of handling. It should be removed from the final solution by successive gentle evaporation (to near dryness) and dilution. It must *never* be taken to dryness directly.

Safety gloves, safety glasses, and laboratory overalls must always be worn when handling acids.

Mineral acids have high heats of dilution and should be diluted by adding to water, slowly and with constant mixing. When other reagents are to be mixed with acids for use in sample dissolution, the mixture should be prepared as required and only in the quantities needed. It is not safe to store the mixed ingredients.

In closed systems, elevated pressures can cause vessels to rupture. This is best counteracted by the use of either a pressure release valve, as is often employed in acid pressure decomposition, or by monitoring and controlling the pressure, as done in many microwave digestions. There are, however, certain drawbacks to the use of these precautions. Pressure release systems permit the loss of volatiles and pressure monitoring may result in incomplete digestion.

When performing Schöniger oxygen flask combustions, the operator must always be protected from possible violent reaction by working behind a safety screen.

An alternative and safer oxidising agent for many purposes is hydrogen peroxide. It has the additional advantages of being highly pure and leaving no residue. It can, however, also react explosively if added to large amounts of easily oxidisable organic matter.

3.1.6 Reference

1. Analytical Methods Committee, 'Notes on Perchloric Acid and its Handling in Analytical Work', *Analyst*, 1959, **84**, 214.

3.2 The Destruction of Organic Matrices

3.2.1 Overview

The analysis of inorganic analytes in an organic matrix usually requires the destruction of the entire matrix. This is in contrast to inorganic matrices, where there may be alternatives to total dissolution of the matrix prior to the analysis. One exception to this is the analysis of metals in organic liquids such as oils. For these samples it is generally more cost effective to 'thin the matrix' to enable an easier aspiration. This would normally be done by diluting the matrix with a solvent such as toluene (methylbenzene).

Where destruction of the matrix is required, this usually involves complete oxidation of the matrix. The oxidant employed may be gaseous oxygen, in combustion or 'dry oxidation'. The alternative, 'wet oxidation', uses an oxidising acid or mixture of oxidising acids, sometimes in combination with further powerful oxidants such as per-acids or their salts or hydrogen peroxide. The most common acids used are sulfuric and nitric.

The choice of technique depends upon three factors:

(i)   the sample matrix;
(ii)   the selected analyte;
(iii)   the end determination method.

During destruction of the matrix, some elements may be lost by volatilisation. More than one method of destruction might therefore be required to determine several elements in a particular matrix. Table 3.2.1 summaries the available methods of destruction.

Table 3.2.1 *Destruction techniques and apparatus used*

Technique	Apparatus type
Dry ashing	
Open vessel	Crucible in muffle furnace
Closed vessel	O_2 bomb
	Schöniger O_2 flask
Low temperature	Flow of activated O_2
Wet digestion	
Open vessel	Kjeldahl/other long-necked flask
	With reflux condenser
	With ultrasonic agitation
Acid pressure decomposition	PTFE-lined bomb
Microwave heated	Closed vessel
	Closed vessel with valve
	Semi-closed vessel

The different techniques will be suitable for different analyses depending on the matrix and the analytes of interest. Generally, the techniques that risk losses due to volatilisation, such as dry-ashing, will be suitable only for non-volatile elements and those present at relatively high concentration (*e.g.* 100 mg kg^{-1}) such as the nutritional elements in foodstuffs. The other forms of ashing such as low-temperature ashing and closed vessel wet digestion are suitable for the more volatile elements and those at lower concentrations (*i.e.* less than 1 mg kg^{-1}).

A comparison of the general features of dry-ashing and wet oxidation is presented in Table 3.2.2. These factors will affect the choice of technique to employ for the selected analytes. In practice, dry-ashing is used for the higher

Table 3.2.2 *Comparison of dry-ashing and wet digestion*

Dry ashing	Wet digestion
Does not need constant monitoring	Needs careful monitoring
Relatively slow	Relatively rapid
Needs high temperatures (*e.g.* >400°C)	Temperatures lower than dry-ashing
– therefore more volatilisation	– therefore less volatilisation
Safer than wet digestion	Potentially very hazardous
Small reagent blank[a]	Reagent blank larger than dry-ashing

[a] If a 10% MgO or Mg(NO$_3$)$_2$ slurry is required to reduce volatilisation (quite common) then the value of the reagent blank is likely to increase.

concentration, non-volatile elements, and wet oxidation is preferred when there is any doubt as to the likely behaviour of an analyte during the destruction process. Indeed many laboratories now favour wet oxidation methods over dry-ashing. Low-temperature ashing is used only occasionally.

As well as the analyte, the other key variable in selecting the destruction technique is the sample matrix. Some generalisations can be made relating the sample type to the technique. For food matrices, where the non-nutritional analytes of interest are likely to be at low concentration, the organic content is usually high. Low concentrations of analyte require a technique that can be applied to a quantity of sample that will provide sufficient analyte for the end-determination. This consideration will usually exclude techniques that can accommodate only small amounts of sample, such as the closed techniques of bomb or microwave digestion. In contrast, the analyte levels in environmental samples are generally higher and the organic content is somewhat lower (sediments, soils, *etc.*). The difference in composition is due to inorganic constituents which are not likely to be decomposed under the conditions used to destroy the organic matter. For these reasons, environmental samples are often more amenable to closed vessel techniques. These considerations are briefly summarised in Table 3.2.3.

Table 3.2.3 *Applications of closed and open vessel techniques*

Closed vessel methods	Open vessel methods
Volatile analytes	Non-volatile analytes
Higher concentrations	Lower concentrations
Low organic content	High organic content

The initial decision on which technique to use will be have to be based on the nature of the sample matrix and the nature and concentration of the analyte. Where more than one analyte is sought, it may be necessary to use different methods of destruction according to the properties of the selected analyte(s).

3.2.2 Comparison of Techniques

This section sets out the options for the analyst when presented with a sample. All the techniques mentioned are set out in tabular form, showing the benefits and problems associated with each. Those most widely used are currently ashing in a furnace and wet digestion in a Kjeldahl flask, although microwave digestion is gaining in popularity.

Dry-ashing in an Open System

The usual dry-ashing conditions involve placing a sample in an inert vessel and destroying the organic matter by combustion in a muffle furnace. Vessels commonly used include silica, porcelain, Pyrex glass, and platinum. The size and shape of the vessel are important; it is recommended that the vessel is uncovered and that it is high-walled compared with the sample depth.

Table 3.2.4 *Dry-ashing in an open system*

Benefits	Problems
Easy to use	Losses due to retention
Low level of supervision	Incomplete combustion of some matrices
	Pre-treatment of oils and fats needed
	May need to add acid or other ashing aid
	Limited to organic matrices
	Loss of volatiles

Table 3.2.4 lists some of the advantages and disadvantages of using an open system. Loss of elements during ashing is the most important problem associated with the technique. Because of their lack of volatility and presence in higher concentrations than many other trace elements, the nutritional elements (*e.g.* Na, K, Ca, Mg, Cu, Fe, P, Mn) can be analysed in this way; it is unsuitable for other trace and ultra-trace analytes. Elements lost from a selection of samples are given in Table 3.2.5. It must be concluded that for ultra-trace analysis, ashing is *unsuitable* for many analytes due to volatilisation.

Table 3.2.5 *Elements lost during dry-ashing*

Ag	As Au	Be	Cd	Cs	Ge	Hg	IrLi
Ni	Pb Pd	Pt	Rb	Sb	Se	Sn	Zn

Underlined elements will volatilise under 500°C.

The elements that are underlined in Table 3.2.5 will volatilise under 500°C. For most elements the rate and extent of volatilisation rises rapidly at and above the melting point. The 'Rubber Handbook' is a useful source of the physical properties and characteristics of elements and their compounds.[1] The formation of chlorides is a particular problem because most elements have volatile chlorides. Nitric acid may be used as an alternative to hydrochloric acid to produce involatile nitrates. More information on volatile species is given in Section 3.2.3.

It has often been claimed that the presence of halides causes losses of certain elements during ashing. Gorsuch investigated losses of antimony, chromium(VI), iron(III), lead, and zinc using radiochemical procedures.[23] Samples were heated to 600°C for 16 h in the presence of ammonium and sodium chlorides. Severe losses of antimony, lead, and zinc occurred in the presence of ammonium chloride but not sodium chloride. Further work showed serious losses of zinc in the presence of magnesium and calcium chlorides, but not in the presence of barium chloride. Losses of chromium and iron were insignificant.

It should also be mentioned that trace elements may be lost by reaction with the materials of the vessels used. It has been found that there are serious losses of some elements when new, glazed silica crucibles are used; this does not happen with old crucibles that have lost their glaze.

A metal salt (usually of Na, K, or Mg) or sulfuric acid may be used as an aid to dry-ashing. These should be used with some reservation and in known quantities as they could sometimes contribute significant blank values. The aids are of two kinds: those that aid the combustion process, such as sulfuric acid; and those that result in an inorganic, alkaline residue, *e.g.* sodium carbonate or magnesium oxide; these aid the retention of some elements. Magnesium nitrate is often used and it serves both purposes.

Dry-ashing in a Closed System

In closed combustion systems, oxygen is used to aid the decomposition of the matrix. The main benefit of this technique is that oxygen, like other gases, can be easily purified. The most common form of dry-ashing in a closed system is the Schöniger flask method which is usually applied to sample weights up to about 1 g. Table 3.2.6 lists some of the advantages and disadvantages of ashing in a closed system.

Table 3.2.6 *Dry-ashing in a closed system*

Benefits	Problems
Speed	Mainly limited to organic matrices
Ease of use	Reactions may be violent
Low level of supervision	Specific apparatus required

Low-temperature Ashing

To avoid the problems caused by elevated temperatures activated oxygen may be used. The oxygen is activated by high frequency induction, then passed over the sample at about 120°C. There are several benefits to this so-called cool plasma ashing and they are summarised, together with associated problems, in Table 3.2.7. It is possible to add fluorine to accelerate the combustion process, but this may increase problems both from an analytical aspect and from increased safety requirements. Specific apparatus may be required but it is possible to modify existing apparatus.

Table 3.2.7 *Low-temperature ashing*

Benefits	Problems
Low degree of volatilisation Low risk of contamination	Some apparatus problems Long decomposition times Conflicting reports of performance Limited samples per batch

Wet Digestion in Open Vessels and Kjeldahl Flasks

Open vessel wet digestion is the most widely used method for destroying organic matrices. Standard glassware is normally used, although for some purposes (usually depending on the acids used) quartz, glassy carbon and PTFE are employed. Heat may be provided by gas burners, though hot-plates and infrared heaters are now more usual. The apparatus is normally covered with a watch glass or glass bubble caps to prevent both ingress of airborne contaminants and losses due to the digest spraying or bubbling. Table 3.2.8 lists some of the advantages and disadvantages of performing acid digestions in Kjeldahl flasks.

Table 3.2.8 *Wet digestion in open vessels and Kjeldahl flasks*

Benefits	Problems
Widely applicable Simple apparatus Suitable for large batches	Risk of reagent contamination Reactions can be violent Needs high level of supervision Labour intensive

Wet Digestion in a Pressurised, Closed Vessel

Table 3.2.9 lists some of the advantages and disadvantages of performing acid digestions in closed vessels. Acid pressure decomposition will normally be carried out in a sealed inert container, such as glass or plastic, surrounded by a strong metal casing to withstand elevated pressures. Operation at elevated pressures, and hence elevated temperatures, assists the decomposition process and avoids the problem of analyte loss by volatilisation.

The pressure increase results from the release of carbon dioxide and steam from the decomposing organic material. It is important to work with samples that are as small as practicable to avoid unnecessary pressure increases.

It must be noted, however, that these bombs take some time to reach the decomposition temperature and to cool down after the digestion.

Table 3.2.9 *Wet digestion in a pressurised, closed vessel*

Benefits	Problems
Retains volatile analytes	Specific apparatus required
Low level of supervision	Can be dangerous
Possibility of automation	Unsuitable for foods and other samples with high moisture content

Microwave Wet Digestion

The reported advantages of a closed microwave system are that elevated temperatures and pressures, which aid the destruction of organic matter, can be achieved in a relatively short time. One of the disadvantages is that, being a recently developed technique, specific methods are not widely available. Some method validation will be necessary when introducing this technique.

Microwave systems, for the purposes of digestion, can be divided into two categories: dedicated digestion systems and home made 'domestic' systems. Home made systems are much cheaper, but are considerably more dangerous (explosive). They usually have to employ enclosed bombs which do not have the same safety features as the reflux tubes usually used in dedicated instruments. Another difference between the two types of instrument is that dedicated instruments tend to be for a single analysis with the option of automation. 'Domestic' ovens tend to be able to take four or five samples with no possibility of automation.

Microwave digestion was first used with open vessels in home made systems and so suffered from the usual problems of volatilisation losses, contamination from the environment, and temperature limitations. Developments have now made it possible to circumvent these difficulties. Small reflux units can be placed on top of the vessels to trap volatiles and permit the use of elevated temperatures. Decompositions in closed vessels are also frequently used. Metallic vessels cannot be used in microwave digestion and so vessels are constructed of polycarbonate, polyethylene, or PTFE, and fitted with screw caps. A PTFE pressure vessel cannot withstand the pressures possible with steel jacketed bombs, but it is nevertheless possible to operate at pressures and temperatures sufficiently elevated to obtain significantly shortened digestion times. Containers must be tightly closed to avoid losses of volatile analytes. Polycarbonate is a useful material for vessel construction since it possesses a high tensile strength, is acid-resistant, and is transparent to microwaves.

The output of a domestic microwave oven is typically 650 W and will often have the capacity to contain several digestion vessels. For dedicated microwave systems the output will be 200 W (but more focused) with the option of having single digests or automated for several. The combination of multi-sample capability and shorter processing times has led to a wide adoption of the technique. Table 3.2.10 lists some of its advantages and disadvantages.

Table 3.2.10 *Microwave wet digestion in closed vessels*

Benefits	Problems
Very efficient	Specific apparatus needed
Can be automated	Sometimes dangerous (domestic ovens)
Low analyst time	Needs high level of expertise
Low risk of volatilisation	Unsuitable for large batches
	Sample types cannot be mixed

Dynamic Closed Vessel Wet Digestion

An alternative to the rather basic open vessel decomposition is to use apparatus fitted with a reflux head to prevent losses by volatilisation.

There are several reflux-based methods. These rely on alternating stages of digestion and the removal of accumulating water, which dilutes the oxidising acids and slows the decomposition process. Sulfuric, nitric, and perchloric acids are used, in different combinations, and sulfuric acid with hydrogen peroxide. These methods are particularly suitable for volatile analytes such as mercury, arsenic, and selenium. Table 3.2.11 lists some of the advantages and disadvantages of this technique.

Table 3.2.11 *Dynamic closed vessel wet digestion*

Benefits	Problems
Improved retention of volatiles	More complex equipment
Avoids high pressures	Labour intensive

3.2.3 General Precautions

Avoiding Losses

The commonly used vessel materials are glass, silica, porcelain, platinum, and glassy carbon. These materials, particularly glass, have a tendency to retain analytes on their surfaces by various mechanisms. Analytes may also be retained by insoluble (usually silica) matter remaining after the decomposition process. Plastics such as PTFE, polyethylene, and polycarbonate are generally less prone to adsorption problems, and are a suitable alternative if the required temperatures are below their melting points (*e.g.* for microwave digestion). However, mercury has been known to migrate through PTFE and the volatility of selenium is always a problem.

It is reported that Teflon PFA (perfluoroalkoxy), which is more highly cross-linked, displays minimal retention. It is, however, more expensive than standard PTFE.

Occlusion in Metal Vessels. A platinum container should not be used when analysing elements that are likely to occlude in its walls. This is most likely to occur when metallic elements are formed as a result of the decomposition (see Table 3.2.12). Platinum crucibles are used rarely, and then only for dry-ashing.

Table 3.2.12 *Elements likely to occlude with platinum*

Iodine
Mercury
Gold
Platinum
Palladium
Iron

Adsorption of Elements by Siliceous Materials. Elements that form oxides in the decomposition process may undergo interaction with silica-based materials such as glass and quartz containers. Indeed most elements can be retained by silica to some degree. The matrix may also react with the analyte metals to cause losses by other mechanisms. If the sample contains a high proportion of NaCl as a result of destruction of a food matrix, interaction between the NaCl and the silica-based container wall may result in retention of analytes by the wall, through reaction with the hydroxyl groups in the silica. Retention by silicates present in the matrix occurs similarly. The elements retained most strongly by silica are lead, copper, and zinc.

To prevent losses by adsorption to active sites, careful washing with less concentrated acids should result in fewer active sites being produced. This will in turn result in reduced losses by adsorption.

Volatilisation. In addition to retention, the other major cause of loss is volatilisation. Volatile species can be formed in a number of ways. Some of the better known examples are given in Table 3.2.13.

The production of these volatile species will depend on the presence of other substances to promote their formation. It should be noted that many elements form volatile chlorides. Generally these will be produced by reaction with hydrochloric acid. Chloride salts in the presence of perchloric acid are unlikely, on thermodynamic grounds, to form these chlorides. To counteract any possible formation of volatile chlorides, sulfuric acid is normally added as it readily produces non-volatile salts. However, it may also form insoluble sulfates. If this is the case, the possibility of using nitric acid as an alternative to hydrochloric acid should be investigated; most nitrates are less volatile than the corresponding chlorides.

Table 3.2.13 *Volatile species formed by ashing*

Oxides	Chlorides	Hydrides	Other
Os	Sb	Sb	Cd
Re	As	As	Hg
Ru	Cr	Se	
	Co	Te	
	Cu		
	Ge		
	Fe		
	Pb		
	Zn		

When analysing for Hg, As, and Se, which may bind to organic ligands, sealed bomb destruction methods should always be used. Organic compounds of these elements are often more volatile than the elements themselves. Compounds of this type are common in marine samples.

Contamination

Contamination may occur from any material coming into contact with the sample. In the case of dry-ashing, the contaminants may be in the walls of the ashing vessel as constituents or as occluded residue from a previous sample. Ashing aids that may be employed (MgO and sulfuric acid) are also a possible source of contamination. The possibility of 'fall-out' when using an open vessel must also be considered.

In wet digestion sources of contamination are the same as those for dry-ashing and, in addition, the possibility of contamination from reagents is greatly increased. It is therefore generally recommended that the least possible quantities of reagents are used to reduce the risk of contamination. This can be achieved through the use of closed digestion systems, such as microwave digestion and pressure decomposition. The conditions used for reagent blanks must be as identical as possible with those used for the samples.

If appropriate, dry ashing could be used to remove some contaminant or interfering elements. For example Cd or Pb could be deliberately volatilised prior to a further digestion sequence.

3.2.4 Practical Advice

General Use of Acid Digestion Techniques

It is common practice for an organic sample to be digested in a Kjeldahl flask with a mixture of nitric and sulfuric acids. After no carbon is visible in the digest and while nitric acid is still present, a distinct yellow coloration indicates the presence of organic nitro compounds. When the nitric acid has been boiled off and the liquid cools, these residual organics cause the solution to appear cloudy. The decomposition must then be continued by heating and adding small volumes of nitric acid until the solution remains clear when hot and cold. The solution may be tested for organics by rapidly cooling the flask in cold water (use borosilicate glass and take care against breakage!) to see if cloudiness results. A pale straw colour remaining in the sulfuric acid is not at all unusual and is caused by nitro-sulfuric acid. This is a strong oxidant and must be removed as it may cause subsequent problems in the analysis, especially if an electrochemical deter-mination step is to be employed. Prolonged boiling after the digestion is complete will achieve this.

A quicker method is to cool the sulfuric acid (this is very important) and then to add, carefully and with constant mixing, an approximately equal volume of water. The strongly exothermic reaction between water and sulfuric acid destroys the nitro compound and expels the resulting nitric acid. It must be stressed that this procedure requires mixing during the addition of water, and it must be conducted in a Kjeldahl flask or in apparatus that will prevent the loss of any digest by spraying.

When analysing water, soil, or certain food samples (for example oyster tissues), it is possible to have a residue of silica. This appears at the end of the digestion as small white grains at the bottom of the flask. Often it is not necessary to dissolve these as it is likely that all the elements of interest would have been leached out from the surface of the silica particles. Alternatively, it may be possible to analyse the sample as a slurry. However, the decision on whether or not to dissolve the silica (by HF treatment) must be left to the judgment of the analyst.

It is recommended that the use of sulfuric acid be restricted for many analyses because it is seldom of sufficient purity for trace level determinations. Useful alternatives are nitric acid and hydrogen peroxide (see suggested method for microwave digestion in this section). They are both safer and cleaner and, when combined with microwave heating, will carry out efficient digestions. With this mixture the analyst should look for a clear green solution in the digestion process. This will indicate sufficient digestion of the sample. If the solution is opaque or a cloudy yellow, it will require more time in the microwave.

Analysis of Difficult Matrices

Most organic matrices can be digested, at least to a large extent, by treatment with nitric acid at 120°C. However, there may be some difficulty with matrices containing large amounts of protein or fat. In these circumstances elevated temperatures may be necessary to destroy the matrix together with sulfuric acid, perchloric acid, or hydrogen peroxide. However, when further acids or oxidation steps are introduced, the possibility of contamination increases and must be monitored using blanks. Alternatively, the digestion could be performed at a higher temperature in a glassy carbon container. Digestions at temperatures up to 220°C are possible in this material and it is also resistant to both nitric and hydrofluoric acids.

Microwave Heated Digestion

When applied to closed vessel decomposition this method of sample digestion requires careful control of both temperature and pressure. All the plastics used to fabricate these vessels have inherent temperature limitations. For example, the melting point of polycarbonate is 135°C and that of PTFE is 303°C (below the boiling point of sulfuric acid) and so reaction temperatures must be kept safely below these levels.

It is important always to have the same number of digests in the oven at any time. If there are fewer cells than normal the energy received by each of them will be higher than normal, resulting in excessive pressures and temperatures. In addition, only similar samples can be digested in the same batch.

It is necessary to be alert to safety considerations when using microwave heating for digestion, particularly for the inherently more hazardous closed vessel version.

It is good practice to have a means of monitoring the temperature as described by Kingston and Jassie.[17] This will not only make the process considerably safer, but also prevent loss of sample. Monitoring of the pressure is advisable for closed systems but it can lead to incomplete digestion through pressure reduction.

An alternative to monitoring parameters in the reaction mixture is to incorporate safety features into the closed unit. A pressure valve is fitted in the screw cap of the vessel so that pressure will be released if a predetermined level is exceeded. In addition, an overflow fitted to an aspirator will prevent a dangerous build-up of pressure. The problem with these safety precautions is that when they operate, volatile analytes will be lost. They are also very expensive.

The working range for the digestion containers is dependent upon their tensile strength at different pressures and temperatures. At approximately 10 atm. (10^6 Pa) cracks begin to show in the caps of PTFE containers. This has been found at temperatures well below the melting point of PTFE and so it is the pressure in PTFE vessels that is the limiting factor. For polycarbonate, the melting point is so low that temperature is most likely to be the limiting factor.

A recommended microwave digestion method for many matrices of environmental interest is: 0.2500 g of sample + 1 ml of H_2O_2 + 3 ml of HNO_3, leave loosely capped for 1–3 hs and then expose to microwave radiation (first at low power for a couple of minutes, allow to cool, release the pressure, and then at medium power for 2 min).

Use of Combustion Techniques

When using a muffle furnace, the temperature should eventually rise to between 450 and 550°C. It must be above 450°C to ensure the complete combustion of organic matter. Below this temperature some of the analyte may remain bound to surviving matrix. At temperatures above 550°C volatilisation losses may occur. It is best to heat up the furnace slowly to prevent the matrix flaming which could also cause losses. It is important not to add samples directly into a hot furnace without pre-charring. The position of the vessel inside the furnace may affect the overall efficiency of the decomposition. It is best if the vessel is not placed in direct contact with the furnace floor as conducted heat is less suitable than radiant heat for achieving complete combustion. Suspending the vessel, for example on a silica triangle, is recommended. It is also best to preheat the digestion vessel at the temperature which will be used for combustion. This will remove water from the crucible and thus prevent misleading results arising from the weight of the crucible plus sample before and after combustion.

To assist the destruction of organic matter, the addition of small amounts of acids or other ashing aids may be employed. Gorsuch has recommended that either 10 ml of 10% sulfuric acid or 10 ml of 7% magnesium nitrate solution be added per 5 g of sample.[2] Alternatively, nitric acid may be used as recommended in the Association of Official Analytical Chemists manual, adding 10 ml of water and 4 ml of 50% nitric acid per 1 g of sample, and evaporating to dryness on a hot-plate before ashing. The choice of ashing aid will depend on the nature of the analyte(s).

Analysis of Slurries

At the beginning of this section it was stated that most inorganic analyses are performed on dissolved solutions with a small percentage performed on solid matrices. Another important area is the analysis of slurries. Slurries of virtually any type of solid material may be prepared. The preparation of slurries tends to decrease the amount of contamination introduced from reagents, and decreases the use of hazardous chemicals (particularly HF). In comparison with some dry-ashing procedures it is relatively quick, and may be calibrated using aqueous standards. Slurries have been introduced successfully into a number of different analytical instruments including ICP-AES, ICP-MS, DCP-AES, and ETAAS (there have been one or two papers introducing them to flame AAS). The important factor in determining the success of the analysis has been found to be particle size. If the transport efficiency of the slurry is to be comparable to that of aqueous solutions, a particle size of less than about 2 μm is required. This however, is dependent on the sample introduction component of the spectrometer. A high solids nebuliser (*e.g.* v-groove) must be used for ICP instruments, because a Meinard nebuliser would quickly block. A maximum level of suspended solids in the slurry is 1% m/v.

Slurries may be prepared in a number of different ways. Each way produces its own contaminants, so the method chosen will depend on the analyte of interest and on the hardness of the sample.

The bottle and bead method has been the most common method of slurry preparation. A 1.00 g sample is weighed into a nalgene bottle, zirconia beads (10 g) and 3–5 ml of dispersant are added, and the bottle placed on a laboratory flask shaker for several hours. The particle size of the resulting slurry may be easily checked using optical microscopy or particle size measuring equipment. This method may lead to contamination by Zr, Hf, and Si, and to a lesser extent Ca, Mg, Al, Fe, and Ti. It is suitable for the large majority of sample types, as the zirconia beads have a hardness of 7–8 on the Mohs (Measure of hardness) scale.

Other methods of slurry preparation include the microniser. This is similar to the bottle and bead method, but here the sample is weighed into a container filled with agate rods which act as the grinding medium. This method may lead to contamination by Si, Al, Na, Fe, K, Ca, and Mg. This method is suitable for many sample types, but the agate rods have a hardness of 6–7 Mohs so more contamination may arise from hard samples. The Tema mill (a tungsten carbide grinder) can be used for extremely hard samples. It has a hardness of 8.5–9.0 Mohs, so does not often produce contaminants. If contamination does occur, it is most likely to be W, Co, Ta, Ti, or Nb.

The choice of dispersant depends upon the sample type. The dispersant is necessary to prevent agglomeration of slurry particles into loose collections of much larger particle size. Some suitable ionic dispersants (also called surfactants) include sodium hexametaphosphate, 'aerosol OT' and sodium pyrophosphate. Solutions (typically 0.1% m/v) may be used to disperse inorganic samples such as firebrick, dolomite, sediments, soils, *etc.* Non-ionic surfactants such as Triton X-100 (0.05–1%) are used to disperse organic materials such as foodstuffs, plant material, blood, *etc.* One problem associated with the use of Triton X- 100 is that it foams. The amount of foaming obtained from a 1% solution is substantial and may lead to problems when diluting solutions in a volumetric flask.

3.2.5 References

General Reading

1. 'CRC Handbook of Chemistry and Physics', ed. D. R. Lide, CRC Press, Boca Raton, FL, published annually.
2. T. T. Gorsuch, 'The Destruction of Organic Matter', Pergammon Press, Oxford, 1970.
3. G. Tölg, 'Role of Sample Decomposition and Preconcentration in Elemental Trace Analysis', *Pure Appl. Chem.*, 1983, **55**, 1989.
4. G. Knapp, 'Mechanised Methods of Sample Decomposition in Trace and Ultra-trace Analysis', *Anal. Proc.*, 1990, **27**, 112.
5. B. Griepink and G. Tölg, 'Sample Digestion for the Determination of Elemental Traces in Matrices of Environmental Concern', *Pure Appl. Chem.*, 1989, **61**, 1139.
6. Analytical Methods Committee, 'Methods for the Destruction of Organic Matter', *Analyst*, 1960, **85**, 643.
7. E. Jackwerth and S. Gomiscek, 'Acid Pressure Decomposition in Trace Element Analysis', *Pure Appl. Chem.*, 1984, **56**, 479.
8. G. Knapp, 'Decomposition Methods in Elemental Trace Analysis', *Trends Anal. Chem.*, 1984, **3**, 182.
9. P. Tschopel, 'Modern Strategies in the Determination of Very Low Concentrations of Elements in Inorganic and Organic Materials', *Pure Appl. Chem.*, 1982, **54**, 913.
10. A. G. Howard and P. J. Statham, 'Inorganic Trace Analysis. Philosophy and Practice', John Wiley, Chichester, UK, 1993.
11. C. Vandecasteele and C. B. Block, 'Modern Methods for Trace Element Determination', John Wiley, Chichester, UK, 1993.
12. R. Anderson, 'Sample Pretreatment and Separation', John Wiley, Chichester, UK, 1987.

13. R. Bock, 'A Handbook of Decomposition Methods in Analytical Chemistry', International Textbook Company, London, 1979.
14. J. Dolezal, P. Povondra, and Z. Sulzek, 'Decomposition Techniques in Inorganic Trace Analysis', Iliffe Books, London, 1966.
15. E. V. Williams, 'Low Temperature Oxygen–Fluorine Radio-Frequency Ashing of Biological Materials in Poly(tetrafluoroethylene) Dishes Prior to Determination of Tin, Iron, Lead and Cadmium by Atomic Absorption Spectroscopy', *Analyst*, 1982, **107**, 1006.
16. G. Knapp, 'Decomposition Methods in Elemental Trace Analysis', *Trends Anal. Chem.*, 1984, **3**, 182.

Microwave Digestion

17. 'Introduction to Microwave Sample Preparation. Theory and Practice' (ACS Professional Reference Book), ed. H. M. Kingston and L. B. Jassie, American Chemical Society, Washington, DC, 1988.

Analysis of Slurries

18. L. Ebdon, M. E. Foulkes, and S. Hill, 'Fundamental and Comparative studies of Aerosol Sample Introduction for Solutions and Slurries in Atomic Spectroscopy', *Microchem. J.*, 1989, **40**, 30.
19. P. Goodall, M. E. Foulkes, and L. Ebdon, 'Slurry Nebulization Inductively Coupled Plasma Spectrometry: The Fundamental Parameters Discussed', *Spectrochim. Acta, Part B*, 1993, **48**, 1563.
20. K. O. Olayinka, S. J. Haswell and R. Grzeskowiak, 'Development of a Slurry Technique for the Determination of Cadmium in Dried Foods by Electrothermal Atomization Atomic Absorption Spectrometry', *J. Anal. At. Spectrom.*, 1986, **1**, 297.

Applications and Specific Problems

21. K. May and M. Stoeppler, 'Wet Digestion of Fatty Biological Samples', *Fresenius' J. Anal. Chem.*, 1978, **293**, 127.
22. H. J. Robberecht, 'Losses of Selenium in Digestion of Biological Samples', *Talanta*, 1982, **29**, 1025.
23. T. T. Gorsuch, 'Losses of Trace Elements during Oxidation of Organic Matter: The Formation of Volatile Chlorides During Dry Ashing in the Presence of Inorganic Chlorides', *Analyst*, 1962, **87**, 112.
24. J. L. Down and T. T. Gorsuch, 'The Recovery of Trace Elements after the Oxidation of Organic Material with Fifty Percent Hydrogen Peroxide', *Analyst*, 1967, **92**, 398.

3.3 Decomposition and Dissolution of Inorganic Matrices and Residual Inorganic Material from Organic Matrices

3.3.1 Overview

The dissolution of inorganic matrices for elemental analysis may be required either when initial destruction of organic matter has left an inorganic residue or when the sample matrix is inorganic, as exemplified by geological samples or metal ores.

This section deals with the dissolution of inorganic matrices and analytes to allow for easy analysis of a solution. However, as described in Section 3.2, there are alternatives to total destruction of the matrix. Analysis of solids, analysis of slurries, and leaching from the matrix can all be used to gain satisfactory information from the sample without the uncertainty associated with total destruction.

The common inorganic matrices that need to be dissolved prior to analysis include: metals and alloys; soils, sands, rocks, glasses, and other silicates; metal ores and concentrates; and any dry ash residues including those of plastics, paper, and paints. These will usually need to be dissolved by an acid or combination of acids and possibly with the addition of other reagents, for example oxidants. Alternatively, fusion followed by leaching is sometimes necessary or desirable to solubilise the analyte.

When an inorganic matrix is to be totally dissolved, the first consideration of the analyst is to choose a medium suitable for dissolving the major constituents. The trace elements will also usually dissolve, but once the bulk has dissolved, further conditions can then be employed to ensure that all the analytes of interest are in solution. If a slurry is to be analysed or the sample to be leached, milder conditions can be used. In addition, if leaching is to be used, the conditions only need to be suitable for dissolving the analytes, not the whole matrix.

Another aspect that needs to be considered is the choice of apparatus (containers and heating device).

3.3.2 Comparison of Techniques: Acid Decomposition

In the decomposition and dissolution of inorganic matrices, several techniques applied to organic samples are frequently used, namely:

(i)   acid decomposition in open vessels, including Kjeldahl flasks;
(ii) acid decomposition in a pressurised closed, vessel;
(iii) acid decomposition with microwave heating.

Each of these techniques has been discussed in Section 3.2, and their operation for inorganic materials is similar. The main difference concerns the choice of acids employed (the choice available for inorganic dissolution is much wider).

The discussion of the selection of acids which follows considers it in the order of the mildest to the harshest conditions. The conditions that are most suited to particular sample types are listed in Table 3.3.1.

Table 3.3.1 *Methods of dissolving inorganic material*

Dissolution method	Examples of applications
Water	Salts
Dilute acids (approx. 0.1 M)	Residues from dry-ashing, electropositive metals, salts, soils, sand
Individual concentrated acids	Electronegative metals and alloys, stainless steel, some metal oxides
Hydrofluoric acid	Silicates, rocks, sand
Concentrated acids / oxidising reagent mixtures	All metals and alloys, refractory minerals, some aluminosilicates, mixed carbon/ silicate particulates, particulates from engine exhausts

Prior to the discussion of the different conditions required for dissolution (which mainly involves acids), another important area deserves some mention, *i.e.* leaching.

Leaching and Partial Dissolution

Leaching is the selective dissolution of one or more analytes from a matrix which can then be removed by filtration or decanting. This method will not have the quantitative robustness of the total dissolution techniques but will give an estimate of the extractable analytes in a matrix such as soil. Leaching is usually performed in one of two ways:

(i)   dissolution in water, aqueous solutions, or dilute acids;
(ii)   complexation in aqueous solutions of organic ligands.

Each of these sets of conditions presents a safe option that may well suffice for the needs of the analyst. This is particularly the case when analysing silicate-containing samples. It may often be better to dissolve easily soluble material and either separate the solid fraction or analyse as a slurry. The other options are to use hydrofluoric acid (HF) or salt fusion, both of which present their own problems.

Total Dissolution

Use of Dilute Acids (~0.1 M).   Dissolution of inorganic material is usually carried out using dilute hydrochloric or nitric acid at room temperature. Most electropositive metals and metal salts will dissolve in any acid, but in practice sulfuric acid is not widely used because many metal sulfates are insufficiently soluble, whereas nitric and hydrochloric acids form soluble salts. These conditions can also be used for leaching.

Some metals do not dissolve, even on heating. This is particularly true of aluminium, which rapidly becomes coated with an impervious layer of insoluble oxide. One way to overcome this problem, and which is always good practice with metallic samples, is to clean the surface prior to dissolution with a non-contaminating abrasive such as emery followed by washing with acetone. In the case of aluminium, however, the oxide layer re-forms quickly so it may be necessary to add a little mercury(II) chloride as a catalyst to aid dissolution.

The solubility of difficult matrices, for example boride salts, can be improved by adding an oxidant. One method is to immerse the sample in dilute hydrochloric acid to which a few drops of hydrogen peroxide have been added.

Matrices suitable for dissolution in dilute acids include residues from dry ashing, soils, sand, electropositive metals, and salts. In the case of salts, it may not be necessary to use dilute acids; water may be sufficient to effect dissolution.

Use of Concentrated Acids, either Individually or In Sequence. The scope for dissolution of matrices is increased when the acid is concentrated and heated. There are obvious safety problems with this procedure (see Section 3.1). It is always important to add the acid to the cold sample and then heat rather than to add the sample to hot acid. This is particularly vital when using sulfuric or perchloric acids.

It may sometimes be desirable to use individual acids in sequence according to their properties. For example, perchloric acid should never be used as the first acid: it should always be preceded by nitric acid to destroy all the more easily oxidisable matter. Another common occasion for using two acids in sequence is in the removal of silicon from a sample by volatilising it as SiF_4. The sample is first treated with hydrofluoric acid and then fumed with either sulfuric or perchloric acid. Alternatively, after HF treatment, boric acid may be added to convert excess fluoride into tetrafluoroborate.

Hot, concentrated mineral acids can be used to dissolve samples such as electronegative metals and alloys, stainless steel, and some metal oxides. Rocks, sands, and other silicates will dissolve in hydrofluoric acid.

Use of Mixtures of Concentrated Acids with Other Reagents. Powerful conditions can be obtained through the combined properties of concentrated acid with other reagents that assist with the breakdown of the sample matrix. Examples include aqua regia, combination of an oxidising agent with an acid, and using hydrofluoric acid with an oxidising acid. Aqua regia (see Section 3.1.4) will dissolve metals such as gold and platinum. Combinations of acids and oxidants are frequently used, *e.g.* sulfuric with a dichromate(VI) ion to form the powerful oxidant chromic(VI) acid, or with hydrogen peroxide to form permonosulfuric [peroxo-sulfuric(VI)] acid, or Caro's acid, which is also a powerful oxidant. Mixing an oxidising acid with hydrofluoric acid combines acid, oxidant, and complexation characteristics.

Such mixtures are much more generally applicable than either dilute acids or single treatments with concentrated acids. They can be used for all metals and alloys, refractory minerals, carbon/silicate particulates, and some aluminosilicates.

Fusion and Sintering

If the matrix cannot be dissolved easily by any of the acid digestion techniques, fusion or sintering may well provide a suitable means of sample decomposition.

The vessels employed for fusion or sintering must withstand attack and are constructed from resistant metals such as platinum, tantalum, zirconium, silver, or nickel, or for some purposes from graphite.

Fusion and sintering are used mainly for matrices with high alumina or silica contents and other refractory oxides. Fusion is often used as an alternative to lengthy decomposition methods involving hydrofluoric, perchloric, and nitric acids for the destruction of aluminosilicate ashes. A summary of the matrices for which fusions or sintering are used is given in Table 3.3.2.

Table 3.3.2 *Matrices for which fusion or sintering are used*

Cement
Sands
Ceramics
Other aluminosilicates
Slags
Rocks (when looking for rare earths)
Al, Be, Fe, Si, Ti, and Zr ores (or mixtures and residues)
Al, Cr, Fe, Si, and W oxides

Fusion (or sintering) is, however, the last method of choice when analytes that could volatilise are to be determined. Also, several of the fusion substances are not available in high purity and could contribute significant blanks in trace analysis. For these reasons, fusion and sintering are not generally considered suitable for sensitive trace analysis, and they will not be considered further here. Excellent accounts of the various fusion and sintering methods and their applications have been given by Dolezal, *et al.*[1]

3.3.3 General Precautions

The majority of decompositions/dissolutions are carried out with acids. The safe handling of acids is covered in Section 3.1.

Plastic apparatus is recommended for most purposes, but the problems of powerful oxidants must be considered. In general, PTFE apparatus is suitable for many digestions and polycarbonate is suitable for storing solutions.

Losses

Losses of analyte are caused by volatilisation during decomposition or by retention on the walls of the vessels used. These matters have been discussed throughout Section 3.2 which deals with organic sample destruction. The advice and information given there apply equally to the dissolution of inorganic samples.

Contamination

Contamination from the reagents, vessels, and laboratory atmosphere have also been covered in Sections 2.2 and 3.2. It is worthy of note that somewhat harsher dissolution conditions are used for impervious inorganic materials than are employed in organic sample digestion. These harsher methods use a wider range of reagents which are often not available in high purity and so the blank values obtained may be high. The reagents are also more likely to attack the surfaces of the vessels used, particularly in fusion methods, therefore adding to possible contamination problems. The particular problem of reagent contamination in fusion and sintering methods has already been mentioned. Although, in principle, this makes the technique unsuitable for inorganic trace analysis, there is some-times no alternative method. It must also be acknowledged that for measurement by X-ray fluorescence spectroscopy, fusion is the recognised standard method of preparation for non-metallic samples. This method is not, however, suitable for measurements at low (*e.g.* sub-parts per million) concentrations.

3.3.4 Critical Aspects

Analytes

The critical factor in the dissolution of an inorganic matrix is the choice of acid or acid mixture to be used. This decision will be based on the properties of the matrix and not the individual analytes. Once the bulk of the sample has dissolved, it is important then to establish, if necessary, conditions under which the individual *trace* analytes will dissolve. These are not necessarily the same conditions as would apply to major amounts of the same substances. The analyte itself and the conditions for its dissolution (remembering that it may be speciated or in various oxidation states) are therefore important.

Table 3.3.3 presents a classification of analytes and some notes on the media normally required for solubilisation. The analytes discussed are only common examples. The other elements cannot easily be classified, but their behaviour can frequently be predicted from their position in the periodic table and the properties of nearby elements. For example, Be behaves similarly to Al, Ba and Sr behave like Ca, Rb and Cs like K, and Co often like Fe, Ni, and Cr.

3.3.5 Uncertainty

The major sources of error in sample dissolution are losses due to volatilisation of analytes, mechanical losses, retention of analytes in undigested matrix, retention

Table 3.3.3 *Dissolution conditions for some common analytes*

Class of analyte	Examples of analyte	Dissolution conditions	Notes
Nutritional elements (*e.g.* 1–100 mg kg^{-1})	Ca, Cr, Cu, Fe, K, Li, Mg, Mn, Na, Ni, Zn	Dilute (*e.g.* 2 M) HCl or HNO$_3$	Have few volatile compounds. Ni is volatile in dry-ashing
Environmental contaminants	Cd	Dilute HCl	
	Se, Pb	Concentrated HNO$_3$	
	As, Sn	Concentrated HCl	
	Ge	Concentrated HNO$_3$ / HF (complexed)	
Noble metals (geological matrices)	Au, Os, Ir, Pt, Rh, Ru	Aqua regia	Os, Rh, and Ru form volatile oxides. Os oxide is toxic. Extra NaCl may be needed to keep Pt in solution.
	Ag	HNO$_3$ (exclude light)	
	Pd	HNO$_3$ and HCl$_4$	
Volatile elements	As, Sn, Tb	Concentrated HCl	Analysed by hydride generation
	Bi, Pb, Se	Concentrated HNO$_3$	Analysed by hydride generation
	Ge	Concentrated HNO$_3$ / HF (complexed)	Analysed by hydride generation
	Hg	Dilute HCl	Analysed as metal vapour
	Sb	Concentrated HCl or tartrate	Analysed by hydride generation
Refractory elements	Al	Concentrated HCl	
	Be	Concentrated HCl or HNO$_3$	
	Mo	Concentrated H$_2$SO$_4$ or H$_3$PO$_4$	
	Si	Alkaline fusion / concentrated HCl	
	Ti	Concentrated H$_2$SO$_4$	
	U	Concentrated HNO$_3$ or H$_3$PO$_4$	
	V	Concentrated HNO$_3$	
	W	H$_2$SO$_4$ or H$_3$PO$_4$	
Elements needing complexation	Ge, Hf, Nb, Ta, Zr	Once in solution require complexing with oxy-ligands, *e.g.* tartrate	

losses on the walls of the digestion vessel, and contamination from laboratory ware, reagents, and laboratory fall-out. These have been discussed in Sections 2.2 and 3.2.

3.3.6 References

General Reading

1. J. Dolezal, P. Povondra, and Z. Sulcek, 'Decomposition Techniques in Inorganic Analysis', Iliffe Books, London, 1966.
2. R. Bock, 'A Handbook of Decomposition Methods in Analytical Chemistry', International Textbook Company, London, 1979.
3. R. Anderson, 'Sample Pretreatment and Separation', John Wiley, Chichester, UK, 1987.
4. A. G. Howard and P. J. Statham, 'Inorganic Trace Analysis. Philosophy and Practice', John Wiley, Chichester, UK, 1993.
5. C. Vandecasteele and C. B. Block, 'Modern Methods for Trace Element Determination', John Wiley, Chichester, UK, 1993.
6. P. Tschopel, 'Modern Strategies in the Determination of Very Low Concentrations of Elements in Inorganic and Organic Materials', *Pure Appl. Chem.*, 1982, **54**, 913.
7. G. Knapp, 'Decomposition Methods in Elemental Trace Analysis', *Trends Anal. Chem.*, 1984, **3**, 182.
8. E. Jackwerth and S. Gomiscek, 'Acid Pressure Decomposition in Trace Element Analysis', *Pure Appl. Chem.*, 1984, **56**, 479.
9. G. Knapp, 'Mechanised Methods of Sample Decomposition in Trace and Ultratrace Analysis', *Anal. Proc.*, 1990, **27**, 112.
10. G. Tolg, 'Role of Sample Decomposition and Preconcentration in Elemental Trace Analysis', *Pure Appl. Chem.*, 1983, **55**, 1989.

Applications

11. G. Tolg, 'Recent Problems and Limitations in the Analytical Characterization of High Purity Material', *Talanta*, 1974, **21**, 327.

3.4 Separation and Preconcentration in Inorganic Analysis

3.4.1 Overview

Once the matrix has been dissolved, inorganic analytes will often be in quite harsh conditions, for instance in concentrated sulfuric acid or hydrofluoric acid. It is usually unnecessary and undesirable for the element to remain in these conditions. It is therefore common practice to transfer the element to conditions more suitable for safe and reliable end-determination.

It is often impossible to directly apply the various measurement techniques of trace analysis, even after decomposition or dissolution because the concentrations of the desired elements are below the detection limit or at too low a concentration to give reliable data, or because interfering substances are present. Preconcentration, also called enrichment, is the generic term for the various processes employed to increase the ratio of determinand to matrix. The change of matrix and the change of concentration, if needed, are performed in the same step, if possible.

The most important factor that will determine which separation and preconcentration techniques are to be used is the end analytical technique. Both the solvent and the speciation of the analyte will need to be compatible with that technique.

Trace metals may be present in one of four forms.

Free Ion in Aqueous Solution

This is the most common case; solutions resulting from dissolutions and digestions as described in Sections 3.2 and 3.3. They are analysed either by using a spectroscopic technique or by electrochemistry.

Chelated Ion in Solution

After dissolution or digestion, the analyte may be chelated. The chelate is analysed using a spectroscopic technique.

Volatile Compound/Element

Analytes in this form will be usually be analysed by hydride generation spectroscopy.

Electrodeposited Metal

This final case applies almost exclusively to the field of stripping voltammetry, although electrodeposition is occasionally used as a concentration technique in other areas of inorganic analysis. In this case the preconcentration technique is an integral part of the end-determination technique. The use of electrochemical techniques is covered in Section 8.2.

3.4.2 Comparison of Techniques

Many methods of preconcentration are available to the analyst including some devised for the solid and gas phases. Methods applicable to liquids, usually sample solutions, include evaporation, precipitation, solvent extraction, volatilisation, ion-exchange, ion chromatography, electrodeposition, adsorption, molecular sieving, flotation, freezing, electrophoresis, and dialysis. Preconcentration is inherent in some techniques: in sample decomposition by dry-ashing or selective dissolution; in hydride generation; in cold vapour mercury determination by AAS; and in stripping voltammetry by electrodeposition. The more widely used of these solution techniques are considered here.

The literature must be consulted for further details of the various methods of separation and preconcentration and a number of suggestions for further reading are listed in Section 3.4.3.

Evaporation and Distillation

This is the most widely used technique because it requires relatively cheap equipment and often avoids the need for transfer of the solution to another vessel. It is, however, often slow and requires a significant amount of supervision. With non-volatile elements, evaporation can be taken to low volume or even dryness to remove unwanted excess acid, and the residue can then be redissolved in an appropriately selected solvent. When evaporating to dryness, it is expedient to add a small quantity of a non-volatile, non-interfering solid, such as an alkali metal compound, to act as a collector for the trace elements sought.

Evaporation is prone to interference, as it must obviously be performed in an open vessel. It may also lead to the concentration of dissolved solids, which can cause 'salting up' of the nebuliser or burner of the spectrometer.

Distillation is used for several elements, particularly boron (as its methyl ester), arsenic (as $AsCl_3$), germanium (as $GeCl_4$), selenium (as $SeBr_4$), rhenium, ruthenium, and osmium (as their volatile covalent oxides).

Volatilisation / Hydride Generation

Volatilisation is used mainly for the analysis of elements that form volatile hydrides, including As, Se, Sn, Ge, Te, Bi, Sb, and Pb. Volatilisation is also used for the analysis of Hg as the monoatomic metal vapour. It is always linked directly with a suitable detector, and so is thought of as a combined preconcentration/end analysis technique. It is useful for several reasons:

(i)   it is linked directly on-line to the end analysis (such as AAS), and so cuts down on potential contamination or losses;
(ii) it is a successful method for the analysis of the above volatile elements which are otherwise rather more difficult to analyse;
(iii) it is a very selective technique which excludes many interferences.

Solvent Extraction

Solvent extraction is another popular technique as it can preconcentrate many analytes simultaneously and, by careful control of pH, unwanted elements can sometimes be selectively separated.

Unless the elements required are already present as neutral, non-polar, extractable compounds, the first step is to convert them into such compounds. The most common method is to form neutral complexes using organic chelating agents such as the dithiocarbamates or 8-hydroxyquinoline. Ion-association systems, *e.g.* the ternary complex formed between Fe^{2+}, 4,7-diphenyl-1,10-phenanthroline, and the perchlorate anion, can also be extracted into a number of non-polar solvents.

Preconcentration using chelating compounds provides the opportunity to choose conditions for the selective extraction and enrichment of particular metals. An example is the determination of trace iron by spectrophotometry by complexing with 1,10-phenanthroline, partition into an organic solvent, and measurement of the absorbance of the organic phase. Chelation can also be used for the general preconcentration of a wide range of metals prior to end-determination by a multi-element technique. An example is the preconcentration of most heavy and transition metals by complexing with diethylammonium diethyldithiocarbamate or ammonium pyrolidinedithiocarbamate, then partition into 1,1,2-trichloro-1,2,2-trifluoroethane followed by acidification and back-extraction into the aqueous phase as a preliminary to simultaneous multi-element analysis by ICP emission spectroscopy.

Extraction into an organic solvent has the advantage that the surface tension and viscosity of the extracting solvent are lower than for aqueous solutions, and the vapour pressure higher. This results in greater nebulisation efficiency in the spectrometer, giving enhanced sensitivity.

Precipitation

Precipitation techniques are widely used for the separation and preconcentration of trace metals. The direct precipitation of trace constituents is not feasible because of the minute quantities involved but co-precipitation on a collector (or carrier) that has been intentionally added to the solution is very useful. The technique has been used particularly for the separation of traces of radioactive isotopes and for the analysis of waters including sea water.

Ion-Exchange and Ion Chromatography

Ion exchange can be used in trace analysis for the two processes of preconcentration and separation from an unwanted or interfering matrix. The resin may be either packed in micro columns and the sample pumped through or added to the sample solution and then filtered out at a later stage.

Provided that suitable operating conditions are chosen, ion-exchange columns may be used to preconcentrate most metals. Separations may also be performed by ion-exchangers; for example in strong HCl solution, many metals form anionic chloro-complexes which may be separated on a Dowex-1 column, progressively diluting the hydrochloric acid so that the metals having the lower distribution coefficients are eluted first.[†]

In some cases, for example where there is a high ratio of alkali metal ions to the required metals, chelating ion-exchange resins are very useful. These resins

---

[†] The distribution coefficient, K, is defined as the theoretical amount of solute held on the ion-exchanger divided by the amount of solute in solution within the column when the column is at equilibrium. (See E. Glueckauf, 'Ion Exchange and its Applications', Society of Chemical Industry, London, 1955, p. 34.) The technique of adjusting selectivity by hydrochloric acid strength is discussed by W. Rieman and H.F. Walton in 'Ion Exchange in Analytical Chemistry', Pergamon Press, Oxford, 1970, p. 154.

contain chelating functional groups, such as iminodiacetic acid, which function like ethylenediaminetetraacetic acid and form chelates with the metal ions. Distribution coefficients are very high and the chelated metals may be recovered by a batch process usually at low pH.

High performance liquid chromatography using ion exchangers offers the advantages of speed, and the possibility of automation, with precise control and hence reproducibility.

Ion-exchange chromatography can be performed on-line, decreasing both the time taken for a method and the risk of airborne contamination. A further advantage of ion-exchange chromatography for trace analysis is the possibility of on-line enrichment. This can be achieved by on-column concentration using large injection volumes, pre-injection dialysis, or concentration on a pre-analytical column followed by back-flushing. The last technique shows most promise, but it is by no means straightforward. One disadvantage of the technique is that it usually requires the use of buffers, which are also preconcentrated and can interfere with the end-determination.

Electrodeposition

A number of interfering metals including copper, lead, cadmium, nickel, and cobalt may be electrolytically deposited at controlled potentials at a mercury cathode leaving many trace analytes quantitatively in solution, including Be, Al, Ti, Zr, Nb, Ta, W, and U.

If the analyte is a member of the first group of elements it may be separated from the second by controlled potential electrolysis at platinum, carbon, or mercury cathodes.

Stripping Voltammetry

This is a technique that combines the preconcentration stage with the end-determination. It is discussed in Section 8.2, Electrochemical Techniques.

3.4.3 References

General Reading

1. 'Atomic Absorption Spectrometry. Theory, Design and Applications', ed. S. J. Haswell, Elsevier, Amsterdam, 1992.
2. A. Mizuike, in 'Separations and Preconcentrations in Trace Analysis', ed. G. Morrison, Interscience Publishers, New York, 1965.
3. J. M. Miller, 'Separation Methods in Chemical Analysis', John Wiley, New York, 1975.
4. Yu A. Zolotov, 'Preconcentration in Trace Analysis', *Pure Appl. Chem.*, 1978, **50**, 129.
5. J. Minczewski, J. Chwastowska, and R. Dybczynski, 'Separation and Preconcentration Methods in Inorganic Trace Analysis', Ellis Horwood, Chichester, UK, 1982.
6. A. Mizuike, 'Enrichment Techniques for Inorganic Trace Analysis', Springer Verlag, Berlin, 1983.

7. A. G. Howard and P. J. Statham, 'Inorganic Trace Analysis: Philosophy and Practice', John Wiley, Chichester, UK, 1993.
8. C. Vandecasteele and C. B. Block, 'Modern Methods for Trace Element Determination', John Wiley, Chichester, UK, 1993.
9. G. Tolg, A. Mizuike, Yu A. Zolotov, M. Hiraide, and N. M. Kuz'min, 'Microscale Preconcentration Techniques for Trace Analysis', *Pure Appl. Chem.*, 1988, **60**, 1417.
10. A. Mizuike, 'Preconcentration Techniques for Inorganic Trace Analysis', *Fresenius' Z. Anal. Chem.*, 1986, **324**, 672.
11. 'Preconcentration for Inorganic Trace Analysis', E. Jackwerth, A. Mizuike, Yu A. Zolotov, H. Berndt, R. Hohn, and N.M. Kuzmin, *Pure Appl. Chem.*, 1979, **51**, 1195.

Solvent Extraction

12. M. S. Cresser, 'Solvent Extraction in Flame Spectroscopic Analysis', Butterworth, London, 1978.

Chelation

13. C. Kantipuly, S. Katragadda, A. Chow, and H. D. Gesser, 'Chelating Polymers and Related Supports for Separation and Preconcentration of Trace Metals', *Talanta*, 1990, **37**, 491.
14. S. Sachsenberg, T. Klenke, W. E. Krumbein, and E. Zeeck, 'Back-extraction Procedure for the Dithiocarbamate Solvent Extraction Method. Rapid Determination of Metals in Sea Water Matrices', *Fresenius' J. Anal. Chem.*, 1992, **342**, 163.
15. M. C. Williams, E. J. Cokal and T. M. Niemczyck, 'Masking, Chelation and Solvent Extraction for the Determination of Sub-parts-per-million Levels in High Iron and Salt Matrices', *Anal. Chem.*, 1986, **58**, 1541.

Precipitation and Coprecipitation

16. A. Townshend and E. Jackwerth, 'Precipitation of Major Constituents for Trace Preconcentration: Potential and Problems', *Pure Appl. Chem.*, 1989, **61**, 1643.
17. A. Mizuike and M. Hiraide, 'Separation and Preconcentration of Trace Substances III – Flotation as a Preconcentration Technique', *Pure Appl. Chem.*, 1982, **54**, 1555.
18. Z. Marczencko, 'New Type of Flotation of Ion Association Compounds of Complexes of Multicharged Anions with Basic Dyes', *Pure Appl. Chem.*, 1985, **57**, 849.

Applications

19. R. K. Skogerboe, W. A. Hanagan, and H. E. Taylor, 'Concentration of Trace Elements in Water Samples by Reductive Precipitation', *Anal. Chem.*, 1985, **57**, 2815.
20. R. Van Grieken, 'Preconcentration Methods for Analysis of Water by X-Ray Spectrometric Techniques', *Anal. Chim. Acta*, 1982, **143**, 3.

CHAPTER 4

Inorganic Analytes: Determination

4.1 Introduction

4.1.1 Overview

This chapter deals with the analysis and determination of trace levels of inorganic analytes after the sample has undergone any necessary preparation. The term 'inorganic analytes' includes metallic and non-metallic species, different oxidation states, and the analysis of speciated metals. Some compounds which are technically inorganic, such as nitrosamines, will be treated as organic compounds for their analysis. The reader should refer to the appropriate organic determination section for the relevant information. This chapter covers those techniques that are most commonly employed in inorganic analysis: it complements Chapter 6 on the analysis of organic analytes, Chapter 7 on speciation, and Chapter 8 which covers techniques applicable to both organic and inorganic analytes. The preparation of samples so that they can be presented for final analysis is covered in Chapter 3.

Many of the techniques discussed in the organic determination modules will, in some way, be applicable to the analysis of inorganic analytes. Indeed, both chromatography and mass spectrometry, which account for much of the organic analyte determinations are also very important in the inorganic field. However, one particular field, atomic spectroscopy, is almost exclusive to the analysis of inorganic analytes. It is also, by some considerable way, the most important trace element technique and is given due emphasis in this chapter.

This introduction presents an overview of each of the techniques in the context of the analytes that need to be tackled since this is the point of view from which the practising analyst will be approaching the analysis.

The methods of trace inorganic analyte determination are:

(i)   atomic spectroscopy;
(ii)   elemental mass spectrometry;
(iii)   electrochemical techniques;
(iv)   chromatography of ions;

(v)   spectrophotometry;
(vi) radiochemical methods.

Separation stages can be linked to some of these determination techniques, particularly atomic spectroscopy and mass spectrometry, to provide information on the speciation of the analyte.

There are other techniques that may be used for the determination of inorganic analytes but, because of the infrequency of their use, the fact that they have been superseded, or there are doubts over their applicability to trace analytes, they are not dealt with in any detail here. Such techniques include:

(i)   neutron activation analysis;
(ii)   scanning electron microscopy;
(iii) gravimetry;
(iv)   electron spectroscopy (*e.g.* Auger);
(v)   X-ray diffraction.

4.1.2 Sample Presentation: the Form of the Matrix

This chapter covers mainly the analysis of (reasonably) clean aqueous solutions by an instrumental technique. However, there are many other matrices and sample conditions. These include: natural solid samples; pelletised solid samples; dilute acid solutions; concentrated acid solutions; salt fusion/acid solutions; slurries; organic solutions/suspensions; and natural viscous fluids (*e.g.* food).

Ideally, the end-determination techniques would require a reasonably clean aqueous solution of the analyte. For most analytes dissolution in an appropriate dilute acid provides a stable solution, suitable for analysis by atomic absorption. In a few cases, and these are discussed in the appropriate section, the analysis of solids is feasible and gives sufficiently accurate results [*e.g.* X-ray fluorescence (XRF) or laser ablation]. There are some instances, however, where it will be advantageous to analyse a matrix as a slurry. This can be successfully achieved with a few modifications to the nebuliser of the atomic spectroscopy instrument [and, by extrapolation, inductively coupled plasma mass spectrometry (ICP-MS)]. Use of slurries is expedient since it may allow the analyst to avoid some difficult (and dangerous) sample preparation with several different acids. With any type of solid sample analysis including, to a certain extent, slurries, suitable standards will be required which are often difficult to obtain. For most atomic spectroscopy techniques, a slurry 'concentration' cannot exceed 1% m/v, if aqueous standards are used as calibrants. The analyst has to tailor the pre-treatment stage to suit the end-determination. Concentrated acids or lithium borate fusion should be used with caution to avoid introducing contaminants. Sample introduction systems will also have to be compatible with these more demanding conditions.

When the matrix is purely organic, such as a lubricating oil, which contains traces of metals, it is inadvisable to try to destroy the matrix. It is preferable to use a dispersant or thinning agent such as toluene (methylbenzene) and the resultant mixture analysed directly (for more information on this topic, the reader is referred to Chapter 7 on speciation which deals with the problems of organic eluents in atomic spectroscopy).

One of the stated advantages of using electrochemical methods, particularly ion-selective electrodes, is that they can be used directly for the analysis of 'dirty' viscous fluids. This is illustrated by the direct analysis of tomato ketchup for trace metals. It is unlikely that any spectroscopic technique could successfully analyse such an unprepared matrix. The ability to analyse analytes directly in their natural matrices is one of the main advantages of ion-selective electrodes and has consequently led to their use in field applications and flowing industrial processes. However, doubts about their reliability in terms of both their accuracy and need for maintenance have resulted in their slow uptake. Ion chromatography, by contrast, requires much cleaner (usually aqueous) matrices. Because it is essentially the HPLC of ions, many of the constraints that apply to the organic analysis version will also apply to the 'ionic' version.

4.1.3 The Analyte

It is possible to group inorganic analytes in terms of their chemical nature and hence in the way they can be analysed. These are listed, but for more detail and examples see Chapter 3.

  (i)   nutritional elements;
  (ii)   contaminant metals;
  (iii)   refractory elements;
  (iv)   volatile elements or derivatives;
  (v)   noble metals;
  (vi)   inorganic compounds;
  (vii)   anions;
  (viii) speciated metals.

Unlike organic chemistry, where the groups are much better defined, *e.g.* alkanes, carboxylic acids, lipids, or steroids, the groups listed above are much more vague with some analytes existing in more than one group. However, there are far fewer analytes in inorganic chemistry than in organic chemistry.

For the analysis of cations [groups (i) to (v) above], atomic spectroscopy and mass spectrometry dominate the field; the remaining techniques have had to restrict themselves to the niches left over. These will normally include the analysis of halides and other anions (sulfates, nitrates, *etc.*).

The method of choice is dependent on the chemical nature of the analyte. Sometimes the inherent difficulty of analysing, say, volatile analytes can be turned to the advantage of the analyst. Atomic spectroscopy is useful for analysing those elements on the border of metals and non-metals such as As, Ge, Se, and P. These elements form volatile hydrides which allows direct analysis, combining preconcentration and analysis in one step through 'Hydride generation AAS'.

Ion chromatography and electrochemical techniques tend to fill the remaining niches (anions and inorganic compounds). Which of these techniques is employed will depend on what other analytes need to be analysed. For example, if a group of three halides had to be analysed in the same sample, ion chromatography would suffice. However, the analysis of fluoride is performed far more successfully by the appropriate ion-selective electrode.

The analysis of organometallic complexes (speciated metals) at trace levels has become a matter of concern in environmental monitoring. The 'speciation' of a metal will determine the toxicological properties and transport of that metal through the ecosystem. A classic example is the analysis of butyltin compounds. The speciation of the tin is important in confirming certain properties; some butyltin compounds are classified as rat poisons whereas inorganic tin is not particularly toxic.

4.1.4 Summary

The information in the following sections will, it is hoped, enable the reader to make informed choices on which determination technique should be used. However, it is important that any analysis is considered as a whole, *i.e.* the determination must be compatible with the pre-treatment and *vice versa*. Therefore the whole analysis should be planned taking this into account.

4.2 Atomic Spectroscopy

4.2.1 Overview

This section concentrates on the application of atomic spectroscopy to trace element analysis. There is a natural bias within the related techniques which favours the analysis of metals, rather than non-metals. This uneven split will be reflected in the space devoted to the trace analysis of metals, with less attention being given to the other inorganic elements, such as the halogens.

To assist the reader there is a glossary of abbreviations in Appendix 1.

The periodic table in Figure 4.2.1 shows the range of elements that may be analysed using atomic spectroscopy techniques. Non-metals such as boron, silicon, and phosphorus are included and it is possible to cover some of the remaining elements with some modifications to the standard instrumentation.

A recent survey showed that atomic spectroscopy accounts for over one-third of the applications in the field of inorganic trace analysis.[1] The relative contributions from the four main modes of atomic spectroscopy were as follows:

Atomic Absorption Spectroscopy (AAS)	63%
Atomic Emission Spectroscopy (AES)	23%
X-Ray Fluorescence Spectroscopy (XRF)	12%
Atomic Fluorescence Spectroscopy (AFS)	2%

Another important technique, directly competing with atomic spectroscopy, is inorganic mass spectrometry, which is discussed in Section 4.3.

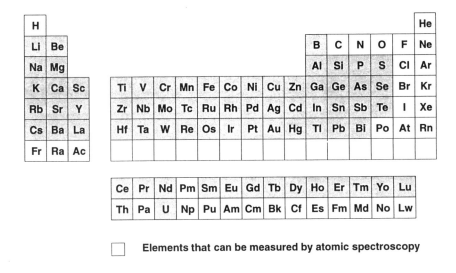

Elements that can be measured by atomic spectroscopy

Figure 4.2.1 *Elements that can be measured by atomic spectroscopy*

Because atomic spectroscopy is based on atomic transitions, it will not, as a rule, provide information on the speciation of the analyte. For example, it will not differentiate between chromium(III) and chromium(VI) oxidation states. Similarly it will not differentiate between, *e.g.* different copper salts, such as the nitrate and the sulfate. Information about the chemical species will usually involve an additional step, such as a chromatographic separation, prior to the atomic spectroscopy measurements. This approach is commonly used for organometallic complexes (see Chapter 7). Ion chromatography does yield information on the oxidation state of the ion.

The outstanding value of atomic spectroscopy in trace analysis is that it provides a means for achieving very specific and quantitative information at trace levels. Typical analyte concentrations can range from 1 ng ml^{-1} to 100 μg ml^{-1}.

Which of the atomic spectroscopy techniques will be employed depends on the nature of the sample and the information that is required from the analysis. One of the first options will be whether to analyse the sample as a solid, slurry, or solution. Section 4.2.2 deals with the form in which different samples can be introduced into the analytical instruments. Section 4.2.3 compares the various forms of atomic spectroscopy, indicating the preferred approach for the different elements.

4.2.2 Comparison of Techniques for Different Sample Types

Much of the work in this section will follow on from the previous chapter on sample preparation. For the purposes of the end-determination, the sample has to be in one of the following forms:

(i)   acid/aqueous solution (*e.g.* sea-water);
(ii) organic solution or dispersion (*e.g.* lubricating oil);
(iii) slurry (aqueous solution containing suspended solids, *e.g.* partially digested sediment);
(iv) solid (*e.g.* rock).

Most atomic spectroscopy systems are designed to accept samples in a solution form. This will usually ensure that samples can be introduced in a reproducible manner. However, vapours, *e.g.* Hg and volatile hydrides, can also be measured. A more detailed comparison of the choice of sample introduction systems is given in Table 4.2.1.

Table 4.2.1 *Characteristics of methods of sample introduction*

Sample mode	Homogeneity	Ease of use	Calibration
Solutions	Good	Good	Easy
Slurries	Depends	Poor	Difficult
Solids	Not good	Good	Difficult

It is possible to measure solids directly. However, most of this section is concerned with the analysis of solutions and slurries. These can be done through the introduction of these matrices into AAS, optical emission spectroscopy (OES), or AFS.

Solid Samples

The first decision to be made is whether the sample can be presented as a solid or if it has to be dissolved using an appropriate digestion procedure. Table 4.2.2 summarises the advantages and disadvantages of analysing solids.

Table 4.2.2 *Measurement of solids by AAS*

Advantages	Disadvantages
Minimal sample preparation	Inhomogeneity of sampling method (*i.e.* localised laser ablation)
Enables depth profiling of sample	Difficult to calibrate accurately
Refractory materials do not need to be dissolved	Generally poorer sensitivity (XRF)

If the sample can be analysed as a solid, then the analysis can be restricted to one of several techniques:

(i)   laser ablation;
(ii)   glow discharge;
(iii)   XRF.

Of the ones listed, XRF is a complete analytical procedure. The other two are methods of introducing a sample into most of the spectroscopic techniques to be discussed in this section and in Section 4.3. Laser ablation and XRF could be considered the most popular, though the former is prohibitively expensive.

X-ray fluorescence can be used for the qualitative, semi-quantitative, or quantitative analysis of elements in solid inorganic samples. It works by the irradiation of a sample using a beam of X-rays with enough energy to displace the electrons in the inner shell of the atoms in the sample. The electrons in the outer shells drop into the vacant 'holes' with a concurrent emission of X-radiation. It is this secondary or fluorescing radiation which, when detected, yields quantitative and qualitative information. The main deficiencies of XRF are relatively high detection limits (when compared with AAs and ICP-MS) and that the analyses are of the surface of the sample. It is particularly suited to monitoring a production process, such as that used in the steel industry. In this context it can provide very rapid analysis, enabling problems with the manufacturing process to be detected at an early stage. The capabilities of XRF are such that it is commonly used as a first screening technique prior to analysis by a confirmatory technique.

Solutions and Slurries

Solutions and slurries are the most common matrices that are analysed by atomic spectroscopy and the remainder of this section will be devoted to these. Tables 4.2.3 and 4.2.4 summarise the merits of solutions and slurries. For more information on the preparation of slurries the reader should refer to Chapter 3.

When analysing a slurry the homogeneity will depend on the particle size. There is a slight contamination risk as a result of the mechanical grinding that is required. These are mostly exhibited by zirconium contamination from the bottle bead method preparation and silicon from an agate rod microniser. These potential sources of contamination are offset by the need for only individual acids for the

Table 4.2.3 *Measurement of solutions by AAS*

Advantages	Disadvantages
Homogeneous	For solid samples the original analyte concentration will be reduced
Easy to dilute	Requires sample digestion
Easy to calibrate	

Table 4.2.4 *Measurement of slurries by AAS*

Advantages	Disadvantages
Easy to prepare	Can cause blockages
Low contamination	Need to be prepared consistently
Calibrated with aqueous solutions	Validity of aqueous standards needs checking

preparation compared with the two or three often required for total decomposition. The reader should refer to Chapter 3 for information on sample preparation and the problems of contamination.

Nebulisers and Spray Chambers

Having identified solution-based systems as the most desirable it is worth comparing the common options available for introducing the solution into the instrument. It is generally thought that all of the current nebulisers and spray chambers are similar in their performance, although some are more suitable for specialised tasks such as slurries. The most popular method is based on producing a fine aerosol which is then allowed to pass through an expansion chamber to remove large droplets. The resulting fine mist is then passed by a carrier gas into the analytical region of the instrument.

The options available for producing an aerosol and removing the larger droplets are summarised in Table 4.2.5.

Table 4.2.5 *Nebulisers*

	Advantages	Disadvantages
Pneumatic		
Cross-flow	Minimal blocking	High sample consumption
Concentric	Block easily	Reduced sample uptake
V-grooved	Good for slurries	More pump noise
Ultrasonic		
Continuous flow	Improved sensitivity	Longer wash out
Discrete volume	Low sample uptake	Less convenient

All the pneumatic nebulisers can be used for ICP systems but only the cross-flow nebuliser would be used for AAS. The ultrasonic nebulisers are increasing in popularity for many analyses.

Expansion Chamber

Usually the expansion chamber is not the most critical part of the sample introduction system. However, the following two points are worth noting:

(i)   a heated spray chamber will generally improve the detection limit of an aqueous system;

(ii) a cooled spray chamber will improve the performance when dealing with organic solvents, *e.g.* for speciation studies, when linking HPLC systems involving organic eluents (see Chapter 7).

4.2.3 Comparison of Techniques

In atomic spectroscopy, one of the first decisions that has to be made is the choice of instrument/technique. This decision will normally be made on the basis of the expected nature of the sample and the analytes of interest. The decision will be between, for the most part, atomic absorption techniques and plasma emission techniques. This is reflected by the prevalence of these two techniques in many analytical laboratories. For single element analysis the method of choice would often be atomic absorption. Where there are several elements to be analysed in any one sample, or if the elements are likely to have formed stable complexes that need to be broken down, as in the analysis of refractories, the method of choice would be plasma emission. However, it must be noted that flame techniques can handle organic matrices better than inductively coupled plasmas, while direct current plasmas are good with any matrix including solids. These generalities are useful in the establishment of the analytical method.

The three main atomic spectroscopy techniques are AAS, OES/AES, and AFS. The configuration of the instruments determines the ease with which the three techniques can be used in a multi-element mode.

General aspects of the three main forms of atomic spectroscopy can be dealt with in broad terms. Table 4.2.6 summarises some of the key features. Some of these factors will be influenced by considering the different options within the three main categories. Other considerations, especially sensitivity, are very dependent on the specific technique used.

Table 4.2.6 *Comparison of atomic spectroscopy modes*

	AAS	OES	AFS
Multi-element	No[a]	Yes	No
Range of elements	Good	Good	Good
Dynamic range	Poor	Good	Good
Specificity	Very good	Medium	Good
Cost	Medium	High	Low

[a] AAS may be multi-element, but this must be done sequentially. Some more advanced instruments have a turret that may hold four hollow cathode lamps, and the software will allow at least these four analytes to be determined sequentially using an autosampler.

Atomic Absorption Spectroscopy (AAS)

A key feature of most AAS procedures is that they are for single element analysis. The three different forms are discussed.

Flame AAS (FAAS).   Flame AAS involves introducing the sample in the form of a fine aerosol which is mixed with the necessary gases to support an air/acetylene flame. The atom concentration is monitored by positioning the flame in an optical light path. Thus the atoms have a limited period in the measuring zone as they are swept upwards by the gas flow supporting the flame. Alternatively, a hydrogen/argon/diffused air flame is used for low wavelength lines (*e.g.* tin at 224.6 nm).

The nitrous oxide flame provides a hotter atomisation zone and also an environment that is 'lean' in oxygen content. Both these factors make it an ideal atomiser for refractory elements, such as aluminium. These type of elements tend to form very stable oxides, which in turn can rapidly reduce the sensitivity of the analytical method.

In Figure 4.2.2, the periodic table shows the elements best dealt with using a nitrous oxide/acetylene flame.

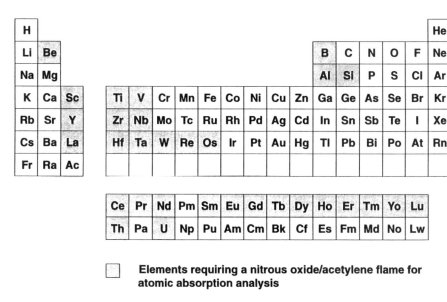

☐ **Elements requiring a nitrous oxide/acetylene flame for atomic absorption analysis**

Figure 4.2.2 *Elements requiring a nitrous oxide/acetylene flame for atomic absorption analysis.*

Electrothermal AAS (ETAAS).   The alternative of ETAAS is most commonly performed using a graphite tube. The tube is aligned longitudinal to the optical light path of the instrument. The sample is introduced into the furnace through a small hole in the side of the tube. Manual or more commonly automated injection of a fixed volume (typically 5–50 μl) is the norm. The furnace is electrically heated in stages, with the final hottest stage releasing the atomic species. Electrothermal methods do need a degree of operator skill.

Because this process is taking place in the contained volume of the graphite tube, the residence time for the atoms in the light beam is much longer than for the flame atomisation system. This factor, and to a lesser extent the poor transport efficiency of the sample introduction system of the FAAS, is responsible for greatly improved sensitivity of the furnace compared with the flame mode. Only 10–15% of the nebulised sample reaches the flame.

These and other aspects of FAAS and ETAAS are compared in Table 4.2.7 along with hydride/cold vapour AAS.

Table 4.2.7 *A comparison of AAS techniques*

| | *Flame* | | *Electrothermal* | *Hydride/cold vapour* |
	Air/acetylene	*Nitrous oxide/ acetylene*		
Cost	Low	Medium	Very high	Medium
Elemental scope	Good	Very good	Good	Poor
Speed/ease of use	Good	Good	Medium[a]	Medium
Sensitivity	Poor	Medium	Very good	Good
Safety risk	Medium	High	Low	Medium

[a]Assumes automated sample introduction system.

There are alternatives to the graphite tube as a means to achieving electro-thermal atomisation; a solid graphite rod has been used in various forms. However, the hollow tube has established itself as the most common form and the comments in Table 4.2.8 are all directed at this device.

Table 4.2.8 *Advantages and disadvantages of electrothermal atomic atomisation*

Advantages	*Disadvantages*
Very sensitive	Requires very clean reagents to avoid blank problems
Modern automated sample introduction makes for ease of use	Instruments with good background correction (*i.e.* Zeeman) are expensive
Can analyse solids	Graphite surface can interact with sample
Low sample volume required	Prone to interferences
	Matrix modifiers may cause contamination

The development of effective background correction (an essential feature) and automated sample introduction has made ETAAS a much more effective technique. These are discussed under the section on interferences. Like all absorption techniques it has a somewhat limited dynamic range.

Hydride/Cold Vapour AAS. The additional option of the hydride generation technique for Ge, As, Sb, Se, Te, and Bi is specific to these elements. The cold vapour technique for Hg can also be added to this category, as an element-specific option. In both cases the hydride (or cold vapour) stage can be viewed as having the advantages of acting as a clean-up stage and as a preconcentration step for the analyte. It is possible that, when analysing for mercury, some of it will exist as organo-mercury complexes. It is good practice to add bromate to the sample to break down these complexes and thus avoid the possible associated fluorescence. Additionally, it is good practice to form complexes of the transition metals, *e.g.* nickel.

The hydride generation method is particularly useful for those elements lying on the boundary between metals and non-metals. These elements have an innate lack of sensitivity, and the hydride generation step provides a neat way of compensating for this limitation.

The earlier hydride systems required samples to be prepared in individual flasks. This reduced the convenience of dealing with large numbers of samples and restricted the sample throughput. Modern flow injection systems are now providing an improvement. The automated valve control can enable batches of samples to be dealt with sequentially, with minimal operator interaction.

Atomic Emission Modes (OES/AES)

The following sections deal with the various forms of atomic emission. The major difference between OES and AAS is that the emission modes lend themselves to multi-element analysis. Table 4.2.9 summarises the features of the main emission techniques.

Of all the emission techniques the ICP ones now dominate the field. They compare well in terms of sensitivity with the various forms of FAAS but offers significant advantages for multi-element analysis. They have a very good dynamic range. The torroidal shape of the plasma, with the sample passing through a central channel, ensures efficient atomisation and excitation, with minimal chemical interferences or problems of self-absorption. Detailed aspects of the various emission systems are described below.

Flame Emission Techniques. Flame OES is a term normally used for low-temperature systems which are specifically set up to analyse the alkali elements (*i.e.* Na, K). They are relatively cheap systems which perform a sensible role for well-defined and limited matrix materials. The following sections deal with the more widespread forms of flame emission.

The flame emission techniques are limited to the range of elements they can cover. This is mainly a function of the flame temperature which can be improved by using nitrous oxide instead of air as the oxidant, see Table 4.2.10.

Table 4.2.9 *Atomic emission modes*

Technique	Advantages	Disadvantages
Flame emission		
Air/acetylene	Cheap	Limited elemental range Usually linked to single element detection system
Nitrous oxide/acetylene	Cheap Reasonable element range	Detection system usually geared to single element
Plasma emission		
ICP	Multi-element Good dynamic range Minimal matrix interferences Sensitive	Expensive
DCP	Cheaper than ICP	More prone than ICP to chemical interferences[a]
MIP[b]	Good for non-metals, halogens, P, S, O, and N	Not robust. Only suited to gaseous samples No solvents allowed
Glow discharge	Good conducting solids Avoids dissolution problems	Difficult to quantify Needs significant amount of sample Sample throughput not slick
Spark/arc	Cheap, photographic-based instruments	Not quantitative Sample preparation is tedious

[a]These can be controlled by adding lithium (but this can cause contamination problems).
[b]Microwave induced plasma.

Table 4.2.10 *Comparison of emission source temperatures*

Emission source	Temperature (K)
Air/acetylene	2200
Nitrous oxide/acetylene	3200
DCP	5000
ICP	7000
MIP[a]	$12000 - 15000$[b]

[a]Microwave induced plasma
[b]Plasma not in thermal equilibrium, kinetic temperature ~8000 K.

Even with the nitrous oxide/acetylene flame there is still a clear necessity to 'tune' the flame conditions to suit individual elements. This automatically detracts from the inherent multi-element advantage of emission techniques.

Invariably flame emission systems are options on an atomic absorption-based system. Because of this the optics and detection system are designed with the normal single element requirements of AAS.

Plasma Emission Spectrometers. Plasma atomic emission spectrometers are based on inert ionised gases which are raised to high temperatures by induction, microwave, or direct current processes. The key characteristics of these plasmas is that they can be operated at atmospheric pressure. A detailed summary of the three main plasma systems is given in Table 4.2.11.

Table 4.2.11 *Comparison of plasma sources*

	MIP[a]	ICP	DCP
Elemental range	Excellent for non-metals	Very good for metals and B, P	Good
Tolerance of solvents	Predominantly gas phase	Fine for aqueous. Organics require care	No problem
Dynamic range	Limited	Excellent	Good
Non-spectral interferences	Prone to problems	Very few[b]	Not as good as ICP
Cost	Running costs are high	High capital cost	Medium
Tolerance to solids	Not good	Particle size needs to be small	No problem

[a]Microwave induced plasma.
[b]Few interferences due to good energy transfer characteristics.

In general the ICP is the favoured plasma source. The microwave induced plasma (MIP) is particularly good for non-metals as shown in Figure 4.2.3. The DCP can provide a simpler and cheaper alternative to ICP sources.

The simplest detection system for a plasma emission source is based on a computer controlled monochromator which is programmed to scan the relevant wavelength regions. This is usually referred to as a *sequential* instrument. The normal alternative is a wavelength dispersive diffraction grating with a battery of photomultipliers sited at the wavelengths of interest. This is usually referred to as a *polychromator* instrument, with a true simultaneous multi-element capability.

A more recent innovation has been the combination of a two-dimensional dispersive system (*i.e.* an echelle spectrometer) with numerous semiconductor strips [charge coupled device (CCD) or charge injection device (CID)] covering all the wavelength regions of interest. This is the closest electronic equivalent to

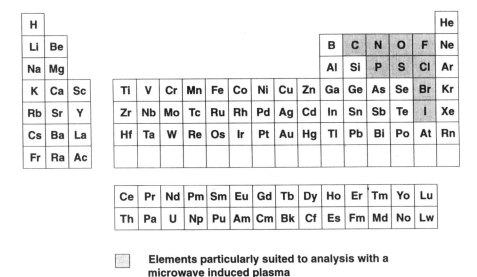

Elements particularly suited to analysis with a
microwave induced plasma

Figure 4.2.3 *Elements particularly suited to analysis with a microwave induced plasma*

the comprehensive wavelength coverage provided by the now outdated photographic plate detection systems. The key features of these different systems are shown in Table 4.2.12.

Table 4.2.12 *Detection systems for plasma emission sources*

	Sequential	Polychromator	CCD/CID echelle
Cost	Medium	High	High
Speed	Slow	Fast	Fast
Flexibility	Good	Poor	Good
Detection limits	Medium	Good	Very good
Background	Good	Poorly defined	Good

The polychromator systems require very stringent control of temperature. This is to ensure the position of slits of the individual detectors coincides exactly with the very specific wavelengths of the atomic emission lines. On modern polychromators this is achieved using a self-contained thermostatting system.

Both the sequential and echelle/CID systems have the potential for dynamic adjustment to locate the exact positions of the atomic emission lines. A further advantage of these two systems is that they permit the measurement of the background, in the regions immediately adjacent to the atomic wavelengths. Most polychromator systems do not have this as a routine feature.

Atomic Fluorescence

Atomic fluorescence can be applied to flame, plasma, and cold vapour systems. It has evolved to play a key role for certain specific elements. In particular, elements suited to hydride generation or cold vapour treatments, have been successfully adapted to instruments based on fluorescence measurements. A summary of fluorescence modes is given in Table 4.2.13.

Table 4.2.13 *Fluorescence modes*

	Flame/hydride	*Plasma*	*Cold vapour*
Range of elements	Limited	Good	Hg only
Cost	Low	Medium	Low
Linearity	Good	Good	Good
Sensitivity	Very good	Medium	Good
Interferences	Poor	Good	Medium

In theory atomic fluorescence using a plasma system should provide a significant saving. Interference filters are adequate to isolate the elemental fluorescence thus obviating the need for a high performance dispersive grating. However, each element requires its own excitation source/interference filter/photomultiplier. Although it has the potential to be a multi-element technique, the practicalities usually result in instruments designed for single element operation.

There are several excitation sources that exist for atomic fluorescence. Normal hollow cathode lamps may be used but the resulting fluorescence is not very intense. Boosted lamps increase the sensitivity and lasers yield more sensitivity. The cost increases as the instrumentation becomes more complex.

The hydride generation mode provides several clear advantages, just as it does for atomic absorption measurements. It results in a combined preconcentration and clean-up stage, which can reduce spectral and chemical interferences. However, it should be noted that the hydride formation step will itself still be affected by matrix and sample digestion factors which will influence the yield of the hydride species.

The advantages of operating hydride systems in the fluorescence rather than absorbance mode are two-fold. Firstly the fluorescence detection system can be cheaper. Secondly the inherent detection limits for the fluorescence system will generally be superior to those achieved by an absorption instrument.

4.2.4 Interferences and How to Deal with Them

Generally, interferences are applicable to both absorption and emission techniques. These can distort the results obtained for a particular analysis and so need to be taken into account. They can be split up into two groups, spectral and chemical interferences.

Spectral Interferences

It is generally acknowledged that atomic absorption measurements are less troubled by spectral interferences than the equivalent emission techniques. Indeed, spectral interferences for atomic absorption are barely worth mentioning. However, both flame emission and argon plasma atomic emission studies will have spectral interference problems needing action.

Absorption Measurements. Atomic absorption measurements have the advantage of being based on the very narrow bandwidth of the light emitted by a hollow cathode lamp. For another element to interfere with this radiation, it must possess an absorption line within 0.01 nm. Some examples of this are given in Table 4.2.14. In addition lines in the low ultraviolet suffer from scatter, which causes a pseudo-absorption signal. This can be a problem with, for example, As at 193.7 nm. Most of these examples are spurious as the Fe and Al lines given are minor and Co as an interfering element is very rare in nature.

Table 4.2.14 *Examples of interfering elements*

Analyte of hollow cathode lamp	Interfering element	Wavelength separation (nm)
Fe 271.902	Pt 271.904	0.002
Al 308.215	V 308.211	0.004
Hg 253.652	Co 253.649	0.003

Another form of spectral interference is due to molecular species. It has been found that CaOH absorbs at the Ba wavelength of 553.55 nm, thus giving enhanced figures for the concentration of barium. Cyanogen emission at 357.4 nm interferes with the 357.4 nm La line. Solutions to these and problems associated with high scatter can be achieved using continuum background correction, Smith Hieftje[2] correction, or Zeeman[3]-based background correction facilities.

Emission Studies. For emission studies, the resolution between potentially interfering wavelengths is determined by the optics of the detection system. This will usually mean that emission lines within 0.5 nm can interfere depending on the relative abundances of the two elements. An example of this is Al 396.15 nm and Ca 396.84 nm. A 20-fold abundance of Ca to Al will be sufficient to influence the emission intensity at the Al wavelength when using a 1 m monochromator.

In both the atomic absorption and atomic emission examples the most common solution is to select an alternative wavelength. Using a weaker absorption/emission wavelength is also a way of avoiding unnecessary dilutions.

Molecular interferences also exist in emission experiments with notable spectral structure from OH⁻ bands at 295–310 nm. The main solution is to look for alternative wavelengths. Many modern emission instruments now supply spectral interference tables as part of their computer software. This may be in the form of a spectral atlas with likely interferences for the individual elements. Wavelength reference books are also available.[4]

Chemical Interferences

Both atomic absorption and atomic emission measurements can suffer from chemical interferences. Premature claims were made for ICP, suggesting negligible chemical interferences. Such claims have not stood the test of time and were considered unrealistic by most experienced analysts.

As a general rule the hotter the 'flame' the fewer the chemical interferences. Thus a progression from air/C_2H_2 to N_2O/C_2H_2 to argon plasma, will generally lead to a reduction but not elimination, of potential chemical interferences. The possibilities of chemical interferences are too numerous to summarise in detail. Some common examples of chemical interferences are given in Table 4.2.15.

Table 4.2.15 *Examples of chemical interferences*

Element	Interference	Comment
Al	Al_2O_3 in air/ C_2H_2 flame	Use N_2O/C_2H_2 flame for AAS
Ca	Suppressed if PO_4^{3-} present	Complex formed. Use hotter flame or add La to bind PO_4^{3-}
Mg	Al, Ti, and Zr	Add 8-hydroxyquinoline

Low analyte concentration may occur because of shifts in ionisation equilibria. This can be eliminated by addition of an ionisation suppressor (*e.g.* Na or K) which provides a relatively high electron concentration. It is common practice therefore to add an excess of one of these elements to both samples and standards to compensate for this.

Some of the solutions that can be considered for dealing with chemical interferences include:

(i)   matrix matching of standards;
(ii) using internal standards;
(iii) standard additions;
(iv) chemical separation/extraction of analyte.

4.2.5 Hints and Tips

Units of Concentration

The commonly used unit abbreviations for concentration, such as parts per million (ppm) or parts per billion (ppb) can cause confusion, especially if it is not specified if they refer to weight/volume or weight/weight. It is preferable to use mg ml^{-1} or mg g^{-1} *etc.*

The tendency is to assume that 1 ml of solution equates exactly to 1 g. This will be a source of error as high solid content or strong acid solutions will significantly alter the density of aqueous solutions. For example 10% HNO$_3$ has a density of 1.039 g ml^{-1} (at 20°C). Thus an error of ~4% will be made if it was assumed that 1 ml = 1 g.

Alternative Methods of Calibration

Always try to match samples and standards for the acid strength and matrix composition.

If the samples have an unpredictable matrix then the use of standard additions should be considered. This involves spiking aliquots of the same sample with increasing amounts of the analyte of interest. It should be remembered that the spike should be a very small volume of a high concentration, *e.g.* if there is 100 ml of sample a spike of 1 ml is more than sufficient. This prevents the dilution of any matrix effects. This provides a perfect match of acid and matrix factors between standards and samples. However, it is very time consuming and still leaves scope for differences due to the chemical form (*e.g.* oxidation state) of the analyte in the original sample.

In atomic emission systems use may be made of an internal standard to aid the calibration and concentration measurements. For this to be useful it is important that the internal standard matches the excitation characteristics of the analyte element.

Accuracy

Avoid preparation steps that involve dispensing very small volumes (*i.e.* less than 1 ml). The potential errors can have a large influence on the final result.

Avoid excessive dilution stages as these will amplify any error when converting the instrumental measurement to a final concentration figure.

Use background correction for all ETAAS work whenever possible.

Extending the Dynamic Range

Rotating the burner head in FAAS shortens the analytical pathlength, enabling more concentrated solutions to be analysed without the need for further sub-dilutions.

Use alternative atomic lines. This can be applied to emission and absorption systems, although the choice of suitable alternative lines is likely to be more extensive in the emission mode. Again this approach can avoid the need for additional dilution steps and avoid the risk of increased errors.

Using an alternative line is usually a very attractive way to extend the dynamic range. However, for some elements this is not feasible. Mercury has no second line, and for zinc the secondary line (307.6 nm) is 3500 times less sensitive than the primary line (213.9 nm).

Indirect Determination of Non-metals

Analytes such as the halogens, ammonia, and cyanide can be measured indirectly using AAS. This works by the addition of a known amount of an atomic analyte which then precipitates out along with the target analyte. The remaining metal concentration, in solution, is then determined by conventional AAS.

Some examples are given in Table 4.2.16; these are reviewed along with some other examples.[5]

Table 4.2.16 *Examples of indirect measurement of non-metallic analytes*

Analyte	Matrix	Atomic analyte
Ammonia	Nitrogen compounds	Ag
Iodide	Seaweeds	Hg
Cyanide	Water	Cu

Single Element and Multi-element Standards

Avoid using commercial atomic *absorption* stock standard solutions for multi-element atomic *emission* work. There is a risk that what would be an insignificant contaminant in a single element AAS standard, could make a significant contribution to the analyte level in a multi-element standard with widely differing elemental concentrations. Use standards specified for multi-element atomic emission analysis.

Preconcentration of Samples

Improved detection limits can be achieved by preconcentrating the sample. Some well-established methods are:

(i)   complexation and solvent extraction;
(ii) pre-columns used with a flow injection analysis system (FIAS);
(iii) electrochemical deposition;
(iv) coprecipitation (see electrochemical techniques).

More information on each of these topics can be found in Chapter 3.

'One-off' Samples

'One-off' samples invariably present special problems. Ensuring that results are reliable from new or unusual samples will entail significant time and effort; often 10 times more work than running additional samples for an established method.

The idea that reliable data can be provided from a 'quick analysis' should be discouraged. It is possible to obtain a quick scan of a sample: however the analyst and the customer must be clear of the limitations on the analytical accuracy of such an approach.

Limited Sample Volume

Small sample volumes, where dilutions are to be avoided, can be dealt with by using discrete sample cups.

Some alternatives to conventional sample introduction can be used to preserve samples of limited volume.

(i)   Discrete sampling cups can be used to limit the volume aspirated (~100 µl) into a flame or plasma.
(ii)   A FIAS system can be used to introduce discrete aliquots.
(iii) Ultrasonic nebulisers can reduce the sample uptake rate.
(iv)   Electrothermal techniques are very economical with sample consumption.

In all cases it is preferable to use chart recorders or software capable of trapping transient signals.

Ultrasonic Agitation

For quick digestion or leaching of suitable samples (usually finely divided particles, air filters), ultrasonic agitation can provide a very convenient alternative to traditional heating or fusion techniques.

Detection Limit

Approximate guidance of adjusting detection limits, as the matrix becomes more complex, is given in Table 4.2.17.

Table 4.2.17 *Detection limit adjustment*

Detection limit adjustment factor[a]	Matrix
×1	Aqueous solutions
×10	Food, body fluids
×100	Metallurgical samples
×1000	Geological samples

[a]If the detection limit for an analyte in an aqueous solution is 2 µg ml^{-1}, it would be expected to be 20 µg ml^{-1} in a body fluid.

4.2.6 Factors Affecting Quality

In addition to the advice on how to obtain more reliable results given in the previous section, there are aspects to do with quality that are simple to carry out and in many cases imperative.

Standards and Samples

Avoid having concentrated standards in close proximity to samples. Where possible keep samples and standards separate prior to analysis. These precautions should help minimise the risk of cross-contamination.

Sample Introduction

Consistent performance of nebulisers throughout a run is critical if the calibration procedures are to be valid for each sample. A trustworthy washing procedure is very important. The flow injection technique will often be very useful for this.

Automated introduction of samples into graphite furnaces has now reduced the errors associated with manual pipetting. It is important to align the introduction mechanism properly.

Analytical Range

Reporting results close to the detection limit or the blank level can lead to very misleading data. There should be set procedures for dealing with results at three times or 10 times the signal-to-noise ratio.

New Data Manipulation Programmes

New versions of software should be validated for each specific application. Extending the scope (*e.g.* to additional elements), should also be checked before reporting results. Aspects such as inter-element corrections should be fully assessed.

Other examples of good practice should include:

(i)   overlapping 'new' and 'old' batches of standards.
(ii) ensuring realistic blanks are included. These should have been through as many of the sample preparation stages as possible.

Finally, it is vital that all results should be reported with realistic uncertainty limits applied to the numerical values.

Uncertainty

Most of the steps contributing to uncertainty of the final result are common to many forms of analysis. The following steps provide guidance on means of measuring uncertainty:

(i)   use certified reference materials to give an overall estimate of uncertainty;
(ii)   carry out 'spiked' experiments to assess recoveries;
(iii)   analyse samples using an alternative technique(s);
(iv)   repeat analysis of selected 'typical' samples;
(v)   use several analysts to assess the operator-dependent factors.

4.2.7 Safety Features

There are a number of safety features that should be observed in carrying out atomic spectroscopy.

Use of Gases

(i)   Before using acetylene flames note the existing legislation and the terms of the licence granted by the Health and Safety Executive or the Guidance document provided by instrument manufacturers.
(ii)   Use flash back arrestors in the gas supply line.
(iii)   Avoid Cu tubing, as this can lead to the formation of explosive acetylide.
(iv)   Acetylene cylinders should not be run 'dry'.
(v)   Modern expansion chambers should have a pressure diaphragm.
(vi)   Care is needed in switching from acetylene/air to acetylene/nitrous oxide flames. On modern instruments this is now an automated facility.

Chemicals

Special care should be taken in handling concentrated acids. In particular hydrofluoric and perchloric acids should be treated with extreme caution. Some analysts suggest avoiding the use of perchloric acid.

Instrumental

High voltages on ICP instruments should have manufacturers safety switches fitted which avoid dangerous access to the instrument. These switches should not be overridden.

Operators should avoid any direct line of sight with UV-emitting sources. Viewing ICP plasmas should only take place with appropriate UV safety spectacles.

Fume cupboards are essential for many digestion procedures. Digestions involving the use of pressurised bombs should be carried out behind an appropriate screen.

4.2.8 Applications

Application reviews appear biannually in *Analytical Chemistry*. Another more regular source of application reviews is in the *Journal of Analytical Atomic Spectroscopy*, which has review sections each year that cover the areas of

Environment, Food, Clinical, and Metallurgy/Chemicals. *ICP Information Newsletter*, is another useful source of current developments.

The applications listed in Table 4.2.18 have been organised under sample type rather than pursuing each element in turn. Following each selected category there are a number of key references that can be used as an entry point to the literature in this area.

Table 4.2.18 *Applications of atomic spectroscopy*

Categories	*References*	
Food and beverages	General	11,12
	AAS	13,14
	Plasma	15,16
Geological	General	17
	AAS	18,19
	ICP	20,21
Metallurgy	General	22
	AAS	23,24
	OES	25
Chemicals (oils and organics)	General	26
Clinical	General	27
	AAS	28, 29
	ICP	30, 31
Environment	AAS	18,19,32
	ICP	33
Environment, water		34,35,36
Environment, soil		37,38,39

4.2.9 References

1. T. Braun and S. Zsindely, 'Comparative Standing of Individual Instrumental Analytical Techniques', *Trends Anal. Chem.*, 1992, **11**, 267.
2. S. B. Smith Jr and G. M. Hieftje, 'A new Background-correction Method for Atomic Absorption Spectrometry', *Appl. Spectrosc.*, 1983, **37**, 419.
3. W. Slavin, G. R. Carnrick, D. C. Manning, and E. Pruszkowska, 'Recent Experience with the Stabilized-temperature Platform-furnace and Zeeman Background Correction', *At. Spectrosc.*, 1983, **4**, 69.
4(a) M. L. Parsons, 'Atlas of Spectral Interferences in ICP Spectroscopy', Plenum Press, New York, 1980.
  (b) R. K. Winge, V. A. Fassel, V. J. Peterson, and M. A. Floyd, 'Inductively Coupled Plasma – Atomic Emission Spectroscopy: an Atlas of Spectral Information', Elsevier, Amsterdam, 1985.
5. K. W. Jackson and H. Qiao, 'Atomic Absorption, Atomic Emission, and Flame-emission Spectrometry', *Anal. Chem.*, 1992, **64**, 50R.

General Reading

6. L. Ebdon, 'An Introduction to Atomic Absorption Spectroscopy', Heyden, London, 1982.
7. G. L. Moore, 'Introduction to Inductively Coupled Plasma–Atomic Emission Spectrometry' (Analyst Spectroscopy Lib. Vol.3), Elsevier, Amsterdam, 1989.
8. J. Anwar, M. I. Farooqi and M. M. Atthar, 'Indirect Determination of Ammonia by Atomic Absorption Spectroscopy', *J. Chem. Soc. Pak.*, 1989, **11**, 117.
9. D. Chakraborty and A. K. Das, 'Indirect Determination of Iodide in Seaweeds by Cold-vapour Atomic Absorption Spectrophotometry', *At. Spectrosc.*, 1988, **9**, 189.
10. S. Chattaraj and A. K. Das, 'Indirect Determination of Free Cyanide in Industrial Waste Effluent by Atomic Absorption Spectrometry', *Analyst*, 1991, **116**, 739.

Applications

11. A. Taylor, S. Branch, H. M. Crews, and D. J. Halls, 'Atomic Spectrometry Update – Clinical and Biological Materials, Foods and Beverages', *J. Anal. At. Spectrom.*, 1993, **8**, 79R.
12. H. M. Kuss, 'Applications of Microwave Digestion Technique for Elemental Analyses', *Fresenius' J. Anal. Chem.*, 1992, **343**, 788.
13. R. Barbers and R. Farre, 'Determination of Cobalt in Foods by Flame and Electrothermal-atomisation Atomic Absorption Spectrometry. A Comparative Study', *At. Spectrosc.*, 1988, **9**, 6.
14. T. C. Rains, 'Application of Atomic Absorption Spectrometry to the Analysis of Foods', *Anal. Spectrosc. Libr.*, 1991, **5**, 191.
15. W. A. Grant and P. C. Ellis, 'Determination of Heavy Metals in Shellfish by Flame Atomic Absorption Spectrometry and Inductively Coupled Plasma Atomic Emission Spectrometry', *J. Anal. At. Spectrom.*, 1988, **3**, 815.
16. J. W. Jones, 'ICP AES: a Realistic Assessment of its Capabilities for Food Analysis', *J. Res. Natl. Bur. Stand.*, 1988, **93**, 358.
17. T. T. Chao and R. F. Sanzolone, 'Decomposition Techniques', *J. Geochem. Explor.*, 1992, **44**, 65.
18. M. S. Cresser, J. Armstrong, J. Dean, M. H. Ramsey, and M. Cave, 'Atomic Spectrometry Update – Environmental Analysis', *J. Anal. At. Spectrom.*, 1991, **6**, 1R.
19. M. S. Cresser, J. Armstrong, J. Dean, P. Watkins, and M. Cave, 'Atomic Spectrometry Update – Environmental Analysis', *J. Anal. At. Spectrom.*, 1992, **7**, 1R.
20. I. Jarvis and K. E. Jarvis, 'Plasma Spectrometry in the Earth Sciences: Techniques, Applications and Future Trends', *Chem. Geol.*, 1992, **95**, 1.
21. G. E. M. Hall, 'Inductively Coupled Plasma Mass-spectrometry in Geoanalysis', *J. Geochem. Explor.*, 1992, **44**, 201.
22. J. Marshall, J. Carroll, and S. T. Sparkes, 'Atomic spectrometry Update – Industrial Analysis: Metals, Chemicals and Advanced Materials', *J. Anal. At. Spectrom.*, 1989, **4**, 251R.
23. K. Ohls and D. Sommer, 'Applications of Atomic Absorption Spectrometry', *Anal. Spectrosc. Libr.*, 1991, **5**, 227.
24. G. Hein, 'Determination of Trace Elements in Steel by Means of Flame Atomic Absorption Spectrometry', *Metalurgija*, 1990, **29**, 29.
25. R. S. Pomeroy, R. D. Jalkian, and M. B. Denton, 'Spark Spectroscopy Using Charge Transfer Devices: Analysis, Automated Systems and Imaging', *Appl. Spectrosc.*, 1991, **45**, 1120.

26. M. R. Todorovic, 'Determination of Trace Metals in Petroleum and Petroleum Products by Atomic-emission Spectroscopy Methods', *J. Serb. Chem. Soc.*, 1987, **52**, 499.

27. A. Taylor, S. Branch, H. M. Crews, and D. J. Halls, 'Atomic Spectrometry Update – Clinical and Biological Materials, Foods and Beverages', *J. Anal. At. Spectrom.*, 1993, **8**, 79R.

28. J. M. Harnly and D. L. Garland, 'Multielement Atomic Absorption Methods of Analysis', *Methods Enzymol.*, 1988, **158**, 145.

29. H. T. Delves and I. L. Shuttler, 'Elemental Analysis of Body Fluids and Tissues by Electrothermal Atomisation and Atomic Absorption Spectrometry', *Anal. Spectrosc. Libr.*, 1991, **5**, 381.

30. K. A. Wolnik, 'Inductively Coupled Plasma-emission Spectrometry', *Methods Enzymol.*, 1988, **158**, 190.

31. P. Schramel, 'ICP and DCP Emission Spectrometry for Trace Element Analysis in Biomedical and Environmental Samples. Review', *Spectrochim. Acta, Part B*, 1988, **43**, 881.

32. M. S. Cresser, J. Armstrong, J. Cook, J. R. Dean, P. Watkins, and M. Cave, 'Atomic Spectrometry Update – Environmental Analysis', *J. Anal. At. Spectrom.*, 1993, **8**, 1R.

33. T. Jickells, M. M. Kane, A. Rendell, T. Davies, M. Tranter, and K. E. Jarvis, 'Applications of Inductively Coupled Plasma Techniques and Pre-concentration to the Analysis of Atmospheric Precipitation', *Anal. Proc. (London)*, 1992, **29**(7), 288.

34. M. Blankley, A. Henson, and K. C. Thompson, 'Waters, Sewage and Effluents (and Atomic Absorption Spectrometry)', *Anal. Spectrosc. Libr.*, 1991, **5**, 79.

35. J. A. C. Broekaert, 'Use of ICP Atomic Spectrometry for Water Analysis', *Tech. Mess.*, 1992, **59**, 147.

36. J. C. Fischer and M. Wartel, 'Trace-metals in Marine Environments', *Analusis*, 1992, **20**, M38.

37. J. Einax, B. Machelett, S. Geiss, and K. Danzer, 'Chemometric Investigations on the Representativity of Soil Sampling', *Fresenius' J. Anal. Chem.*, 1992, **342**, 267.

38. M. Cresser, 'Analysis of Polluted Soils (by Atomic Absorption Spectrometry)', *Anal. Spectrosc. Libr.*, 1991, **5**, 515.

39. A. M. Ure, 'Trace Elements in Soil: their Determination and Speciation', *Fresenius' J. Anal. Chem.*, 1990, **337**, 577.

4.3 Elemental Mass Spectrometry

4.3.1 Overview

Inductively coupled plasma mass spectrometry (ICP-MS) represents the state of the art in inorganic analytical chemistry. The technique is capable of producing semi-quantitative data for most of the elements in the periodic table, both in aqueous and in organic matrices, in under 2 min. The technique can be used for quantitative determinations with detection limits in the parts per trillion range (ng kg^{-1}) with typically a relative standard deviation of 5%.

There are several different elemental mass spectrometry techniques. However, the majority of these are now infrequently used. Inductively coupled plasma mass spectrometry is the one elemental technique that is widely used and increasingly so. The other techniques are discussed briefly in Section 4.3.2, where they are summarised in relation to ICP-MS. Most of this section deals with ICP-MS.

The first mass spectrum of a plasma was obtained in the early 1980s, and although these first spectra gave poor sensitivities, they were relatively simple and free from interferences. The ICP-MS technique was born out of ICP-optical emission spectrometry (OES), which is a well-established technique used in a wide range of disciplines, from the steel industry to clinical trials. Although ICP-OES is capable of producing detection limits to nearly the low ng g^{-1} range, it does suffer from spectral overlap from other elements in the sample. The ICP-MS technique, although itself prone to interferences, gives a much simpler spectrum than ICP-OES, and yields detection limits in the sub-nanogram g^{-1} concentration range. The ICP-MS technique is typically two to three orders of magnitude more sensitive than ICP-OES. Both are multi-element techniques, ICP-MS being able to determine the majority of the elements in the periodic table in one acquisition, while ICP-OES instruments typically are limited to between 35 to 45 elements per analysis. The ICP-MS technique is tolerant of total dissolved solids only to around 1% w/v, whereas ICP-OES is generally tolerant to around 2% w/v. The detection limits achievable by ICP-MS are generally one order lower than those obtained by Graphite Furnace Atomic Absorption Spectroscopy (GFAAS), although GFAAS is a single element technique.

The ICP-MS technique is still relatively new; the first commercially available instruments appeared in 1983. The vast number of papers in the scientific literature relating both to fundamental aspects of ICP-MS and applications of the technique pay testimony to its role in a modern analytical laboratory. The ICP-MS instruments may now be used for a wide variety of sample types from water to oils, soils to plastics, foods to wastes.

The basic advantages of ICP-MS over existing techniques are:

(i)   lower limits of detection;
(ii)   wide linear dynamic range (six to seven orders of magnitude if the extended dynamic range is used);
(iii)   multi-element capability (over 70 elements in one analysis compared with 35 – 45 elements by ICP-OES);
(iv)   semi-quantitative analysis within 2 min;
(v)   isotopic information.

Instrumentation

A schematic outline of an ICP-MS instrument is shown in Figure 4.3.1. An argon plasma provides the ion source. The high thermal temperature of the plasma (approximately 7000 K) provides an excellent source of excited state atoms. The ICP is also a very good ion source producing predominantly singly charged positive ions. Indeed many of the most useful analytical emission lines used in ICP-OES are ion lines.

To operate effectively, the instrument has to extract ions from the ICP, at atmospheric pressure, and transmit them to the MS, operating at $\leq 10^{-5}$ torr (10^{-3} Pa). Ions are extracted from the ICP through a sampling cone, into the first vacuum chamber, through a skimmer cone, into the second vacuum chamber. The ion beam is then directed via ion lenses into the MS.

The mass separation of the ions can be achieved using a magnetic sector (high

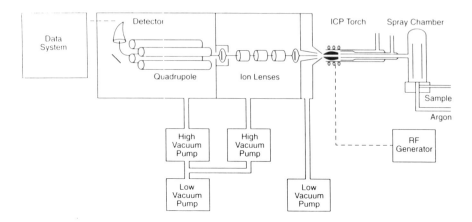

Figure 4.3.1 *Schematic outline of an ICP-MS instrument*

resolution) or a quadrupole (low resolution). As high resolution instruments are not yet commonplace, the discussion is limited to quadrupole-based instruments. Although there are some differences in design (pumps, ion lenses) of the instrument according to the manufacturer, the basic components and principles remain the same.

The multi-element data are relatively free from interference. Figure 4.3.2 shows the elements that may be determined by the technique with an indication of the detection limits achievable. Analysis of the less sensitive elements such as the halides and sulfur and the elements for which there are no data such as the noble gases, oxygen, and nitrogen could be achieved by the complementary technique of Microwave Induced Plasma MS. The technique is also rapid, producing semi-quantitative data on approximately 70 of the elements in around 1 min, and quantitative data on several elements typically in under 5 min. Most instruments use a quadrupole mass spectrometer for detecting and measuring the ions. There are less common, more expensive instruments, which use a high resolution, magnetic sector, mass spectrometer. Such a detector enables potential interfering masses to be resolved.

4.3.2 Comparison of Techniques

As mentioned in the overview, there are a variety of ways that trace elements can be converted into ions so that they can be measured with a mass spectrometer. The main techniques are:

(i)   ICP-MS;
(ii) glow discharge MS;
(iii) secondary ion MS;
(iv) thermal ionisation MS.

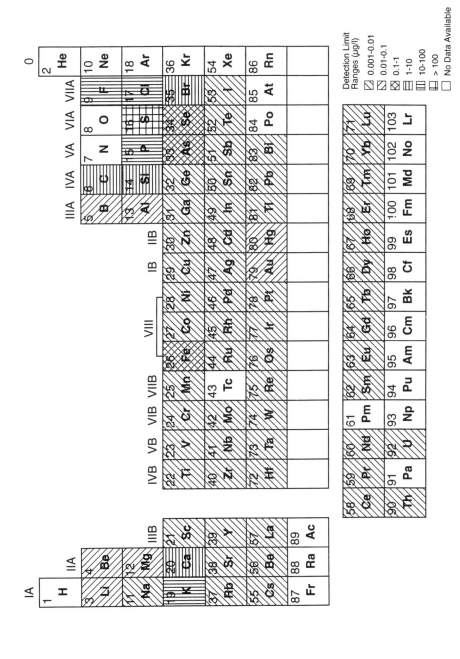

Figure 4.3.2 *Periodic table showing elements that can be analysed by ICP-MS*

It should be noted that spark source mass spectrometry was used for transition element analysis, but it has now been largely superseded by the above techniques.

The following sections highlight the advantages and disadvantages associated with each of the techniques.

Inductively Coupled Plasma MS (ICP-MS)

As outlined in Section 4.3.1, this is usually a solution technique and will therefore require the digestion of solid samples. Exceptions to this include laser ablation, which enables volatilised solids to be analysed, and Electrothermal Vaporisation (ETV), which can introduce solutions, slurries, and solids into the ICP-MS. The latter has the advantage that many interferences may be removed using a suitable temperature programme, *i.e.* water is removed, which decreases the amount of MO^+ interference. The main characteristics of this technique are shown in Table 4.3.1.

Table 4.3.1 *Advantages and disadvantages of ICP-MS*

Advantages	Disadvantages
High sensitivity for a wide range of analytes	Requires sample pre-treatment to remove some matrix interferences
Medium cost of quadrupole instruments	Resolution of quadrupole MS insufficient to separate isobaric overlaps, *e.g.* ArO^+ and ^{56}Fe
Wide linear dynamic range	
Multi-element calibration readily achieved	
Easy to use for isotope dilution analysis	Requires destruction of sample (at least in part)
High sample throughput	

Glow Discharge MS (GDMS)

This form of analysis is characterised by its suitability to measure trace components directly in solid samples. With high resolution MS this forms a very sensitive trace analysis option. Table 4.3.2 lists the advantages and disadvantages.

Table 4.3.2 *Advantages and disadvantages of GDMS*

Advantages	Disadvantages
Multi-element, with high sensitivity	Expensive, as high resolution MS required to avoid isobaric interferences
Low sample consumption (1–100 mg)	Not suitable for high sample throughput
No digestion problems for refractory materials	Accurate calibration requires matrix matched standards
Can be used for depth profiling	Does not lend itself to isotope dilution MS experiments
	Sample preparation can be problematic

Secondary Ion MS (SIMS)

This technique displays similar detection limits to GDMS, with excellent multi-element capabilities. Unlike ICP-MS and GDMS, it does show a much wider range of sensitivities across the periodic table. The sensitivities are also very matrix-dependent. It is, however, the only method that can provide spatial distribution analysis, analogous to the scanning electron microscope with an energy dispersive analyser. Table 4.3.3 lists the advantages and disadvantages.

Table 4.3.3 *Advantages and disadvantages of SIMS*

Advantages	Disadvantages
Multi-element sensitivity high	Two ion sources required to cover periodic table (oxygen and caesium)
Spatial and depth distributions can be acquired	Long analysis time (hours)
Minimal sample preparation, provided the surface is clean	Matrix matched standards for quantification
Very low sample consumption (non-destructive)	Very expensive (instruments and expert staff)
	Does not lend itself to isotope dilution-MS.

Thermal Ionisation MS (TIMS)

This is a solution technique. The samples are dried onto a filament, which is then used as a resistively heated ion source in the mass spectrometer. It is a technique which has been used extensively in elemental isotope dilution mass spectrometry (IDMS) to achieve the ultimate in terms of accuracy and precision. It is particularly useful for providing certified reference values. Table 4.3.4 lists the advantages and disadvantages.

Table 4.3.4 *Advantages and disadvantages of TIMS*

Advantages	Disadvantages
High accuracy and precision when used for isotope dilution measurements	Not generally simultaneous multi-element analysis
Small sample volumes	Low sample throughput
Good sensitivity	Matrix- and analyte-specific sampling considerations are usually required
Ideal for certified reference values (IDMS mode)	

Comparison Summary

Table 4.3.5 summarises the differences, in relative terms, between the different elemental mass spectrometric techniques.

Table 4.3.5 *A summary of the different techniques*

	ICP-MS	GDMS	TIMS	SIMS
Cost	Low[a]	Medium	Medium	High
Accuracy	Good	Adequate	Very good	Poor
IDMS	Yes	Difficult	Yes	No
Sample preparation	Long	Short	Medium	Short
Sample throughput	Fast	Slow	Slow	Slow
Solid sampling	No[b]	Yes	No	Yes
Spatial/depth profiling	No	Yes	No	Yes

[a]This would mean approx. £150 000. Some benchtops can be purchased for £60 000.
[b]Can be achieved with laser ablation, slurries, or ETV.

4.3.3 Physical and Spectral Interferences in ICP-MS

Prior to considering the types of sample acquisition used in ICP-MS it is useful to consider the sources and magnitude of interferences, and hence possible errors observed in ICP-MS. Interferences may be divided into two categories, physical and spectral.

Physical Interferences

Physical interferences are associated with properties of the sample which affect the sample introduction method and ionisation of the sample in the plasma. Changes in viscosity of the sample will affect the nebulisation efficiency and lead to variations in the aerosol that is produced. This can lead to overall suppression or enhancement of the ion signal observed and hence produce errors in the observed elemental concentrations. Sample volatility also affects the aerosol formation. Volatile species in the sample will yield a higher vapour pressure in the spray chamber, leading to improved transport efficiency to the plasma for these elements, which results in an enhancement of the signal from the volatile species and a concomitant suppression of the signal from the other analytes. Because of the difficulty in matching the matrix of standards to the matrix of samples, and because samples will vary themselves, these effects lead to spurious results if not corrected for.

Internal standardisation is a method used in a variety of analytical techniques to overcome these problems, and is used widely in ICP-MS. The method involves adding (spiking) an element(s), to all the solutions to be analysed, standards and samples, at the same concentration. The internal standard may also be used to compensate for instrumental drift. If slurries are being introduced, care must be taken not to assume that the internal standard will have the same transport efficiency as the slurry particles. In some instances the element may be present naturally in the samples, in which case its concentration must be determined beforehand, and be known not to vary significantly between samples. The element(s) selected as internal standards should be free from spectral interferences, not be of analytical interest in the samples, be close in mass to the elements of interest, and ideally be mono-isotopic. It should also have a similar ionisation energy to the analyte of interest. This last point yields high ion count rates and results in better precision. When the standards and samples are analysed, the intensities for the elements of interest are ratioed against the intensities obtained for the internal standard(s). Only once the ratios have been obtained is the elemental concentration calculated. The method assumes that the internal standard(s) will behave in a similar manner to the elements of interest, so that if there is a suppression or enhancement of the signal, the internal standards will also be suppressed or enhanced, the effect being cancelled once the ratios have been taken. Indium or rhodium tend to be good internal standards lying approximately mid-way in the mass range, although for some applications several internal standards across the mass range are required. If specific analytes are being analysed, for instance gold or lead, then a close mass like tellurium would be ideal. The use of internal standardisation in ICP-MS is essential to obtain valid analytical data, especially when sample matrix effects are likely to be present.

Spectral Interferences

Spectral interferences observed in ICP-MS may be conveniently divided into two groups, isobaric interferences and polyatomic interferences, both of which arise as a consequence of the low resolution (unit mass) of the quadrupole mass filters used in the instruments. A knowledge of the possible interferences and the sample type is essential for correct interpretation of results. Each of these interferences is considered in turn.

Isobaric Interferences. These interferences arise when an isotope of an element overlaps with an isotope of another element with the same nominal mass. For example ^{116}Sn overlaps with ^{116}Cd. The problem is worse when the abundance of the interfering isotope is high and the abundance of the analyte isotope is low. In most cases suitable isotopes may be selected which do not suffer from isobaric interferences. The occurrence of isobaric interferences is easy to predict given some basic knowledge of the sample, and can then be corrected for. Table 4.3.6 lists some of the possible isobaric overlaps. The list is by no means exhaustive, but does give some indication of the main isobaric interferences experienced in ICP-MS. It is recommended that IUPAC values are used for natural abundances.[1]

The software now available on many instruments will automatically correct for interferences from elemental species. However, care must be taken when interpreting the results, since there will be no correction for polyatomic interferences.

Polyatomic Interferences. Polyatomic interferences arise from molecular ions which are formed either in the high-temperature plasma or in the interface region between the plasma and the mass filter. These polyatomic ions may then interfere

Table 4.3.6 *Isobaric interferences in ICP-MS (% abundance)*[a]

Mass	Analytes with same nominal mass (natural abundances)		
50	Ti(5.34)	V(0.24)	Cr(4.31)
54	Fe(5.82)		Cr(2.38)
58	Fe(0.33)	Ni(67.84)	
64	Zn(48.89)	Ni(1.08)	
70	Zn(0.62)	Ge(20.52)	
76		Ge(7.76)	Se(9.02)
87	Sr(7.02)	Rb(27.85)	
92	Zr(17.11)	Mo(15.84)	
94	Zr(17.4)	Mo(9.04)	
96	Zr(2.8)	Mo(16.53)	
98		Mo(23.78)	Ru(1.87)
100		Mo(9.63)	Ru(12.62)
102	Pd(0.96)		Ru(31.61)
104	Pd(10.97)		Ru(18.58)
106	Pd(27.33)	Cd(1.22)	
108	Pd(26.71)	Cd(0.88)	
110	Pd(11.81)	Cd(12.39)	
113		Cd(12.26)	In(4.28)
115	Sn(0.35)		In(95.72)
116	Sn(14.3)	Cd(7.58)	
122	Sn(4.72)		Te(2.46)
124	Sn(5.94)	Xe(0.096)	Te(4.61)
130	Ba(0.101)	Xe(4.08)	Te(34.48)
134	Ba(2.42)	Xe(10.44)	
136	Ba(7.81)	Xe(8.87)	Ce(0.193)
138	Ba(71.66)	La(0.089)	Ce(0.25)
142	Nd(27.11)		Ce(11.07)
144	Nd(23.85)	Sm(3.09)	
148	Nd(5.73)	Sm(11.24)	
150	Nd(5.62)	Sm(7.44)	
154		Sm(22.71)	Gd(2.15)
160	Dy(2.29)		Gd(21.9)
164	Dy(28.18)	Er(1.56)	
170		Er(14.88)	Yb(3.03)
176	Lu(2.59)	Hf(5.2)	Yb(12.73)
186	Os(1.59)	W(28.41)	
187	Os(1.64)		Re(62.93)
198	Hg(10.02)	Pt(7.21)	
204	Hg(6.85)		Pb(1.48)

[a]As supplied with Perkin-Elmer Elan 5000 ICP-MS.

with isotopes of the same nominal mass. The polyatomic species generally arise from the plasma support gas (argon) and entrained gases (oxygen, nitrogen), reagents used during sample preparation (acids), and the sample matrix (matrix ions or salts). Table 4.3.7 lists some of the more common polyatomic species and the isotopes they interfere with. The list is by no means exhaustive but does give an indication of some of the interferences observed in the mass spectrum below

Table 4.3.7 *Polyatomic interferences in ICP-MS*

Mass	Analyte	Possible polyatomic species			
24	Mg	$^{12}C^{12}C$			
27	Al	$^{12}C^{14}NH$			
28	Si	$^{14}N^{14}N$ $^{12}C^{16}O$			
29	Si	$^{14}N^{15}N$ $^{14}N^{14}NH$	$^{13}C^{16}O$	$^{12}C^{17}O$	$^{12}C^{16}OH$
30	Si	$^{15}N^{15}N$ $^{14}N^{15}NH$	$^{14}N^{16}O$	$^{13}C^{17}O$	
40	Ca	^{40}Ar			
42	Ca	$^{40}ArH_2$			
44	Ca	$^{12}C^{16}O^{16}O$			
45	Sc	$^{12}C^{16}O^{16}OH$			
46	Ti	$^{14}N^{16}O^{16}O$			
50	Ti,Cr,V	$^{36}Ar^{14}N$			
51	V	$^{36}Ar^{15}N$ $^{36}Ar^{14}NH$	$^{35}Cl^{16}O$		
52	Cr	$^{38}Ar^{14}N$ $^{36}Ar^{15}NH$	$^{40}Ar^{12}C$		
53	Cr	$^{38}Ar^{15}N$ $^{37}Cl^{16}O$	$^{35}Cl^{18}O$		
54	Fe	$^{40}Ar^{14}N$			
55	Mn	$^{40}Ar^{15}N$ $^{40}Ar^{14}NH$			
56	Fe	$^{40}Ar^{16}O$			
57	Fe	$^{40}Ar^{16}OH$			
58	Ni	$^{40}Ar^{18}O$			
59	Co	$^{40}Ar^{18}OH$			
63	Cu	$^{23}Na^{40}Ar$			
64	Zn	$^{32}P^{16}O_2$ $^{32}S^{16}O_2$	$^{32}S_2$		
68	Zn	$^{40}Ar^{14}N^{14}N$			
70	Ge	$^{40}Ar^{14}N^{16}O$			
72	Ge	$^{36}Ar^{36}Ar$			
73	Ge	$^{36}Ar^{36}ArH$			
74	Ge	$^{36}Ar^{38}Ar$			
75	As	$^{36}Ar^{38}ArH$	$^{40}Ar^{35}Cl$		
76	Ge, Se	$^{36}Ar^{40}Ar$	$^{38}Ar^{38}Ar$		
77	Se	$^{36}Ar^{40}ArH$	$^{38}Ar^{38}ArH$		
78	Se	$^{38}Ar^{40}Ar$			
79	Br	$^{38}Ar^{40}ArH$			
80	Se	$^{40}Ar^{40}Ar$			
81	Br	$^{40}Ar^{40}ArH$			
82	Se	$^{40}Ar^{40}ArH_2$			

mass 82. Interferences from S species (H_2SO_4) have not been mentioned but need to be considered if this acid is used in sample preparation. The same is true for HCl and HF. Oxide formation can also be a problem especially for the rare earth elements; the oxides of one rare earth tend to interfere with other members of the rare earth series. Doubly charged ions are also formed to some extent in the plasma, and give rise to an interference at exactly half the true mass. For example $^{138}Ba^{2+}$ interferes with ^{69}Ga. The extent to which doubly charged and polyatomic species are present in the plasma may be controlled to some extent by the operating conditions used, and the sample preparation step, the use of acids *etc.*, although total removal of these species is rarely achieved. Again, an understanding of the interferences and the sample is essential when interpreting the spectrum.

4.3.4 Modes of Sample Introduction

The modes of sample introduction are analogous to that for ICP-OES (see Section 4.2). They include:

(i)   conventional nebulisation;
(ii)   flow injection analysis;
(iii)   hydride generation;
(iv)   laser ablation for solids;
(v)   electrothermal vaporisation with a graphite furnace;
(vi)   chromatography coupling.

The major difference between ICP-OES and ICP-MS with regard to sample introduction is that ICP-MS has a lower tolerance of solids (1% w/v). This is simply because there is direct physical contact between the MS detector and the argon plasma in ICP-MS. In ICP-OES, the optical detection process provides a natural physical 'buffer' between the plasma source and the detection system.

4.3.5 Modes of Data Acquisition in ICP-MS

There are several different modes of acquiring data depending on the type of information required. The mode of acquisition used will depend upon the sample type, the analytes of interest, the quantity of sample, and the precision and accuracy demanded by the customer. There are basically two different types of analysis that may be carried out using ICP-MS, semi-quantitative analysis (SQ) and quantitative analysis. Each of these is considered in turn.

Semi-quantitative Analysis (SQ)

Semi-quantitative analysis provides a very rapid screening of the sample covering the entire mass range from lithium at mass 6 to uranium at mass 238, or any pre-defined section of this mass range. Certain areas of the spectrum are routinely 'skipped' to prevent saturation of the detector. For example, mass 40 ($^{40}Ar^+$), mass 80 ($^{40}Ar^{40}Ar^+$). It is also usual to skip elements that are known to be present in high concentrations in the sample (matrix elements), *e.g.* calcium and magnesium in sea waters. The SQ acquisition covers nearly all the elements of the periodic table, and is typically carried out in under 2 min. The elements that may be determined in a SQ analysis have already been identified in Figure 4.3.2. In this acquisition mode the instrument is not calibrated for each mass or element, rather a response curve of intensity against mass is plotted from data obtained from a multi-element standard. This standard typically contains six to ten elements that are generally chosen to encompass the entire mass range. The instrument software plots the intensities obtained from these elements against their masses and fits a line through these points, Figure 4.3.3. When an unknown sample is analysed, the software compares the intensities obtained for all the elements against this response curve to calculate the concentration of the elements in the sample. The

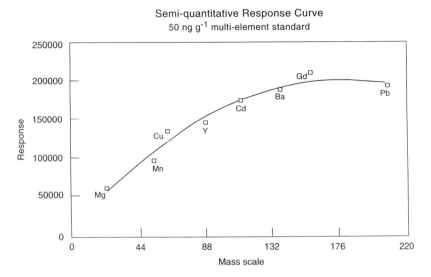

Figure 4.3.3 *Semi-quantitative analysis response curve*, 50 ng g^{-1} *multi-element standard*

advantage of this type of analysis is that you do not need to know what you are looking for prior to the analysis as all the elements are determined. The analysis is very rapid, and is often used as a means of assessing which elements are present in the sample and at approximately what concentration prior to a quantitative determination. It is also useful in highlighting possible interferences from other elements in the sample. The results obtained by this type of analysis tend to be within a factor of 2 of those obtained when the analysis is carried out quantitatively, although if the concentrations of the elements to be determined are similar to the concentration of the standard used to generate the response curve then the accuracy may be somewhat better. Samples are usually diluted 100-fold prior to being analysed by this method. This ensures that no element is likely to be present at a concentration that saturates the detector. If necessary, further dilution may be carried out based on the results of the first analysis.

It is usual to include an internal standard in all standards and samples. This procedure helps correct for matrix effects and instrumental drift. If a multi-element standard of known concentration is used as a sample, then the relative accuracy of the calibration may be assessed.

Quantitative Analysis

Quantitative analysis follows the usual format for any analytical determination. External calibration curves are plotted from standards containing the elements of interest, over the expected concentration range. Once the calibration has been performed the unknown samples may then be analysed and the intensities obtained are read off the calibration curve. It is usual to include as many different isotopes of each element as possible in the analysis, although those known to suffer from an isobaric or polyatomic interference are usually avoided if possible. The use of an internal standard is also necessary for accurate results. Matrix-matched standards, where the standards are prepared in a similar matrix to the samples, are used when complex matrices, such as strong acids, or highly saline media, are present. In order to obtain the precision and accuracy required in quantitative analysis, it is necessary to dwell on each mass for longer, leading to typical analysis times of around 5 – 10 min per sample depending upon the number of elements being determined.

The method of standard additions is used when there is known to be a major matrix problem. The technique is more time consuming and generally requires larger sample volumes than the procedures outlined above. Using this method an aliquot of the sample is analysed in the conventional manner. A second aliquot of the sample is then spiked with a known concentration of the element to be determined, at approximately the same concentration as the element present in the sample. Clearly this requires some previous knowledge of the sample in order to make the spike at the right concentration (*i.e.* SQ analysis). This spiked solution is analysed along with a third aliquot of the sample which is spiked with approximately twice the concentration of the element present in the sample. These three points are then plotted and the line extrapolated to cross the concentration axis. It is this negative intercept that yields the elemental concentration in the original sample, Figure 4.3.4. To a large extent the method overcomes matrix effects, as the calibration is performed in the sample matrix itself. However, as each sample essentially requires its own calibration curve the analysis time is longer. For greater certainty a fourth aliquot of sample can be measured along with a concentration of spike which gives a reading between the second and third point. This increases the analysis time further.

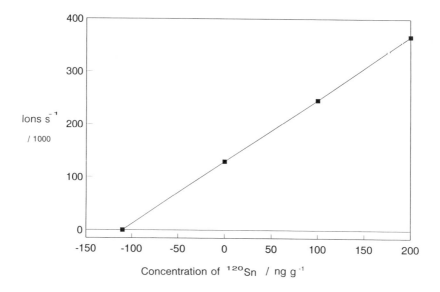

Figure 4.3.4 *Graph showing the analysis of tin by standard addition*

4.3.6 Isotope Ratio and Isotope Dilution Analysis

One of the main advantages of ICP-MS over ICP-OES is that isotopic information may be obtained. This allows the acquisition of isotope ratio data and the use of isotope dilution analysis.

Isotope Ratio Analysis

As the name suggests this involves measuring the ratio of two or more isotopes of the same element (or in some cases different elements). Such work is especially important in clinical trials where, for example, isotopes are used to track metabolic pathways of drugs. Other uses include geological ageing and environmental monitoring. The isotopic composition of lead varies in nature depending on the origin, because U and Th decay to ^{206}Pb, ^{207}Pb, and ^{208}Pb. The ratio of the isotopes may be used to determine the source of the lead in the environment, *i.e.* whether it is as a consequence of pollution or present naturally. Although the isotope ratios may be measured in a short time and with very good precision (0.1 – 0.2% RSD) the accuracy of these measurements is affected by the mass bias, or mass discrimination of the instrument. The causes of mass bias are not well understood in ICP-MS, although a major explanation involves space charge effects in the ion beam during transfer from the plasma to the quadrupole mass analyser. This results in mutual repulsion of positive ions and the concomitant defocusing of the lighter isotopes compared with the heavier isotopes. Thus an apparent increase in signal of the heavier isotopes is observed which distorts the isotopic ratio. This mass bias needs to be measured and corrected for if accurate isotopic ratios are to be measured. This is carried out using a Certified Reference Material (CRM) which has been certified for isotopic composition. The CRM used must be of the same element and isotopes as those to be determined in the samples. Using the CRM the observed isotope ratios are compared with the certificate values and a correction applied. This correction is then applied to the isotope ratios measured for the samples.

The mass bias factor is calculated using Equation (1).

$$CF = t/e \qquad (1)$$

where:
  CF = mass bias correction factor
  t = true (certificate value) ratio
  e = experimental ratio

This CF value is then applied to the raw ratio data obtained from the samples.

Example of Mass Bias Correction Calculation

A blend is normally prepared so that the isotopic ratio and the concentration matches as closely as possible the sample being measured. A known weight of a solution of known concentration of a reference material is added to a known weight of a solution of the isotopically enriched material. Isotope counts are normally made before and after the unknown sample being measured.

Isotope enriched sample

Weight of solution	0.6035 g
Concentration of solution	30.7604 μg g^{-1} of Pb
Lead 208 (atom %)	0.03
Lead 206 (atom %)	99.76
A_r	205.97

Reference material

Weight of solution	7.1235 g
Concentration	10.0417 μg g^{-1} of Pb
Lead 208 (atom %)	52.347
Lead 206 (atom %)	24.144
A_r	207.215

Theoretical ratio of mass bias solution $^{208}Pb/^{206}Pb$:

$$\text{atoms } ^{208}Pb = \frac{0.6035 \times 30.7604 \times 0.03}{205.97 \times 100} + \frac{7.1235 \times 10.0417 \times 52.347}{207.215 \times 100}$$

$$= 0.0000270 + 0.1807055 = 0.1807325$$

$$\text{atoms } ^{206}Pb = \frac{0.6035 \times 30.7604 \times 99.76}{205.97 \times 100} + \frac{7.1235 \times 10.0417 \times 24.144}{207.215 \times 100}$$

$$= 0.0899128 + 0.833468 = 0.173259$$

True ratio $^{208}Pb/^{206}Pb$ = 1.04313

	Isotope counts	
	Leading solution	*Trailing solution*
^{208}Pb	279 211	268 098
^{206}Pb	270 490	259 462
Measured ratio	1.032 24	1.033 28
Mass bias factor	1.0105	1.0095
Mean mass bias correction factor = 1.0100		

The mass bias factor will vary depending upon the isotopic composition of the standard used to calculate it.

The detector has a response that is dependent on the ions per second, or count rate, for each individual isotope. This is why it is essential to correct for the mass bias effect with a certified standard or blend of standards with similar isotopic composition to that in the samples.

Table 4.3.8(a) and (b) illustrate the consequence of using an inappropriate reference to correct a sample containing the natural abundances of ^{208}Pb and ^{206}Pb. The natural abundance of ^{208}Pb = 52.4% and ^{206}Pb = 24.1%.

Table 4.3.8 *The importance of appropriate mass bias correction*

(a) Certified abundances

SRM	Certified abundance		Ratio
	206	208	208:206
SRM 981	24.14	52.35	2.169
SRM 983	92.15	1.255	0.0136

(b) Mass bias corrections to a natural sample

SRM used	208:206 ratio corrected for 'natural' sample
SRM 981	2.086
SRM 983	1.776

There is a difference of 14.9% between the mass bias corrected ratios for the same 'natural' sample. The difference in isotope abundances between standard reference material (SRM) 983 and the 'natural' sample make it an inappropriate choice for mass bias correction.

Isotope Dilution

Isotope dilution is generally regarded as being a definitive method. The technique is well understood, and the possible sources of error may be controlled and minimised. The technique relies upon the measurement of a ratio of two isotopes of the same element, the reference isotope and the spike, or enriched isotope. The reference isotope is generally the isotope of highest natural abundance while the spike isotope is generally taken as one of the minor isotopes in nature, and is usually selected to be close in mass to the reference isotope. As only isotope ratios are measured, and no external calibration curve is required, once the spike isotope has been added to the sample, and equilibrium reached, no further quantitative handling of the sample is required. This minimises possible errors due to loss of sample. The use of the spike isotope is the ideal form of internal standardisation, compensating for any matrix effects and/or instrumental drift. These two factors combine to produce both highly accurate and precise data. Most of the elements in the periodic table are amenable to isotope dilution, the only prerequisite being that the element has two or more stable isotopes, or long-lived radioisotopes. The elemental concentration in the samples can then be calculated as shown in Section 4.3.7.

4.3.7 Calculation of Elemental Concentration from Isotope Dilution Measurements

Since the ICP-MS measures atoms, the concentration of the isotopic species must also be in terms of atoms (number of moles of atoms).

Sample

Weight of sample solution	$= w$
Concentration of sample solution (w/w)	$= c$
Relative atomic mass of the element being determined	$= A_r$
Isotope abundance (atom%) isotope 1	$= X$
Isotope abundance (atom%) isotope 2	$= Y$

Spike

Weight of spike solution	$= w'$
Concentration of spike solution	$= c'$
Relative atomic mass of the element in the spike	$= A'_r$
Isotope abundance isotope 1	$= X'$
Isotope abundance isotope 2	$= Y'$

Note it is the relative atomic mass of the *element* not the isotope which is used to calculate the mole of atoms. This is because the isotope abundances tabulated are atom % (*not weight %*).

When w' g of the spiked solution is added to w g of sample solution, the number of moles of each isotope, n_1 and n_2, is given by Equations (2) and (3).

$$n_1 = \frac{wcX}{A_r} + \frac{w'c'X'}{A'_r} \tag{2}$$

$$n_2 = \frac{wcY}{A_r} + \frac{w'c'Y'}{A'_r} \tag{3}$$

The measured ratio R is given by Equation (4), which is obtained by dividing Equation (2) by Equation (3).

$$R = \frac{n_1}{n_2} = \frac{wcXA_r' + w'c'X'A_r}{wcYA_r' + w'c'Y'A_r} \tag{4}$$

The concentration c of the sample solution is then given by Equation (5).

$$c = \frac{w'c'A_r(RY' - X')}{wA_r'(X - RY)} \tag{5}$$

Example

Enriched spike
Abundance ^{206}Pb (205.97) = 99.76% = Y'
Abundance ^{208}Pb (207.98) = 0.03% = X'
A_r = 205.97
Concentration of spiked solution = 30.7604 μg g^{-1} = c'
Weight of spiked solution added to sample = 0.0991 g = w'

Sample
Abundance ^{206}Pb = 24.51% = Y
Abundance ^{208}Pb = 52.70% = X
A_r = 207.218
Concentration of sample = c
Weight of sample solution = 0.2499 g = w
Measured ratio ^{208}Pb/^{206}Pb (after mass bias correction) = 1.06965

Substituting the values into Equation (5) gives the concentration of the Pb in the sample solution.

$$c = \frac{0.0991 \times 30.7604 \times 207.218 \, (\, 1.06965 \times 99.76 - 0.03 \,)}{0.2499 \times 205.97 \, (\, 52.70 - 1.06965 \times 24.51)}$$

$$= 49.435 \; \mu\text{g g}^{-1} \text{ of Pb}$$

If the isotope abundance (atom%) of the sample is not known this must be determined experimentally. Otherwise, the only information required is the weight of the enriched isotope added (and hence the elemental concentration of the spike solution), the weight of the sample, and the measured ratios. Modern analytical balances are capable of routinely weighing to 1 part in 50 000, so that errors arising from the weights of the sample and spike should be minimal. In practice this means using weights of at least 0.5 g and avoiding the use of weights of less than 0.2 g.

The main parameter affecting the result in isotope dilution is measuring and correcting for mass bias effects. This has been dealt with in the section on isotope ratios. However, it is as well to stress that as the isotopic composition (ratio) of the sample varies, so too will the true mass bias factor. Therefore, as indicated earlier, it is necessary to measure and correct for mass bias using a standard solution with a similar isotopic composition to the sample. For most accurate work it is preferable to have signal ratios between the reference and spike isotopes as close to unity as possible. The standard used to correct for the mass bias should match the composition of the spiked sample.

The other parameter to affect the result is the detector dead-time. The channel electron multipliers, used as detectors in most instruments, have a finite time after detecting an ion during which they are unable to detect other ions striking their surface. This is known as the dead-time. When two isotopes are being measured that have quite different abundances, the proportion of non-detected ions is greater for the more abundant isotope than it is for the minor isotope, so distorting the observed ratio. To achieve accurate results this must be corrected for by using a suitable reference solution which has comparable isotopic concentrations.

Isotope dilution techniques, although capable of highly accurate and precise measurements (with a relative standard deviation typically less than 0.5%) do tend to be significantly more time consuming than external calibration methods, and hence more expensive. To date, isotope dilution has not found use in routine analysis, rather it is used as a tool for checking the accuracy of techniques used within laboratories and for establishing a chain of traceability. The technique is also used in the characterisation of standard reference materials, which requires a high degree of precision.

4.3.8 Analysis of Organic Samples by ICP-MS

Although ICP-MS is purely an inorganic measurement technique, this does not preclude the determination of inorganic species in organic matrices. The instrument conditions required are somewhat different from those used for the measurement of aqueous solutions. The first problem encountered with introducing organic materials into the ICP is the extremely high reflected power, which may adverseley affect the generator if it cannot be lowered by changing the load, *etc.* Another problem encountered is due to the large concentration of carbon in the sample, which results in deposits of carbon on the mass spectrometer interface leading to eventual blockage of the sampling orifice. Depending on the sample, the orifice may become completely blocked in under 1 min, clearly precluding any analysis. This deposition may be overcome to a large extent by the addition of oxygen to the plasma gas. The oxygen is usually added to the nebuliser gas, although some workers have added it to the coolant and auxiliary gases. The oxygen forms CO species in the plasma, so preventing carbon deposition on the sampling cone. The amount of oxygen added to the plasma is crucial, as too little will still lead to soot deposition, and too much will lead to serious deterioration of the sampling cone, with the orifice being 'burnt out', and the cone eventually having to be replaced.

The second main problem with introducing organic samples into the plasma is due to the volatility of some of the solvents used. This leads to a large vapour loading in the plasma, which may put the plasma out, and at the least lead to suppression of the signal for most of the elements. This is overcome by cooling the spray chamber. A spray chamber fitted with a cooling jacket, containing circulating isopropanol, may be routinely chilled to $-15°C$. This is sufficient to allow the direct introduction of solvents such as methanol. In addition to cooling the spray chamber the sample pump rate is usually reduced to around 0.5 ml min^{-1} (typical pump rates for aqueous samples are 1–1.5 ml min^{-1}). When introducing organic solvents into the ICP it is usual to initially light the plasma while introducing an aqueous solution. The organic solvent is then introduced. The plasma is seen to glow a bright green from the C_2 emission band. This may then be effectively 'titrated out' by the gradual addition of oxygen to the nebuliser gas via a second mass flow controller. Once the green emission has been removed from the bulk of the plasma, the oxygen level is usually correct. This prevents possible damage to the cone by introducing oxygen to the plasma in the absence of organic material. A typical flow rate of 0.1 l min^{-1} of oxygen is usually found to be optimum. It is worth noting that while the oxygen is gradually being added, carbon will already have been deposited on the sample cone and it is sometimes necessary to increase the oxygen flow slightly above the optimum flow rate in order to burn this deposit off, reducing the oxygen flow once more after this deposit has been removed. It is possible to tell if carbon is being slowly deposited because the vacuum in the expansion stage slowly improves as the orifice blocks. The signal will also decrease.

Both semi-quantitative and quantitative analysis may be performed in organic matrices, using standards prepared in a similar matrix. Toluene (methylbenzene) has been found to be an ideal diluent for most organic samples, diluting the samples 10-fold, or 100-fold as required in methylbenzene, and preparing all standards in methylbenzene. As a general rule, the absolute detection limits when analysing organic solvents tend to be an order of magnitude worse than for aqueous samples.

4.3.9 Uncertainty (Errors)

The main sources of errors arise from preparation of the samples/standards and from spectral interpretation, both of these leading to uncertainties in the data. Considering sample preparation first, contamination and/or analyte losses during the sample dissolution or digestion will lead to uncertainties. Due to the high sensitivities obtained with ICP-MS, the reagents used in sample preparation (acids, water) need to be scrupulously clean, as do all vessels used during the preparation steps. Whereas for techniques such as FAAS levels of contaminants below 10 ng g^{-1} are suitable for routine applications, in ICP-MS this must be reduced to around 0.01 ng g^{-1}. If high blank values are recorded, these should be checked and if possible reduced prior to continuing with the analysis. High purity acids and reagents should be used throughout, and glassware should be avoided whenever possible as this can be a major source of contamination at the extreme low levels often encountered with ICP-MS. It is convenient to use disposable plastic vessels and to prepare all samples and standards by weight, which is itself inherently more accurate than preparing solutions by volume, this limits the uncertainties further. If properly calibrated and maintained balances are used throughout, the errors arising from these procedures should be around 1 in 50 000. Elements such as Ca, Mg, Al, Na, Si, which are ubiquitous in the laboratory environment, are particularly difficult to control, and care should always be taken in an analysis to control the blanks obtained.

Careful choice of digestion procedures, with particular reference to the elements of interest, will limit any analyte losses. Volatile elements, such as Hg, are best dealt with in closed systems such as microwave digestion vessels. The use of internal standards in the digestion procedure will yield information on recoveries (sample losses) and possibly contamination. The internal standard will also correct any long-term drift of the instrument during the measurement step. Temperature control of the laboratory housing the instrument is also important in maintaining a steady signal.

Spectral interpretation has been dealt with in Section 4.3.3, under the sub-heading of spectral interferences. If there are any residual interferences in the mass spectrum then this will lead to errors in the assigned intensities for the elements of interest and hence errors in the observed concentrations. If these uncertainties are to be kept to a minimum, then a thorough knowledge of the sample and the possible interferences is required to interpret the spectrum correctly.

The software, which comes as standard with these types of instruments, is in general very good and uncertainties arising from poor calculations of, for example, peak areas or regression analysis are negligible. Areas of the software that deal with more specialised applications such as isotope dilution tend not to be as well developed and these functions are better carried out separately using spreadsheets. However, the use of spreadsheets can lead to problems. Occasionally, a sample should be calculated manually and the results compared with that given by the spreadsheet. This should prevent errors arising from mistakes made by the software.

4.3.10 Applications

Elements	Matrix	Reference
Pb	Biological	6
Ag Cr Fe Mn Ni Cu Zn Sn As Cd Hg Pb	Biological materials	7
Ni Cu Zn Ag Cd Pb	Sea-water Solvent extraction	8
Fe	Human faecal matter	9
Re Os	Geological materials	10
Rare Earth elements	Geological materials	11
Alkaline earths	Concentrated brines	12
Mn Co Cu Pb U	Riverine water	13
Mn Co Ni Cu Zn Cd Pb	Sea-water	14

4.3.11 References

1. IUPAC, 'Isotopic Composition of the Elements 1989', *Pure Appl. Chem.*, 1991, **63**, 991.

General Reading

2. D. Colodner, V. Salters, and D. C. Duckworth, 'Ion Sources for Analysis of Inorganic Solids and Liquids by MS', *Anal. Chem.*, 1994, **66**, 1079A.
3. E. H. Evans and J. J. Giglio, 'Interferences in Inductively Coupled Plasma Mass Spectrometry', *J. Anal. At. Spectrom.*, 1993, **8**, 1.
4. J. M. Carey, F. A. Byrdy, and J. A. Caruso, 'Alternative Methods of Sample Introduction for Plasma Mass Spectrometry', *J. Chromatogr. Sci.*, 1993, **31**, 330.
5. S. F. Durrant, 'Inductively Coupled Plasma Mass Spectrometry for Biological Analysis', *Trends Anal. Chem.*, 1992, **11**, 68.

Applications

6. R. M. Barnes, A. Lasztity, M. Viczian, and X. Wang, 'Analysis of Environmental and Biological Materials Inductively Coupled Plasma Emission and Mass Spectrometry', *Chem. Anal.*, 1990, **35**, 91.
7. D. Beauchemin, J. W. McLaren, S. N. Willie, and S. S. Berman, 'Determination of Trace Metals In Marine Biological Reference Materials by Inductively Coupled Plasma Mass Spectrometry', *Anal. Chem.*, 1988, **60**, 687.
8. T. Kato, S. Nakamura, and M. Morita, 'Determination of Nickel, Copper, Zinc, Silver, Cadmium and Lead in Seawater by Isotope-dilution Inductively Coupled Plasma Mass Spectrometry', *Anal. Sci.*, 1990, **6**, 623.
9. B. T. G. Ting and M. Janghorbani, 'Inductively Coupled Plasma Mass Spectrometry Applied to Isotopic Analysis of Iron in Human Faecal Matter', *Anal. Chem.*, 1986, **58**, 1334.
10. S. B. Beneteau and J. M. Richardson, 'Rhenium-osmium Isotope Ratio Determination by ICP-MS', *At. Spectrosc.*, 1992, **13**, 118.
11. F. E. Lichte, A. L. Meier, and J. G. Crock, 'Determination of Rare-earth Elements in Geological Materials by Inductively Coupled Plasma Mass Spectrometry', *Anal. Chem.*, 1987, **59**, 1150.
12. L. Ebdon, A. Fisher, H. Handley, and P. Jones, 'Determination of Trace Metals in Concentrated Brines using Inductively Coupled Plasma Mass Spectrometry Online Pre-concentration and Matrix Elimination with Flow Injection', *J. Anal. At. Spectrom.*, 1993, **8**, 979.
13. D. Beauchemin, and S. S. Berman, 'Determination of Trace Metals in Reference Water Standards by Inductively Coupled Plasma Mass Spectrometry with Online Pre-concentration,' *Anal. Chem.*, 1989, **61**, 1857.
14. J. W. McLaren, A. P. Mykytiuk, S. N. Willie, and S. S. Berman, 'Determination of Trace Metals in Seawater by Inductively Coupled Plasma Mass Spectrometry with Preconcentration on Silica-immobilized 8-Hydroxyquinoline (quinolin-8-ol)', *Anal. Chem.*, 1985, **57**, 2907.

4.4 The Chromatography of Ions

4.4.1 Introduction and Overview

Section 6.3, High Performance Liquid Chromatography, describes the chromatographic separation of molecular compounds. There are also many trace analytical applications for the chromatographic separation of ions, both inorganic ions and organic acids and bases. The techniques involved are described in this section.

As may be expected, there is a large degree of overlap between Ion Chromatography (IC) and High Performance Liquid Chromatography (HPLC). Initially the technology of IC, both in separation and detection, was distinct from that of HPLC, and so IC has traditionally been considered as a technique in its own right. However, in recent years distinctions have become somewhat grey and arbitrary. Methods such as ion-pair chromatography and ion-suppression chromatography could equally be described under either heading. This section therefore concentrates on methods of separating ions, and on detection systems that are specific to IC. Common hardware, detection methods, and other techniques are covered in Section 6.3.

Ion chromatography can be used for the analysis of any ion. There are several adequate methods for the determination of trace levels of cations (*e.g.* AAS, ICP-MS), which are frequently used instead of IC. However, there are many anionic analytes for which no other suitable determination method exists. Examples include halides, nitrate, sulfate, and cyanide. Ion chromatography has become the standard technique for the analysis of such ions. Other recent applications include weak organic acids such as carboxylic acids and sulfonic acids, and bases such as amines.

4.4.2 Ion Separation Mechanisms

There are four main mechanisms of chromatographically separating ions, based upon different stationary phase/mobile phase combinations (Table 4.4.1). In many cases the choice of stationary phase/mobile phase also dictates which detection system is appropriate.

Ion-exchange Chromatography

Ion-exchange resins are probably the most widely used stationary phase. They are usually polymer-based, often derivatives of cross-linked polystyrene.

The ionic analytes are retained by an equilibrium whereby they displace an ion of the same sign (positive or negative) which is bonded onto the column. The mobile phase is an aqueous buffer solution. Separation is dependent upon the degree of interaction between the ions in the sample and the bonded phase, and upon the competition between buffer ions and solute ions for the ion-exchange sites. This is obviously dependent upon the ionic strength of the buffer, which can also be adjusted by changing the pH of the mobile phase or by altering the concentration of any organic modifier in the mobile phase (as in HPLC). In addition, the buffer itself can be changed to one that provides a different amount of competition for the ion-exchange sites on the column.

Table 4.4.1 *Methods of chromatographically separating ions*

Separation method	Range of applications	Comments	Stationary phase	Mobile phase
Ion exchange	Used for strong or weak acids/bases. Only method for many analytes	Very robust. Adjust eluent strength to change retention	Anion or cation exchange resin, *e.g.* $R-SO_{3-}$ H_+, $R-COO^-$ H^+, $R-N(CH_3)^+$ OH^-	Aqueous buffer, with or without organic modifier
Ion exclusion	Used for hydrophylic organics/moderate to weak acids	Little variation/optimisation possible	Sulfonated resin	Water or dilute acid
Ion pair	Used for strong or weak acids/bases	Can use C-18 column. Must modify mobile phase	Any HPLC reversed phase	Aqueous buffer and ion-pair reagent, with or without organic modifier
Ion suppression	Used only for weak acids/bases	Can use C-18 column. Treat as standard HPLC analysis	Any HPLC reversed phase	Aqueous buffer and organic modifier

Ion-exclusion Chromatography

Ion-exclusion chromatography is a selective technique for separating moderate to weak acids. It is similar to gel-permeation chromatography in that molecules with short elution times are unable to enter the stationary phase, while molecules with long elution times penetrate the stationary phase and therefore must traverse a much larger volume inside the column before eluting. Unlike gel-permeation chromatography, the mechanism of exclusion of molecules from the stationary phase depends not on size but on charge.

The stationary phase used for ion-exclusion chromatography is high capacity cation-exchange resin, generally sulfonated. The mobile phase is a dilute, strong acid such as hydrochloric acid, generally in the 1–10 mM concentration range. Because the negative sulfonate functional groups are fixed to the resin, whereas the positive hydronium counterions are mobile, a small fraction of the hydronium counterions diffuse out of the resin, leaving behind a net negative charge, called *the Donnan potential*. As result of this negative charge, anions are unable to penetrate the resin, while neutral molecules can. Therefore, anions of strong acids such as sulfate and nitrate are not retained and are co-eluted in the void volume, which is also called the totally excluded volume. Weak acids that are undissociated at the eluent pH are able to penetrate the resin and elute much later. Partially dissociated acids elute in-between. Separation is therefore dependent on eluent pH, with acids eluting roughly in order of increasing pK_a. There is also some selectivity as a result of interaction of the acid with the resin.

Detection can be by low-wavelength UV absorbance or by suppressed conductivity detection, which produces a superior signal-to-noise ratio and less baseline drift.

Ion-pair Chromatography

Ion-pair chromatography is again predominantly used for the analysis of weak acids and bases, although it has also been applied to more strongly ionised compounds. The mobile phase is modified to include an ion of opposite charge to the analyte ion. Although the exact mechanism is not fully understood, the two oppositely charged ions can be thought of as forming a neutral complex. This can then be separated using standard HPLC equipment. An advantage of the technique is that normal reversed phase stationary phases, such as C-18, can be used. In addition, columns packed with neutral macroporous resin can be used.

Ion-suppression Chromatography

Ion-suppression chromatography is only used for the analysis of weak acids and bases. The mobile phase is either acidified or made alkaline to force the equilibrium between the ionic and molecular forms of the analyte to favour the molecular form, *e.g.* amines are ionic in acidic media, but neutral in an alkaline mobile phase. As in ion-pairing, the analysis is then performed on standard HPLC equipment.

4.4.3 Methods of Detection

The limiting factor for most trace-ion chromatographic analyses is the sensitivity of the detector. All detectors used in HPLC can be used in ion chromatography. However, unlike HPLC, where the majority of analytes are UV absorbers, few ions have such a convenient property which can be measured at very low levels. The majority of ion chromatographic detectors rely upon conductance measurements (conductivity detectors), but where applicable amperometric or photometric detectors are used (Table 4.4.2).

Table 4.4.2 *Detection methods in ion chromatography*

Detection method	Range of application	Comments
Suppressed conductivity	Most common method. High sensitivity for strongly to moderately ionised molecules	Can use gradient elution. High sensitivity
Direct conductivity	Standard method. Near universal application	Measuring small change in sometimes large signal: prone to noise. Not selective. Cannot use gradient elution
Amperometry	Optimum method for aromatic amines, phenols, catecholamines	Limited applications. Sensitive and selective where applicable. Can use gradient elution
Pulsed amperometry	Alcohols, aldehydes, carbohydrates and amines	Sensitive and selective for these analytes
Direct photometry	Used for iodide, nitrate, sulfide. Suitable for soil samples, sea water, *etc.*	Limited applications. Highly conductive eluents can be used. Can use gradient elution
Indirect photometry	Near universal application	Nearly as sensitive as direct conductivity. Measuring small change in large signal. Prone to mobile phase fluctuations
Post-column derivatisation photometry	Only method for some transition metal ions	Can use gradient elution. Increased pump noise and dead-volume

Conductivity Detection

The obvious method, and in many cases the only method, of detecting ions is to measure the change in the conductance of the mobile phase as the ions elute. As the mobile phase contains salts, this entails measuring a small signal change against a large background. Sensitivity is therefore poor, particularly if the difference in conductivity between the analyte and the eluent ions is not maximised. However, sensitivity is greatly increased by the use of a suppressor. In order to use conductance detectors for trace analysis, there are two possible methods of compensating for the high background conductance.

Suppressed Conductivity Detection. In the original design of the suppressed conductivity detector, the eluent was passed through a second ion-exchange column before detection. The resulting weakly ionised species lead to reduced conductance of the eluent. Because of the undesirable increase in dead-volume, ion-suppression columns have now been replaced with membrane suppressors, which occupy very little volume and do not require regular regeneration.

The conditions required for sensitivity is a highly conducting analyte in a poorly conducting eluent. Carbonate and hydroxide are easily suppressed to produce low conducting eluents. The sensitivity is then maximised for highly conducting analyte ions. The analyte conductance is in turn dependent upon the degree of dissociation, the charge, and the mobility of the analyte ion. Mobilities are analyte-dependent; generally the smaller the ion (including the size of its hydrated sphere) the more mobile it is.

Table 4.4.3 shows the ionic conductivities of some common ions and mobile phase components.

Table 4.4.3 *Limiting ionic conductivities* (λ) *in aqueous solutions at 25 °C (in units of 10^{-4} S m^2 mol^{-1})*

Anions	λ	Cations	λ
OH^-	198	H^+	350
F^-	55.4	Li^+	38.7
Cl^-	76.3	Na^+	50
Br^-	78.1	K^+	74
I^-	76.8	NH_4^+	73.5
CN^-	78	$1/2\ Cd^{2+}$	54
NO_3^-	71.4	$1/2\ Mg^{2+}$	53
NO_2^-	71.8	$1/2\ Ni^{2+}$	50
HCO_3^-	44.5	$1/2\ Ca^{2+}$	59.5
$1/2\ SO_4^{2-}$	80	$1/2\ Sr^{2+}$	59.4
$1/2\ SO_3^{2-}$	79.9	$1/2\ VO^{2+}$	32
Mobile phase buffers			
Acetate	40.9	$CH_3NH_3^+$	58.7
Benzoate	32.4	Tetraethylammonium	32.6
Formate	54.6	Tetrapropylammonium	23.4
$1/2\ HPO_4^{2-}$	33		
$H_2PO_4^-$	33		

λ is the molar (equivalent) conductivity at infinite dilution.

Non-suppressed Conductivity Detection. If a suppressor is not used, the only method of improving the sensitivity is to maximise the difference between the conductance of the mobile phase and that of the analyte. If the analyte has a high conductance the eluent should have a low conductance (*e.g.* a benzoate or phthalate mobile phase), so that a measurable increase in signal occurs (direct detection). If the analyte has a low conductance, the decrease in signal as the mobile phase ions are displaced can be measured providing that the mobile phase has a high background conductance (indirect detection). In either case it is the *difference* between analyte and background that must be maximised. In practice, if using direct detection a suppression system is invariably used.

Indirect detection requires the high background signal to be as steady as possible; any degree of noise markedly reduces sensitivity. It is vital that the detector cell is precisely thermostatted and that there is no pump noise. In order to obtain a background signal low enough to be offset, eluents are generally weaker than for suppressed systems. In turn, this means that columns often have a lower ion-exchange capacity. Indirect systems are therefore unsuitable for the separation of ions with a large variation in concentration, *e.g.* trace analytes in sea water samples.

Conductivity detection is used for the analysis of many inorganic and organic ions that do not absorb UV light. This includes all strong-acid anions (*e.g.* Cl^-, NO_3^-, SO_4^{2-}, CF_3COO^-) and inorganic cations (*e.g.* alkali metals and alkali earths). Because transition metals form oxyanions in aqueous solution (except at low pH) they cannot be analysed using conductivity detection.

A major limitation of non-suppressed conductivity detection is that gradient systems cannot be used; the background conductivity of the mobile phase must remain constant. However, gradient elution can be performed with suppressed conductivity detection using hydroxide eluent, since hydroxide is suppressed to water, which has very low background conductivity.

Amperometric Detection

Certain analytes can be detected by their oxidation or reduction at an electrode. As with conductivity, there is an inherent background from the mobile phase. The technique is therefore only useful for trace analysis if the analyte is oxidised or reduced at a lower potential than the components of the mobile phase. For oxidations, the analyte must also be oxidised at a lower potential than the working electrode.

Analytes with a sufficiently low redox potential include aromatic compounds with a delocalised charge, phenols, and aromatic amines. The most common use of the technique is the analysis of catecholamines.

The material from which the electrode is constructed is important. Carbon electrodes are generally used, as they are slow to oxidise and foul. Some electrode materials, however, also act as catalysts for the oxidation of certain analytes. Platinum electrodes aid the oxidation of iodide, and silver electrodes are similarly used for cyanide, sulfide, and the halides. Pulsed amperometry is also used to detect certain species that foul the surface of the electrode. These include carbohydrates, alcohols, aldehydes, and amines.

Photometric Detection

There are certain ions that do absorb UV or visible radiation, and can be detected photometrically. Table 4.4.4 lists some of the ions most commonly detected by UV absorbance. The molar absorption coefficients are given for species with an absorbance maximum at wavelengths above 200 nm. As detection at or below 200 – 210 nm is impractical due to mobile phase absorbance, the molar absorption coefficient for the others is given at both 200 nm and 210 nm (the lowest practical detection wavelengths, depending upon the mobile phase). Those most often encountered are iodide, bromide, the nitrate/nitrite ions, and thiocyanide, as well as certain organic acids and amines.

Table 4.4.4 *Optimum detection wavelengths and extinction coefficients (ε) of UV absorbing ions*[1]

Ions With max. absorbance >200 nm		λ_{max} (nm)	$10^{-3} \varepsilon$ (l mol^{-1} cm^{-1} at λ_{max})
$AuCl_4^-$ [a]	(i)	227	38
	(ii)	313	5.5
CrO_4^{2-}	(i)	273	3.6
	(ii)	372	4.7
$Fe(CN)_6^{3-}$	(i)	260	1.0
	(ii)	303	1.6
	(iii)	420	1.0
I^-		226	12.1
MoO_4^{2-}	(i)	207	10.0
	(ii)[b]	227	5.1
NO_2^-		209	5.2
HS^- [c]		230	8.0
SCN^-		215	3.1
$S_2O_3^-$		215	3.8
VO_3^-		266	3.4

Ions with λ_{max} <200 nm	$10^{-3} \varepsilon$ (l mol^{-1} cm^{-1})	
	at 200 nm	*at* 210 nm
AsO_2^-	9.5	5.0
Br^-	8.9	1.9
BrO_3^-	2.0	1.1
IO_3^-	17	3.9
N_3^-	6.0	1.7
NO_3^-	9.3	7.8
SeO_3^{2-}	5.1	2.3
WO_4^{2-}	6.7	2.7

[a] In 100 mM HCl. [b]Shoulder. [c]In 1 mM NaOH.
Reproduced by permission from R.D. Rocklin, *J. Chromatogr.*, 1991, **546**, 175.

The major advantage of photometric detection is that there is no interference from mobile phase buffers and salts. Highly conducting eluents can be used, and there is also the scope for gradient elution if required. It is particularly appropriate for the determination of UV absorbing ions in environmental extracts that have high conductivities, such as soil and sea-water samples.

Post-column derivatisation is possible in ion chromatography as in any other forms of HPLC. The usual problems of increased dead-volume and noise due to pressure fluctuations apply. The most common applications are the derivatisation of transition metal or lanthanide ions with 4-(2-pyridylazo)resorcinol to give complexes that absorb at 520 nm. The addition of *o*-phthalaldehyde and 2-mercaptoethanol to primary amines also produces fluorescent derivatives. Fluorescence detection without derivatisation is rare because few ions fluoresce. The exceptions are Ce(IV) and U(VI).

Ions that do not absorb UV radiation can be detected by the drop in signal as they displace ions in a UV-absorbing mobile phase (*e.g.* mobile phases containing nitrate, benzoate, or phthalate). This indirect detection is analogous to indirect conductivity detection. Again, it is measuring a change of signal against a high background and so is susceptible to pump fluctuations and other changes in the mobile phase. Despite this, it can still rival the sensitivity of conductivity detection.

4.4.4 Critical Aspects

Ion chromatography, particularly ion-exchange chromatography, is a very robust technique. However, there are many variables of which small changes can have marked effects upon both separations and sensitivity. If there is the time and resources, method development should ideally include comprehensive robustness testing: investigating the effect of small changes in temperature, pH of the mobile phase, buffer strength, injection volume, the use of different but nominally identical columns, and all of the other parameters that may affect an analysis. If the amount of work involved is impractical, analysts should at least endeavour to keep chromatographic conditions constant. Reproducible results should never be assumed.

Temperature

As with HPLC, increasing the temperature of the column will decrease retention times and improve separation efficiency. However, it may also alter the selectivity of the separation. The column temperature should therefore be kept constant. If it cannot be thermostatted it should at least be insulated. This is particularly important for automated, overnight analyses in laboratories where air conditioning may be switched off outside working hours.

pH

Controlling the pH of the mobile phase is essential. The pH controls the degree of dissociation of the weakly ionic analyte, and hence both its retention characteristics and its conductance. Buffers must be prepared accurately.

When developing a method the other effects of varying the pH must also be remembered. Resin stationary phases are stable over a pH range of 0 to 14, but silica columns are unstable above pH 8. Acids below pH 3 can slowly dissolve materials used in certain pump seals; it is best to check the exact working range in the manufacturer's equipment manual.

Mobile Phase Organic Modifier

Increasing the percentage of organic modifier in the mobile phase generally decreases retention times. As is the case in HPLC, the smaller the percentage of modifier the more accurately it must be measured. Pipettes should be used for measuring mobile phase components where a volume of 20 ml or less is added to a volume of 1 l.

4.4.5 Practical Advice

Equilibrating the System

If the column is in routine use, it is best never to shut down the system. Rather than turning the pump off, either leave it on at a low flow rate or leave it on and recycle the mobile phase.

Columns do not need to equilibrate under the exact analytical conditions. Equilibration may be speeded up by altering any of a number of variables. Increasing the temperature generally reduces the time required, as does increasing the flow rate. Using a mobile phase with a higher (*e.g.* 50%) proportion of organic modifier than the analytical mobile phase can also speed up equilibration. A gradient can then be programmed to change over to the analytical mobile phase.

Table 4.4.5 gives a summary of the minimum detection levels for the various analytes in reagent water for given loop sizes used with specific columns and detectors.

Method Development: Predicting Relative Retentions

As there are so many variables in ion chromatography, relative retentions are often difficult to predict under different conditions. However, some general principles do apply.

Relative Retentions under Identical Conditions. For the same buffered solution, ion-exchange elution order is generally:

  larger ions before smaller ions;
  monovalent ions before divalent ions;
  divalent ions before polyvalent ions;
  weak acids (high pK_a) before strong acids (low pK_a) (anion exchange).

Effect of Varying the Buffer pH. Varying the buffer pH affects the degree of dissociation of the substances to be separated. The greater the degree of

Table 4.4.5 *Minimum detection limits for various analytes in reagent water*

Analyte	Column type	Loop size (μl)	Detector type	MDL (μg l^{-1})
Fluoride	Anion exchange	100	Suppressed conductivity	0.6
Chloride				1
Nitrite				1
Nitrate				1
Bromide				8
Phosphate				5
Sulfate				3
Chlorate(III) (chlorite)				10
Chlorate(V) (chlorate)				10
Bromate(V) (bromate)				5
Potassium	Cation exchange			0.25
Calcium				0.5
Magnesium				0.25
Sodium				0.08
Lithium				0.05
Ammonium				0.18
Cyclohexylamine		25		50
Borate	Ion exclusion	50		20
Heptaoxodiphosphate(V) (pyrophosphate, $P_2O_7{}^{4-}$)	Combined reversed phase/ anion exchange	15		10–20
Tripolyphosphate				10–20
Iron(II)	Mixed anion/cation exchange	100	UV/visible	6
Iron(III)				2
Manganese(II)				3
Cobalt(II)				2
Cadmium				15
Lead(II)				15
Copper(I)				2
Zinc				3
Chromium(III)		250		100
Chromium(VI)				1
Aluminium	Cation exchange	50		50
Silicate	Anion exchange	50		200
EDTA		50		50
NTA				50
HEDP				50
Sulfite	Ion exclusion		DC amperometry	100

EDTA = Ethylenediamine tetraacetic acid, NTA = Nitrilotrisacetic acid, HEDP = 1-Hydroxyethane-1,1-diphosphonic acid dissociation, the longer the retention. Therefore, increasing the pH of the buffer will increase the retention of acids while decreasing the retention of bases. Conversely, acidifying the buffer decreases the retention of acids, while increasing that of bases.

Effect of Varying the Buffer Concentration. The effect of varying the buffer concentration can be predicted by visualising a competition between buffer ions and analyte ions for the active sites on the stationary phase. Therefore, if the buffer concentration is increased the analyte retention decreases. An exception is that retention generally increases as the mobile phase concentration increases during ion-exclusion chromatography.

4.4.6 References

1. R. D. Rocklin, 'Detection in Ion Chromatography', *J. Chromatogr.*, 1991, **546**, 175.

General Reading

2. D. M. Radzik and S. M. Lunte, 'Application of Liquid Chromatography – Electrochemistry in Pharmaceutical and Biochemical Analysis: A Critical Review', *CRC Crit. Rev. Anal. Chem.*, 1989, **20**, 317.
3. J. W. Dolan, 'LC Troubleshooting: Questions about Ion Pairing', *LC.GC Int.*, 1994, **7**, 691.
4. D. T. Gjerde, 'Eluent Selection for the Determination of Cations in Ion Chromatography', *J. Chromatogr.*, 1988, **439**, 49.
5. D. R. Jenke, 'Effect of some Operational Variables on the Efficiency of Ion Chromatographic Separations', *J. Chromatogr.*, 1989, **479**, 387.
6. J. S. Fritz, 'A Look at Contemporary Ion Chromatography', *J. Chromatogr.*, 1988, **439**, 3.
7. H. F. Walton and R. D. Rocklin, 'Ion Exchange in Analytical Chemistry', CRC Press, Boca Raton, FL, 1990.

CHAPTER 5

Organic Analytes: Sample Preparation

5.1 Introduction to Planning and Preparation for Organic Analyses

5.1.1 Overview

Organic trace analysis is a much more recent discipline than inorganic trace analysis. It is generally considered to have developed since the early 1950s, whereas its inorganic equivalent has been a challenge to the analyst since the beginning of the 20th century.

Since the early 1950s the levels of trace organic compounds that can be determined have fallen substantially. The need to determine these low levels has led to the development of more sensitive techniques. However, in determining these small amounts, the problems of the work-up of the sample remain difficult if the analyst is to avoid contamination and/or losses.

Organic trace analysis certainly moved into a new era with the invention of the gas chromatograph, which enabled analysts to detect compounds that had previously only been detectable at much higher levels. This coincided with the growing concern about the level of chemical contaminants in the environment; there was little restriction on pollutants such as industrial effluents or the use of the many novel pesticides available at the time. The cycle of increasing analytical sensitivity has been self-perpetuating until the present day: the detection of previously undetected compounds leads to the demand for techniques that can measure lower and lower levels, which leads to the detection of previously undetected compounds and so on. From the detection of low $\mu g\ g^{-1}$ levels of DDT (dichlorodiphenyltrichloroethane) by colorimetry in the 1950s, science has now moved to a stage where fg g^{-1} levels of aflatoxin M1 in milk can be determined, *i.e.* an increase in sensitivity of some nine orders of magnitude.

5.1.2 Potential Difficulties with Organic Analyses

Although there are many parallels between trace organic and inorganic analyses, there are also many problems that are either unique or compounded in the organic field. Many of these difficulties also apply to the analysis of speciated compounds; these are dealt with separately in Chapter 7. The types of problem faced by the organic analyst are dealt with in turn.

Range of Compounds for Which Methods are Required

There are around 100 elements and 10–15 inorganic anions that are of interest to the inorganic analyst, but there are, in theory, a million or more organic compounds of potential interest!

Legislation already exists or is proposed for the permitted levels of a wide range of chemical types. For example, pesticides include more than 30 classes of chemical compounds, residues of which are subject to statutory control. Veterinary drug residues include at least seven classes of antimicrobial compounds, as well as a range of anabolic compounds. There are many other compounds such as phthalates, polycyclic aromatic hydrocarbons, polychlorinated biphenyls, dioxins, and mycotoxins, which are subject to legislative control.

The substrates that have to be analysed for trace organic compounds include:

(i)   gases, *e.g.* air, stack gases, gaseous reaction mixtures, factory atmospheres;
(ii)   liquids, *e.g.* water, body fluids, oils, effluents, industrial process mixtures;
(iii)   solids, *e.g.* food, soil, sludge, human and animal tissues, plant materials, plastics, polymers, packaging materials.

The complexity of analytes to be determined ranges from simple molecules such as carbon disulfide to compounds having a relative molecular mass of thousands, such as the macrolide antibiotics.

The Need for Multi-analyte Methods Covering Many Compounds Within a Group

There are areas of legislation (*e.g.* pesticide residues) that require the measurement of increasing numbers of analytes in a given substrate. Purely on a cost basis there is a need to determine as many compounds as possible in a single analysis. Thus an ideal method for pesticide residues would be capable of determining more than 200 individual compounds. This is obviously going to include analytes with a wide range of physical and chemical properties, which means that it is unlikely that such a method will be the optimum approach for every single analyte that has to be determined.

The Need to Determine Both the Parent Compound and Metabolites

Metabolites may have considerably different properties from their parent compound. By their very nature, they are usually more water-soluble. It is often therefore extremely difficult to determine both types using the same method.

Analysis of metabolites is of particular importance when they have different toxicological properties from the parent and they are included in lists for which there are legal limits.

Problems with Co-extracted Interferents

In inorganic analyses it is often the case that the sample matrix can be completely destroyed (ashing). In organic analyses such fierce conditions would, in most cases, also destroy the analyte. Other separation techniques must therefore be employed.

The Similarities in Properties of Many of the Compounds to be Determined in a Single Sample

When similar compounds exist in a sample, but need to be individually quantified, the specificity of the method becomes crucial. Examples of such analyses are:

(i)   tetrachlorodibenzodioxin, where some 25 isomers exist but only one, the 2,3,7,8-isomer, is of toxicological interest.

(ii)   the increasing importance of separating D- and L-isomers of a particular compound. Many pharmaceuticals, agrochemicals, and industrial intermediates exist as racemic mixtures, although it may only be one enantiomer that is biologically active.

The Instability of Many Organic Compounds to Heat and/or Light and their Susceptibility to Hydrolysis and Oxidation

Examples are mycotoxins, which are unstable to UV light, and many sulfur-containing fungicides, which are readily oxidised.

The Toxicity of the Analytes and, to a Lesser Extent, of the Solvents and Reagents

Some examples of toxic analytes are the *N*-nitrosamines, which are carcinogenic, some mycotoxins, which are mutagenic, and chloropropanols, which are believed to be mutagenic.

Commonly used solvents such as dimethyl formamide, dimethyl sulfoxide, and acetonitrile are acutely toxic and there are very few other organic solvents that do not have some form of acute or chronic hazard associated with their use.

5.1.3 Planning a Trace Organic Analysis

Despite the large variations in type of substrate and compound to be analysed, there is a commonality in many of the techniques employed. All methods of trace organic analysis can be considered to consist of two or more stages from the following: sampling, sample preparation, extraction, clean-up, separation, and determination (Figure 5.1.1).

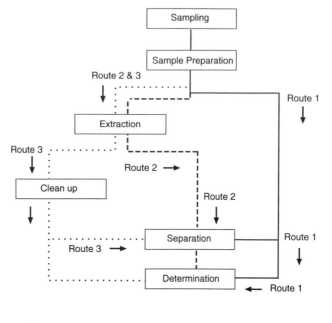

──── Route 1 example: Enzyme Linked Immunosorbent Assays of aqueous samples
– – – Route 2 example: Mass spectrometric screening of crude extracts
· · · · Route 3 example: Traditional organic analysis route for chromatographic analysis

Figure 5.1.1 *Stages in organic analysis*

The first two of these stages are common to all analytical determinations and, in many ways, are the most critical. If the initial sample is not representative or the sample preparation leads to significant losses of the analyte to be determined or to significant contamination, then the final result is meaningless. These potential problems in sampling and sample preparation apply equally well to both organic and inorganic trace analysis, and are dealt with in detail in Chapter 2.

Having prepared a satisfactory test sample, the aim of the method developer is to use the least number of steps possible. Analyte losses and method variability both tend to increase with the number and complexity of the stages used. The desire for minimalism must be weighed against the requirement that the analyte in the final extract is in a suitable form to be determined. This is obviously dependent upon the determination technique used; generally the less selective the technique, the greater the degree of clean-up required and therefore the more stages needed. Thus at one extreme are techniques such as immunoassays (Route 1 in Figure 5.1.1), which are so selective that the crude extract can often be examined directly. At the other extreme are more traditional techniques such as gas chromatography with flame ionisation detection (Route 3 in Figure 5.1.1), where multiple clean-ups are often required before the sample extract is sufficiently free from interferents. The former approach is preferable, both from the aspect of reducing the variability of more complicated methods and because rapid analyses are more economic; it is often precluded by the analyte type, matrix type, and detection systems. This situation is gradually changing, with the widespread use of gas chromatography with mass-selective detection leading a drive towards rapid minimalist methods that rely heavily on the extremely selective end-determination techniques. If the test extract is not clean enough to ensure reliable quantitative results, the preferred procedure is to screen the samples using a crude but rapid method, and restrict use of the full traditional methodology for samples which test positive during the screen.

In addition, in selecting and developing suitable method(s), the analyst must bear in mind the level of analyte to be determined, the availability of equipment and reagents, the expertise of the staff, the purpose for which the results are required, and the timescale and budget available to produce the results.

The purpose for which the results are required must always be borne in mind. The onus is upon the analyst to discuss this with the customer commissioning the work. There will be cases when a quick 'yes/no' at a relatively high level is all that is necessary; for example, whether an animal carcass contains a banned steroid. In such cases a high degree of clean-up is not required: the test extract does not have to be clean enough for quantitation, but only sufficient for a quick screen. Conversely, the analysis of an athlete's urine sample for a drug that has a maximum permissible concentration must be performed with a high degree of quantitative certainty. In such cases a more protracted method is usually required to produce an extract in which the analyte can be measured with a greater degree of confidence.

Sections 5.2, 5.3, and 5.4 follow sequentially through the extraction, clean-up, and concentration stages of the more traditional analytical process (Route 3 of Figure 5.1.1), which is still used in the great majority of cases. Section 5.2 deals with extracting the analyte from solid and gaseous samples. The extract produced is then regarded as analogous to a liquid sample. Techniques to extract the analyte from the resulting liquid, or from liquid samples, and to remove interferents (clean-up techniques) are described in Section 5.3. Section 5.4 compares methods of reducing the final extract volume, both at various stages of the method when it may be necessary and finally to bring the cleaned extract to a suitable concentration for determination of the analyte.

5.2 Methods of Extraction of Organic Analytes from Solid Samples

5.2.1 Overview

There are a number of sample matrices that may be presented for analysis of the trace organic component, the majority of which are solids. Being able to deal with this almost infinite range of sample types is likely to be necessary for the foreseeable future. Heightened concern over the toxicological effects of environmental pollutants, in particular, has driven the need for valid extraction procedures from samples such as foods, soils, sludges, and plant and animal tissues. Other examples of solid samples encountered are the determination of minor impurities in pharmaceutical formulations and colorants in foods and cosmetics.

Extracting analytes from solid samples can give rise to many more problems than the analysis of liquids and gases. The difficulties arise from matrix effects. These can include the binding or entrapment of the analyte by the matrix, or co-extraction of interfering compounds along with the determinand. These may be difficult to remove at a later stage of the analysis. Both of these effects can have profound implications for the validity of an analytical method. Co-extractives can mask the presence of an analyte or, in extreme cases, give rise to misidentifications and false positive results being reported. The problem of bound analytes can be harder to deal with, mainly because very often it is not recognised. Commonly used quality control procedures such as measured recoveries from fortified samples or use of standard addition techniques make no allowance for matrix binding or entrapment. Without expensive field trials involving treatment with radiolabelled analyte isomers, outside the scope of most organisations, the occurrence and degree of binding can only be estimated.

The prerequisite of an extraction technique is that it removes the analyte of interest; preferably all of it, but at least sufficient in a reproducible manner that an estimate can be made of the remaining unextracted material. It is usually fairly easy to find conditions for the optimum extraction yield for a single analyte, but in many organic analyses the situation is much more complex. Multi-analyte extractions are common, particularly for environmental contaminants such as pesticide residues or polyaromatic hydrocarbons. In other cases, such as in the tracing of pharmaceutical metabolic pathways, the analyst is also concerned with the extraction of metabolites of the parent compound. These, by their very nature, are usually of a radically different polarity to the parent and so require different extraction conditions.

Apart from extraction efficiency, the degree of co-extraction has to be considered. Thus, the technique that gives the best extraction of the desired analyte from the sample matrix is not always the method of choice. It is unfortunate that techniques that are generally good for removal of the determinand, *e.g.* Soxhlet extraction, are also good for the removal of everything else. Co-extractives, often including analyte binding, described as 'matrix effects', are the bane of the organic analyst's life!

Extraction conditions are therefore usually a compromise. The yield is still the first priority, but in multi-analyte extractions (including those where metabolites must be extracted) it is impossible for the conditions to be the optimum for each compound. It may then be necessary to further reduce the yield in order to minimise co-extractives. The simplest method of controlling this balance between co-extraction and yield is usually either by the choice of solvent(s) or by varying the physical extraction conditions (*e.g.* temperature, blending time, and speed).

There is an alternative approach to the minimisation of co-extractives at the extraction stage of a method. This is to rely upon a determination technique that is sufficiently selective so that other compounds present in the extract can be safely ignored. For example, extracts to be presented for gas chromatography with electron capture detection must be extremely clean, as most co-extractives will interfere. More latitude is possible with extracts intended for gas chromatography with sulfur-mode flame photometric detection, as only sulfur-containing co-extractives will interfere. The increasing use of gas chromatography with mass selective detection (GC–MS), in particular, is having far reaching ramifications on method development as less and less stringent clean-ups are required. Some techniques, such as Enzyme Linked ImmunoSorbent Assay (ELISA), are so selective that there are virtually no interference problems, whatever the extract contains. Applications are currently rare, but where used, such methods present convenient short cuts to the extensive sample preparation and clean-up usually required in organic analyses.

5.2.2 Choice of Extraction Technique

Available Extraction Methods

A comparison of the more common techniques available is given in Table 5.2.1. The majority of methods are based on the permeation of an extracting solvent through the sample matrix. This solvent can either be an organic solvent or a supercritical fluid. The exception to this rule is in the analysis of volatile analytes. These can be released by methods such as distillation, static headspace, dynamic headspace, and sublimation.

The questions that must be asked before selecting an extraction method are the following:

(i)   Is the extraction exhaustive?
(ii)   Are there undesirable co-extractives?
(iii)   Is the unit time taken for each extraction economical?
(iv)   If the method is time consuming, is there a stage at which the extract can safely be left overnight?
(v)   Can the extraction be automated and/or coupled with the end-determination?
(vi)   What is the maximum amount of sample with which the technique can cope?
(vii)   Will the extract need to be concentrated? How difficult will this be?
(viii) Will the extract be compatible with the end-determination technique?

How these considerations are weighted will depend upon the purpose of the analysis.

Soxhlet extraction is regarded as the most exhaustive method of solvent extraction available for most solid sample types, and is used as a yardstick to measure the efficiency of other techniques. However, because of the long percolation time required and the risk of extracting co-extractives, alternative methods are preferred if applicable. Those such as blending, sonication, and simple shake-flask methods are quick and easy and therefore the methods of choice, with the proviso that the sample matrix must be permeable to the solvent.

Table 5.2.1 *Techniques for extraction from solias*

	Sample type	Restrictions on use	Examples of use	Yield	Operator skill	Time taken	Max. sample size	Extract volume	Cost
Soxhlet	Most solids, some liquids	Analyte must be thermally stable	Fats, phenols, aromatics, hydrocarbons	***** unless bound to matrix	**	> 8 h	~ 20 g	< 50 ml	**
Soxtec	As Soxhlet	As Soxhlet	As Soxhlet	As Soxhlet	As Soxhlet	< 2 h	~ 10 g	As Soxhlet	***
Blending	Solids, tissue, plant material	Sample must be fairly friable	OC in tissue,[1] OP in grain[2] aflatoxins	****	*	~ 2 min	~ 100 g	< 300 ml	***
Cold column extraction	Solids, tissues	Matrix must be solvent-permeable	Fats, hydrocarbons, aromatics	****	***	~ 30 min	~ 50 g	< 1 l	**
Sonication	Soils, granules, feeds	Matrix must be solvent-permeable	Environmental pollutants	****	*	~ 30 min	~ 100 g	< 500 ml	**

Method	Sample	Comments	Applications			Time	Sample size	Solvent	
Shaking	Soils, granules, liquids	Permeable matrix, soluble analytes	Veterinary drugs, vitamins	***	*	~ 5 min	~ 5 l	< 5 l	*
SFE	Solids	< 20% moisture	Food additives,[3] plastic additives[4,5,6]	***	****	~ 1 h	~ 20 ml	< 50 ml	*****
Microwave	Soils, granules	Analyte must be thermally stable	OP in soil[7]	****	***	< 5 min	~ 50 g	< 100 ml	****
Static headspace	Volatiles in liquids and solids	Poor quantitation, low sensitivity	Food flavours, taints	****	***	40 per day	~ 100 g	Minimal	****
Dynamic headspace	Volatiles in liquids and solids	Liquids better by 'Purge and trap'	Spices, perfumes	*****	****	Easily automated	~ 100 g	Minimal	*****
Sublimation	Volatiles in solids	Hard to optimise, limited applications	PAH in soil	***	*****	< 30 min	~ 100 g	Minimal	****
Distillation	Volatiles	Thermal stability	N-Nitrosamines	****	***	< 30 min	~ 100 g	< 50 ml	***

OC, organochlorine pesticides; OP, organophosphate pesticides and nerve gases; SFE, Supercritical fluid extraction; PAH, polynuclear aromatic hydrocarbons.

* – low
***** – high

In these cases, recoveries are often comparable to those obtained using Soxhlet extraction and supercritical fluid extraction (SFE).

If the sample is not permeable to the solvent, it can often be made to be so. The physical form of the sample, particularly its particle size and moisture content, has a major effect on solvent permeation. Dry samples do not allow full penetration, whilst those too wet are immiscible with non-polar solvents. Freeze dried samples can exacerbate these problems. A recent inter-laboratory trial found that the extraction of organo-phosphorus pesticides from freeze dried bread could be reduced by up to 100% if the sample was not first reconstituted with water.[8]

Methods of improving permeability to solvents are listed in Table 5.2.2. Granular material such as sand or anhydrous sodium sulfate is often ground with plant or tissue samples to break down the physical form of the sample or to dry it before the addition of a non-polar solvent. Samples can even be ground with C-18 adsorbents (the matrix dispersion technique) to ensure contact between the solvent and the analyte.

Supercritical fluid extraction (using CO_2) is an attractive alternative to solvent extraction as it provides maximum solvent permeation, and therefore an efficiency comparable to Soxhlet extraction, in a fraction of the time. However, it cannot be regarded as a panacea. As with any other solvent extraction technique there are the same problems with extracted interferents and incomplete extractions. Solvents are not widely available for the extraction of polar analytes, and it cannot be used for wet sample matrices, as this will not allow permeation of the supercritical fluid. However, as the drive towards less toxic and environmentally damaging solvents gains force, SFE is likely to be developed as the preferred technique.

Volatile compounds in solids are the exception to the problem of matrix effects. They can be easily removed by heating, and the resulting vapour treated as a gaseous sample. The usual techniques are distillation and dynamic or static headspace analysis, with sublimation also occasionally used. Table 5.2.3 shows the relative merits of these techniques when used in sample analysis. These are generally coupled with on-line chromatography to give a complete automated analysis. Volatilised compounds are held on an adsorbent or in a cryogenic trap, then purged directly into the top of a gas chromatographic column. The ease of use of these techniques, and the lack of necessary clean-up, make them the method of choice for extracting volatiles.

One final factor in extraction selectivity that is frequently overlooked is pH, of both the solvent and the sample. If the desired analytes are susceptible to speciation effects (*i.e.* an acid or a base) their chemical form and thus polarity will be sensitive to pH changes. The chemical nature of the analyte should be born in mind at all times.

Pre-extraction Clean-up (Defatting)

With any of the solvent extraction methods listed, there is scope to reduce the degree of clean-up required in the subsequent analytical method, by removing some interferents at source. This approach relies upon extracting as many compounds as possible from the sample using conditions that leave the analyte

Table 5.2.2 Aids to permeation of solvent when blending

Action	Use	Advantages	Disadvantages
Add anhydrous sodium sulfate to make sample friable before adding solvent	Non-polar solvents with wet samples	Easy. Available with adequate purity. Does not affect chromatographic detectors	Does not break down tissue cells or micells in sample. Sometimes cannot extract from dried, crystallised substrate
Acid-washed sand added followed by grinding	Mainly tissue samples	Releases analyte from tissue matrices	Can cause contamination
Add acetone and an alcohol, e.g. ethanol	Breaks down cellulose, micelles and tissue cells, etc.	Easy. Releases analyte from matrix. Easy to remove by washing with water	Not effective with all sample types
Add water	Dried (especially freeze dried) samples	Easy. Aids release of analyte by bulking sample	Does not break down tissues or cells in the sample. Lowers mg kg^{-1} detection limit unless wet sample size increased
Heat	To aid solubility of analyte	Easy. No contamination problems	Not applicable to thermally unstable or volatile analytes
Matrix dispersion (grinding with a sorbent)	To improve extraction of analytes caged in sample matrix. Can be put into thimbles, etc.	Often better recoveries than grinding with an inert material	More expensive. Must ensure that analyte desorbs easily from sorbent

Table 5.2.3 *Methods of sampling volatiles*

	Sensitivity	Quantitation	Ease of automation	Use for field sampling	Ease of use
Static headspace	**	**	*****	No	*****
Dynamic headspace	****	***	****	No	****
On-line purge and trap	*****	***	****	No	****
Off-line purge and trap	*****	*****	*	Yes	**
Sublimation	**	***	*****	No	*

* – low
***** – high

intact, then changing the conditions to extract the analyte. Depending upon the analyte, this could involve either changing the extraction solvent or changing the pH of the system to control the extraction selectivity. The most common use of the technique is defatting food and animal tissue samples before the extraction of polar or moderately polar analytes. Typically, the sample is washed with a solvent such as hexane, which will remove non-polar lipids but is insufficiently polar to extract a significant amount of the analyte. This extract is discarded. The analyte can then be extracted using a more polar solvent such as acetone (propanone) or methanol. This procedure is pH-dependent, *e.g.* fatty acids will not be removed in an alkaline medium.

Examples of pH-selective pre-washing are rarer, but can be useful. The polarity of, *e.g.* phenolic analytes, or those containing amine groups, is very easy to control. For example, if samples containing the wood preservative pentachlorophenol are made alkaline the analyte is ionised and the sample can be safely washed with a non-polar solvent. If it is then acidified, a second wash with the same solvent will extract the analyte.

Matrix Binding Effects

As already mentioned, bound analytes can cause major difficulties when analysing solids. The problem is not in dealing with the binding, but recognising that there is a problem. If presented with a sample where binding is likely, *e.g.* protein-containing animal tissue, it is safest to assume that this is the case, and act accordingly.

Table 5.2.4 categorises some of the more common types of matrix binding, and suggests methods of releasing the analyte. By far the greatest incidence of matrix

Table 5.2.4 *Methods of releasing matrix-bound analytes*

Type of binding	Methods of releasing analytes	Examples
Chemical binding (usually to proteins)	Addition of a protease enzyme, *e.g.* papain, phospholipase	Drugs of abuse in body tissue[9]
	Equilibrium dialysis to determine extent of binding, then extrapolation of the concentration of the free analyte[10]	Food colorants[11]
	Use of polar solvents to adjust the dielectric constant or to denature proteins, *e.g.* dichloromethane to release sulfonamides from meat	Analysis of vitamins in meat
pH-dependent binding (*e.g.* hydrogen bonding of phenols)	pH control of the extraction system Saponification	Pentachlorophenol in wood
Encapsulation	Mechanical action, *e.g.* grinding Use of polar solvents to break down matrix	Pesticide residues in plant cellulose

binding occurs in the analysis of meat and body tissues. These samples contain proteins, whose main function is to complex with exactly the type of compound the analyst is trying to release. It is virtually inevitable, in fields such as the forensic analysis of drugs of abuse or residues of feed additives in farm animals, that the problem will have to be addressed. The addition of enzymes to break down the proteins is standard procedure, but the only method of validating the extraction is to empirically try different methods on a sample of known incurred residues and compare the results. Even this is unsatisfactory, as it can never be known if even the most effective technique has given 100% extraction.

Bound analytes also include those physically encapsulated within the sample matrix. These are usually easier to deal with, as release can be effected by the physical destruction of the sample matrix.

5.2.3 Choice of Solvent

The solvent(s) chosen for extraction is obviously highly critical. A list of common solvents, their properties and possible uses is given in Table 5.2.5. Properties of solvents are also covered in Section 5.3, including miscibilities of solvent mixtures (Table 5.3.4). Factors to consider when choosing an extraction solvent include the following:

(i)   miscibility with the sample matrix;
(ii)   solubility of the analytes and their metabolites;
(iii)   solubility of unwanted co-extractives;
(iv)   possible solvent impurities and their significance to the analysis;
(v)   toxicity, possibility of a safer alternative;
(vi)   cost, adequacy of a cheaper alternative;
(vii)   disposal costs and/or difficulties;
(viii) ease of concentrating;
(ix)   compatibility with the detection system to be used. If not compatible, can it be easily removed at a later stage of the analysis?

It is more often than not the case that a combination of two or more solvents will be used to fulfil these criteria, particularly in the case of multi-analyte assays.

Generally, solutes will dissolve in solvents that possess similar intermolecular attraction properties. Non-polar analytes, such as polychlorinated biphenyls, are best extracted with a non-polar solvent such as an aliphatic hydrocarbon. For more polar analytes acetone (propanone), acetonitrile, or a chlorinated solvent tend to be used. Acetone is used frequently: it is non-toxic, and its miscibility with water facilitates permeation into wet samples, such as animal tissues. Aromatic analytes are usually extracted with toluene (methylbenzene) or, if a lower boiling solvent is required, a chlorinated hydrocarbon.

Non-polar solvents will not fully permeate a wet sample, such as foods or tissues. To counter this, a mixture of solvents is often used, *e.g.* hexane to extract the analyte and propanone or an alcohol to increase the hexane/sample miscibility.

Some samples require degradation before full extraction can occur. Acetone will break down plant cellulose and it may be used to destroy micelles in milk and thereby release the entrapped organic components. The advantage of using acetone for this purpose is that if it needs to be removed (depending upon the subsequent clean-up) it can easily be washed out with water or other polar solvent. Alternatively, the sample can be dried before the application of a non-polar solvent as described in Section 5.2.2, 'Available extraction methods'.

Not all solvent combinations are compatible, even when initial results suggest that they may be miscible. Some, such as dichloromethane and methanol, are only miscible in certain proportions. Others, such as methanol and esters, will react. Further information on inter-solvent properties can be found in the literature.[12]

Safety considerations also play a role in the choice of solvents for analysis. Many are acutely or chronically toxic, and must be handled with extreme care. Some are flammable; in particular, diethyl ether (ethoxyethane) vaporises at room temperature and can explode if ignited. Operations involving ethers should ideally be fitted with an ice-cooled condenser.

As safety considerations gain more and more emphasis, an increasing number of solvents are either banned from general use or are in the process of being phased out (see Table 5.2.5). At present, this list includes benzene and carbon tetrachloride (tetrachloromethane), with other chlorinated solvents soon to follow. This can cause problems if a person is following a published or statutory method that specifies the use of a solvent that has since been banned. In the case of benzene, toluene (methylbenzene) is usually used as an alternative, although its higher boiling point can present difficulties if the extract solution requires concentration. As chlorinated solvents are phased out, they are being replaced by a variety of solvents, usually oxygenated, most notably, for example, ethyl acetate or acetone. It may be necessary to use a mixture of solvents to mimic the action of the original solvent. When following a previously validated method, where a solvent change has to be made, a thorough investigation of extraction efficiencies of the new solvent(s) is required.

The toxicity of many organic solvents and the restrictions on the use of chlorinated solvents are currently one of the driving forces for the development of carbon dioxide SFE. The use of a solvent which is in virtually inexhaustible supply, is completely non-toxic, and is self-disposing at the end of the analysis is certainly attractive. Many more analytical methods will need to be modified as an increasing number of solvents are withdrawn from use, and in many cases SFE will be a logical alternative.

When choosing an extraction solvent, the ramifications for the entire analytical method must be considered. Some solvents are incompatible with certain element-selective detectors. Traces of chlorinated or oxygenated solvents can interfere strongly with electron capture detectors. Acetonitrile and nitrated solvents will give a signal on any nitrogen-selective detector. There are many more examples that can be cited. If such solvents must be used for the extraction, it is imperative to ensure they are completely removed before the final determination.

Table 5.2.5 *Solvents for organic extraction*

Solvent	Uses (examples)	Frequency of use	Selectivity	Ease of evaporation (boiling point)	Possible impurities	Cost	Toxicity	Ease of disposal
Dimethyl formamide (dimethyl methanamide)	'Tough' matrices, polar species, heterocycles (drugs)	*	*	* (153°C)	Water	****	*****	*
Dimethyl sulfoxide (DMSO)	π-electron, H-bonding groups (dioxins)	**	*	** (110°C)	Water	****	*****	*
Tetrahydrofuran	'Tough matrices', polar species (plastics)	**	*	**** (60°C)	Peroxides, stabilisers, water	***	*****	*
1,4-Dioxan	'Tough' matrices, polar species	**	*	** (101°C)	Peroxides, water	****	***** (Teratogen)	*
Acetonitrile	Polar species	***	***	*** (82°C)	Water	***	****	*
Carbon tetrachloride[a] (tetrachloromethane)	General	*	**	*** (77°C)	Other chlorinated hydrocarbons	**	CCCC	*
Chloroform[a] (trichloromethane)	General (amphetamines)	*	**	** (61°C)	Ethanol as preservative	**	CCC	*
Dichloromethane[a]	General	***	**	***** (40°C)	Phosphate stabilisers	**	CC	*
Dichloroethane[a]	General	**	**	*** (83°C)	Other chlorinated hydrocarbons	**	CC	*
Trichloroethane[a]	General	***	***	*** (75°C)	Other chlorinated hydrocarbons	**	C	*

Solvent	Use			Boiling point	Impurities to watch for			
Acetone (propanone)	Polar and water-soluble species	*****	**	**** (56°C)	Water	*	*	****
Ethyl acetate	General	****	***	*** (65°C)	Water	*	*	***
Methanol	Polar species	****	***	*** (66°C)	Water	*	*	****
Ethanol	Polar species	*	***	** (78°C)	Water	*	*	*****
Isopropyl alcohol (propan-2-ol)	General	**	**	** (82°C)	Water	*	*	*****
Benzene[a]	Aromatics	*	***	** (80°C)	Sulfur compounds, thiophene	**	CCCC	***
Toluene (methylbenzene)	Aromatics, alternative to DMSO/benzene	***	***	* (110°C)	Benzene	**	CC	**
Hexane	Non-polar solutes (vitamins)	*****	****	*** (69°C)	Very pure	*	**	**
Iso-octane (2,2,4-trimethylpentane)	Non-polar solutes	***	****	** (99°C)	Very pure	*	*	**
Ethers	Non-polar solutes (alkaline amines)	**	***	* (−35°C)	Peroxides	*	*	** Flammable
Light petroleum	Non-polar solutes	*	***	* (40–60°C)	Peroxides	*	*	** Flammable
Supercritical CO₂	Non-polar solutes	***** for SFE	**	*	Very pure	*	*	*
Supercritical modified CO₂	Slightly polar analytes	**** for SFE	Variable	*	Very pure	*	*	* (Methanol modifier)

[a] Solvents that are either banned or being phased out.
CCCC, very high carcinogenic risk; CCC, high carcinogenic risk; CC, carcinogenic risk; C, low carcinogenic risk.
* – low; **** – high.

5.2.4 Hints and Tips

Liquid Extraction from Solids

Blending and chopping is best done in short pulses rather than continuously. This reduces frictional heating and prevents the blades becoming entangled. It also reduces nebulisation and splashing.

A mixture of toluene (methylbenzene) and methoxyethanol gives a high boiling solvent, suitable for Soxhlet extractions, whereas toluene on its own is inadequate.

Iso-octane (2,2,4-trimethylpentane) is a solvent of similar properties to hexane, but with a higher boiling point. It can be used to replace hexane if a slower evaporation is required at any stage of the analysis, *e.g.* removing a lower boiling solvent from a hexane mix by evaporation. If hexane is found to be co-extracting too many interferents from fatty samples, cyclohexane can be used which has similar properties but is less fat solubilizing.

Extraction of non-polar compounds from soils of high organic carbon content, *e.g.* peat, often gives low yields. If this is the case, SFE has been recommended as an alternative.[13]

With all extraction techniques involving hot solvents (*e.g.* Soxhlet extraction), fats and oils may dissolve in the hot solvent, only to precipitate as the extract solution cools. This may cause consequent losses by coprecipitation and entrapment. Any precipitates should not be ignored, but redissolved. There can also be problems at a later stage in the method if the extract solution has to be concentrated; solutions high in oils and fats cannot be reduced by evaporation. If the need for concentration cannot be avoided, the only way around this is to design less severe extraction conditions that do not strip out so much fat. It should be remembered that 'hot solvent' methods can also include cold blending and sonication, as the sample is heated by friction. Heating facilitates chemical reactions, this has to be borne in mind when refluxing to ensure that the analyte remains chemically inert.

Most manufacturers now supply solvents of sufficient purity for use in trace analysis, and any problems with impurities can usually be resolved simply by changing suppliers. If it is necessary to further purify solvents (most commercial chemicals should be of sufficient purity without further purification), distillation is the easiest method. Adsorbents should not be used, as they can introduce more impurities than they remove.[7] Traces of chlorinated solvents can be removed from aliphatic hydrocarbons by reaction with lithium aluminium hydride or sodium metal. Trichloroethene, an impurity in some chlorinated solvents, can be photolytically degraded using UV irradiation. Non-polar contaminants can be removed from acetone (propanone) by adding water, extracting into an immiscible solvent, then freezing out the water, although this is a rather tedious procedure. Sulfur compounds can be removed from aromatic solvents by shaking with sulfuric acid. Use of recovered solvents is not recommended as these are seldom of sufficient purity for trace analysis.

Supercritical Fluid Extraction (SFE) from Solids

There are fewer data available on supercritical fluid solubilities than conventional solvents, as solvent–solute interactions are poorly understood. A rough guide can be obtained from supercritical fluid chromatography retention times, but there is no substitute for empirical results. However, as the rate of SFE decreases exponentially with time it is not always necessary to take an extraction to completion, as the extraction curve can be modelled and the proportion of analyte extracted approximated.[14] Obviously the uncertainty in any given result will increase if this technique is used, but extraction times can be significantly reduced. The time used for the extraction is governed, chiefly, by the confidence required in the result.

The selectivity of a SFE can be controlled by varying the amount of polar modifier added to the fluid. Differences of the order of 1% in the methanol content of supercritical carbon dioxide can have a marked effect on the polarity of the solvent.

Supercritical fluid extraction is unsuitable for the analysis of wet samples as the solvent cannot permeate the sample, it merely compresses against the extraction cell frit. Sample moisture may be controlled by freeze drying (although there is the risk of incomplete extraction if the sample becomes too dry), adding acetone (propanone), or adding inert glass beads to bulk the wet sample. However, one of the more successful ways of dealing with high moisture content is to mix the sample with twice its weight of anhydrous sodium sulfate, which results in a cleaner extract.

In SFE the occasional problem of the restrictor blocking can be avoided by heating the back-pressure restrictor to temperatures between 50 and 200°C. To avoid sample loss by evaporation, the last 3 cm of the restrictor should be kept cool, and the eluent allowed to decompress directly into the collection solvent.

The main cause of analyte losses in SFE is during the decompression of the solvent. As there is a large pressure drop beyond the sample cell, there is the risk that the fluid will decompress prematurely and the extract crystallise-out inside the pipework. As can be seen from Table 5.2.6, the only failsafe way of overcoming this problem is to use a back-pressure regulator.

The SFE eluent can also be adsorbed then desorbed from non-solvent traps. This facilitates on-line coupling with a chromatographic end determination (usually GC). Although this improves sample throughput, coupled techniques are unsuitable for accurate quantitative work because of the band broadening that occurs on desorption.

Extraction of Volatiles

To cope with the frequent problem of band broadening in on-line desorption, cryogenic focusing can be used. The vapour is concentrated on a cold trap just before the analytical column, then the trap is rapidly warmed. This releases the analyte in one focused plug. It is generally more successful for analytes of low volatility, as volatile analytes adsorb poorly onto the trap and tend to leach off before controlled desorption.

Table 5.2.6 *Methods of decompression in SFE*

Method	Reasons for use	Reasons against use
Solvent collection trap	Simple and easy to use. Little concentration needed. Solvent can be chosen to be compatible with end-determination	Can get premature decompression and analyte crystallisation. Losses of volatile analytes reported.
Adsorbent trap	Suitable for on-line automation. Low detection levels	Can get premature decompression. Band broadening on desorption. Not good quantitatively.[6] Limited capacity
Cryogenic trap	Suitable for on-line automation. Low detection levels	Can get premature decompression. Band broadening on desorption. Not good quantitatively[6]
Back-pressure regulator	Obviates premature decompression problem	Expensive. Complex to operate. Can get nebulisation and sample loss due to aerosol formation.

To aid the adsorption of vapours onto a trap, a non-volatile derivative can be formed. Pentachlorophenol, for example, has been successfully extracted by the *in situ* formation of its sodium salt.[15]

5.2.5 Uncertainty Associated with Extraction

As there is no measure of the 'true' amount of an analyte in the matrix, which does not involve extraction, it is impossible to accurately quantify the optimum extraction yield. There are, however, methods of estimating the extraction uncertainty. Fortified samples (spikes) are widely used, and although they do not allow for matrix binding, spiking recoveries will flag problems such as poor analyte solubility in the extraction solvent. Matrix reference materials are available for a narrow range of analytes and matrices, and should be utilised if possible to validate extraction procedures. Results from inter-laboratory collaborative trials are also very useful for this purpose. One other method is to try a further, exhaustive, extraction (*i.e.* Soxhlet or similar) on the remains of a sample which has already been found to contain high levels of determinand. However, even this assumes that Soxhlet extraction is indeed exhaustive.

There is a lower uncertainty in sampling volatiles than most other extraction techniques, which is why it is the preferred option for analytes that have a sufficiently high vapour pressure. During headspace sampling, interference effects are reduced to a minimum as the vaporised analyte is completely free of the matrix. This means that minimal, or no, sample clean-up is required, reducing the number of stages in the analysis and therefore the overall uncertainty. However, when coupled with GC, as in the majority of cases, uncertainty is increased again by the effects of desorption band broadening. The less volatile and more strongly adsorbed the analyte, the greater this effect.

The SFE technique is relatively new, and so there are fewer data available on precision and robustness than for classical extraction methods. Most studies have shown that, for suitable analytes and sample types (usually granular samples), SFE shows a reproducibility and a reliability comparable to other extraction methods. The major source of uncertainty, apart from extraction efficiency, is the decompression of the supercritical eluent beyond the sample cell. With careful use of a back-pressure regulator, this can be reduced to a minimum. Depending upon the volatility of the analyte, losses can be considerable when using simple open vessel collection techniques. Also, automated on-line adsorption and desorption methods introduce a considerable degree of doubt into quantitative results.

As with all other extraction techniques, in order to obtain batch-to-batch reproducibility conditions must remain constant. As it is usually unknown whether extraction is theoretically complete or not, variations in temperature, pressure, flow rate, and extraction time can all lead to poor replication.

5.2.6 References

1. D. E. Wells, 'Extraction, Clean-up and Group Separation Techniques in Organochlorine Trace Analysis', *Pure Appl. Chem.*, 1988, **60**, 1437.
2. Ministry of Agriculture, Fisheries and Food Committee for Analytical Methods, 'Determination of a Range of Organophosphorus Pesticide Residues in Grain', *Analyst*, 1985, **110**, 765.
3. J. W. King, 'Fundamentals and Applications of Supercritical Fluid Extraction in Chromatographic Science', *J. Chromatogr. Sci.*, 1989, **27**, 355.
4. J. H. Braybrook and G. A. MacKay, 'Supercritical Fluid Extraction of Polymer Additives for Use in Biocompatibility Testing', *Polymer Int.*, 1992, **27**, 157.
5. T. P. Hunt, C. J. Dowle, and G. Greenway, 'Analysis of Poly(vinylchloride) Additives by Supercritical Fluid Extraction and Supercritical Fluid Chromatography', *Analyst*, 1991, **116**, 1299.
6. K. D. Bartle, T. Boddington, A. A. Clifford, N. J. Cotton and C. J. Dowle, 'Supercritical Fluid Extraction and Chromatography for the Determination of Oligomers in Poly(ethylene terepthalate) Films', *Anal. Chem*, 1991, **63**, 2371.
7. K. Ganzler, A. Salgó, and K. Valcó, 'Microwave Extraction: A Novel Sample Preparation Method for Chromatography', *J. Chromatogr.*, 1986, **371**, 299.
8. Ministry of Agriculture, Fisheries and Food, 'Food Analysis Performance Assessment Scheme', Series IX (OP Pesticides) Round 3, MAFF, Norwich, UK, September 1992, Report 0903.
9. M. D. Osselton, I. C. Shaw and H. M. Stevens, 'Enzymic Digestion of Liver Tissue to Release Barbiturates, Salicylic Acid and Other Acidic Compounds in Cases of Human Poisoning', *Analyst*, 1978, **103**, 1160.
10. W. E. Lindup, 'Drug–Albumin Binding', *Biochem. Soc. Trans.*, 1975, **3**, 635.
11. N. P. Boley, N. T. Crosby, and P. Roper, 'Use of Enzymes for the Release of Synthetic Colours in Foods', *Analyst*, 1979, **104**, 472.
12. 'CRC Handbook of Chemistry and Physics', ed. D. R. Lide, CRC Press, Boca Raton, FL, 75th edn, 1994.
13. E. G. van der Velde, W. de Haan, and A. K. D. Liem, 'Supercritical Fluid Extraction of Polychlorinated Biphenyls and Pesticides from Soil: Comparison with Other Extraction Methods', *J. Chromatogr.*, 1992, **626**, 135.
14. J. S. Ho and P. H. Tang, 'Optimization of Supercritical Fluid Extraction of Environ-

mental Pollutants from a Liquid–Solid Extraction Cartridge', *J. Chromatogr. Sci.*, 1992, **30**, 344.

15. K. Ballschmiter, 'Sample Treatment Techniques for Organic Trace Analysis', *Pure Appl. Chem.*, 1983, **55**, 1943.

General Reading

16. S. K. Poole, T. A. Dean, J. W. Oudsema, and C. F. Poole, 'Sample Preparation for Chromatographic Separations: An Overview', *Anal. Chim. Acta*, 1990, **236**, 3.
17. J. N. Miller, 'Matrix and Sensitivity Constraints in Trace Organic Analysis', *Anal. Proc.*, 1982, **19**, 114.
18. M. J. Lichon, 'Sample Preparation for the Chromatographic Analysis of Food', *J. Chromatogr.*, 1992, **624**, 3.

Sampling Volatiles

19. Health and Safety Executive, 'Analytical Quality in Workplace Air Monitoring', HSE, Sudbury, UK, March 1991, Document MDHS 71.
20. T. Jursík, K. Stránsky, and K. Ubik, 'Trapping System for Trace Organic Volatiles', *J. Chromatogr.*, 1991, **586**, 315.
21. X.-L. Cao and C. N. Hewitt, 'Trapping Efficiencies of Capillary Cold Traps for C_2–C_{10} Hydrocarbons', *J. Chromatogr.*, 1992, **627**, 219.
22. W. M. Coleman III and B. M. Lawrence, 'Comparative Automated Static and Dynamic Quantitative Headspace Analyses of Coriander Oil', *J. Chromatogr. Sci.*, 1992, **30**, 396.
23. L. J. Mulcahey, J. L. Hedrick, and L. T. Taylor, 'Collection Efficiency of Various Solid-Phase Traps for Off-Line Supercritical Fluid Extraction', *Anal. Chem.*, 1991, **63**, 2225.

Purification of Solvents

24. K. Ballschmiter, 'Sample Treatment Techniques for Organic Trace Analysis', *Pure Appl. Chem.*, 1983, **55**, 1943.

Supercritical Fluid Extraction

25. K. D. Bartle and A. A. Clifford, 'Can Supercritical Fluid Extraction Live Up to its Promise?', *LC.GC*, 1991, **4**, 10.
26. J. W. King and M. L. Hopper, 'Analytical Supercritical Fluid Extraction: Current Trends and Future Vistas', *J. Assoc. Off. Anal. Chem. Int.*, 1992, **75**, 375.
27. S. B. Hawthorne, 'Analytical Scale Supercritical Fluid Extraction', *Anal. Chem.*, 1990, **62**, 633A.
28. T. Greibrokk, 'Recent Developments in the Use of Supercritical Fluids in Coupled Systems', *J. Chromatogr.*, 1992, **626**, 33.
29. M. R. Andersen, J. T. Swanson, N. L. Porter, and B. E. Richter, 'Supercritical Fluid Extraction as a Sample Introduction Method for Chromatography', *J. Chromatogr. Sci.*, 1989, **27**, 371.
30. M. D. Burford, S. B. Hawthorne, D. J. Miller, and T. Braggins, 'Comparison of Methods to Prevent Restrictor Plugging During Off-Line Supercritical Fluid Extraction', *J. Chromatogr.*, 1992, **609**, 321.
31. M. Ashraf-Khorassani, R. K. Houck, and J. M. Levy, 'Cryogenically Cooled Adsorbent Trap for Off-Line Supercritical Fluid Extraction', *J. Chromatogr. Sci.*, 1992, **30**, 361.

Critical Comparisons of Techniques

32. M. Richards and R. M. Campbell, 'Comparison of Supercritical Fluid Extraction, Soxhlet and Sonication Methods for the Determination of Priority Pollutants in Soil', *LC.GC Int.*, 1991, **4**, 33.
33. E. G. van der Velde, W. de Haan and A. K. D. Liem, 'Supercritical Fluid Extraction of Polychlorinated Biphenyls and Pesticides from Soil: Comparison with Other Extraction Methods', *J. Chromatogr.*, 1992, **626**, 135.

Specific Application

34. F. Mangani, A. Cappiello, G. Crescentini, F. Bruner, and L. Bonfanti, 'Extraction of Low Molecular Weight Polynuclear Aromatic Hydrocarbons from Ashes of Coal-Operated Power Plants', *Anal. Chem.*, 1987, **59**, 2066.

5.3 Methods of Extraction of Organic Analytes from Liquid Samples and Clean-up of Liquid Extracts

5.3.1 Overview

This section deals with the extraction of analytes from a liquid matrix. As the techniques involved in extracting from a liquid sample are similar to those used for cleaning up a liquid extract from a solid sample, the two are treated together in this section. The extraction of volatile analytes from liquids by purge and trap methods is also included.

Liquid (usually aqueous) samples generally present less of a challenge than solids, in terms of analyte extraction and subsequent clean-up. Matrix effects are minimal or non-existent, this greatly simplifies extraction and often means higher efficiencies than when extracting from solids. Also, as the majority of liquid samples are aqueous, any trace organic determinands are easily removed, either by partitioning into a non-polar solvent or adsorption onto a non-polar stationary phase. It is mainly in the case of non-aqueous samples, such as oils, paraffins, and fats that complications can arise.

A greater problem can be presented by sample extracts, as they are generally in non-polar solvents and can contain many co-extractives. In order to achieve the extreme detection limits required in trace level analyses, the final extract ideally must be completely free from interferents. This separation of the analyte (or group of analytes) from other compounds that are present in the sample, usually at far higher levels, is the most challenging and demanding stage of most analyses. There are an infinite number of possible co-extractives, and the likelihood is that at least some of them will be chemically similar to the analyte(s). If this is the case, there are two possible routes: to rely upon an end-determination technique in which the co-extractives do not interfere, or to use separation techniques to isolate the determinand. Of the two, the former is preferable, as with fewer stages in the analysis the associated uncertainty is greatly reduced. In practice, however, such specific end-determination techniques are seldom available, and extensive clean-up is the only remaining option. This situation is gradually changing with the advent of techniques such as immunoassays, which are sufficiently selective to determine the analyte with little or no clean-up, but as yet such techniques still have extremely limited applications.

5.3.2 Comparison of Techniques

Liquid extraction and other separation methods are generally very labour intensive. Tables 5.3.1 (extraction from liquid samples) and 5.3.2 (sample extract clean up) list these techniques. It is common, particularly with extracts from complex sample matrices such as foodstuffs, to use a combination of two or more of these techniques to achieve the necessary degree of clean-up. Adsorbent chromatography, liquid partition, and chemical reaction are the most established methods, and as such tend to be better documented and tested for a given application. They are also the most tedious. Because of this, the newer techniques, particularly solid-phase extraction, are rapidly gaining in popularity.

Liquid/Liquid Partition

The traditional method of dealing with liquid samples has been liquid/liquid extraction. It is easy to comprehend, simple to perform, and requires no specialised equipment or training. The properties of the different solvent combinations that can be used are shown in Table 5.3.3, and their miscibilities in Table 5.3.4. Table 5.3.3 is not exhaustive: any pair of immiscible solvents as shown in Table 5.3.4 can be used to effect a partition, although many of the more toxic or environmentally damaging solvents listed are now being phased out of use. Aqueous samples are immiscible with most organic solvents. The denser chlorinated solvents, prior to being restricted, were particularly popular for aqueous partitions as the density difference between phases meant there was less risk of emulsion formation. There is a much narrower choice of extraction solvent when extracting from extracts that are already in an organic medium. The most widely used combination in such cases is hexane/water-modified acetonitrile; these two phases are particularly useful for separating fatty material from analytes of a slightly higher polarity.

The disadvantages associated with liquid partitioning are well-documented. The technique is time consuming and labour intensive; sample volumes are limited by the available glassware, and there is the risk of emulsion formation. These problems can now largely be overcome by the use of automated, continuous extractors (for a comparison of available types, see Table 5.3.5).[2] These require a large capital outlay and have not, as yet, come into widespread use. Alternatives are to use a partition column, with one of the phases bonded to a stationary support whilst the second is allowed to flow past, or to use celite as a liquid partition suspension.

As the partition between two phases is an equilibrium, there can never be complete extraction with one washing. The minimum requirement for a single partition is approximately 60% extraction. It is then usual for at least three

Table 5.3.1 Techniques for extraction from liquids

	Sample type	Restrictions on use	Examples of use	Yield	Operator skill	Time taken	Max. sample volume	Extract volume	Cost
Liquid/liquid extraction	Liquids (usually aqueous)	Liquids must not react or emulsify	Drugs in body fluids	*****	***	~5 min	Limited by flask (~500 ml)	Up to 500 ml	*
Continuous extractors	Liquids (usually aqueous)	Some liquid combinations foam	Beverages	****	**	18–24 h	Virtually unlimited	~100 ml	****
Solid-phase extraction	Liquids (usually aqueous)	Clogged by sediment. Can be overloaded	Trace organics in water	****	****	~1 h	~1 l water	~10 ml	*
Adsorbent traps	Liquids, gases, vapours, smokes	Can be overloaded. Needs rapid desorption	Flue gases	*****	***	Continuous monitoring	~20 l water virtually unlimited for gas	Direct onto instrument	**
Cryogenic traps	Gases, vapours, smokes, exhausts	Analyte must be high boiling	Flue gases	****	***	Continuous monitoring	Virtually unlimited	Direct onto instrument	***
Purge and trap	Volatiles in liquids	On-line coupling problems. Limited sample loading	Flavours in foods	*****	******	Easily automated	~200 ml	Direct onto instrument	*****

* – low
***** – high

Table 5.3.2 *Organic group separation techniques (clean-up)*

Technique	Uses	Maximum sample size	Problems	Advantages
Adsorbents	Environmental samples. Food samples	Depends on adsorbent and nature of sample	Labour intensive. Requires careful control of adsorbent activity. Analyte must be of different polarity to interferents. Unsuitable for multi-residue analyses if analytes have a range of polarities	Effective. Well tested and documented. Few contamination problems
On-line chromatography	Variety of applications	Depends on loading of column	Columns can be overloaded. Relatively expensive	Can be carefully controlled. Easy to automate. Can be coupled to determination. Can be miniaturised
Gel permeation chromatography (Size exclusion)	Multi-analyte analyses with a range of polarities. Biochemical analyses	~1.5 g fat per column	Size difference must be over 25%. Does not remove all lipids so not often used alone	Applicable to all polarities
Solid phase extraction	Variety of applications. Aqueous extracts. Removal of polar species	Typical capacity 500 mg. Smaller capacity than classical techniques	Limited capacity. Leaching of plasticisers can occur. Easily overloaded by 'dirty' samples	Good range of selectivities. Fast and cheap. Can be automated. Can couple to determination

Method	Applications	Sample size	Disadvantages	Advantages
Sweep co-distillation	Routine environmental samples. Pesticides in foods	Maximum 1 g fat[1]	Losses through degradation and adsorption. Analytes must be thermally stable. Analytes must be volatile relative to sample. High capital cost	Fully automated
Liquid partition	Triglycerides in organic samples. Drugs of abuse	~500 ml extract	Rare that all the fat can be removed. Interferents must have different polarity from analytes. Emulsions are readily formed. Requires large volumes of solvents. Used alone, is inadequate for trace analysis.	Little skill required. Cheap. Quick. Concentration possible.
Chemical reaction (acidic or basic)	Lipids in organics. Saponification, then aqueous partition	~100 ml extract	Many oxidisable organics are destroyed	Lipids easy to wash out after saponification. Simple, economic, rapid
Distillation	N-nitrosamines	~100 g	Complicated and expensive glassware. Time consuming	Easy once in routine use. Good for unstable analytes
Low-temperature precipitation	Not widely used	~100 ml extract	Losses by co-precipitation	Easy. No contamination problems

Table 5.3.3 *Common liquid partition phases*

Typical phases[a]	Examples of applications
Hexane/water-modified acetonitrile	Removal of fats from fairly non-polar analytes
Hexane/water-modified dimethyl sulfoxide	Removal of fats from aromatic and π-bonded analytes
Aqueous/chlorinated solvent	Any organics from aqueous extracts, *e.g.* drugs from body fluids
Aqueous/petroleum or diethyl ether	Non-polar analytes from aqueous extracts
Aqueous/ethyl acetate	Medium polarity analytes from aqueous extracts (alternative to chlorinated solvents)
Aqueous/toluene	Aromatic analytes from aqueous extracts

[a] To control selectivity, pH control can be used with all systems that have an aqueous phase.

consecutive extractions to be combined, which, depending upon the system, will result in efficiencies of about 95%.

Gravity-fed Adsorption Columns

Adsorption columns were one of the first group separation techniques employed in organic trace analysis and have retained their popularity. Many statutory and other recommended methods specify the use of adsorption chromatography, ensuring its continued use.

Sorbents are usually, although not invariably, polar materials, facilitating the separation of non-polar analytes from polar interferents. An example of an application is the removal of fats or other polar materials from vegetable extracts, animal products, and similar samples. Some of the more common sorbents are listed in Table 5.3.6. Silica gel is probably the most widely used, with alumina a close second; florisil has particular application in pesticide residue analysis. Alumina is available in three forms: neutral, acid washed, and basic. The application of each of these is summarised in Table 5.3.7; neutral alumina is by far the most widely used. National preferences also play a large role in choosing a suitable sorbent; *e.g.* florisil is used widely in North America but is of only minor interest in Europe. Other materials, such as activated carbon or magnesia, do find occasional application.

The major disadvantage of adsorbent columns is their limited use for multi-analyte determinations where the analytes have a range of polarities. They are better suited to the analysis of groups of related compounds; in order to elute compounds with a range of properties, conditions usually have to be used that also elute most of the undesirable interferents. In some cases the collection of more than one elution fraction, possibly using a graduation of different solvents, can overcome this problem. However, when collecting a range of fractions it is likely that one or more of them will contain the very interferents the technique was intended to remove. An example of this is the separation of organochlorine and

Table 5.3.4 *Miscibilities and reactivity of some common solvents*

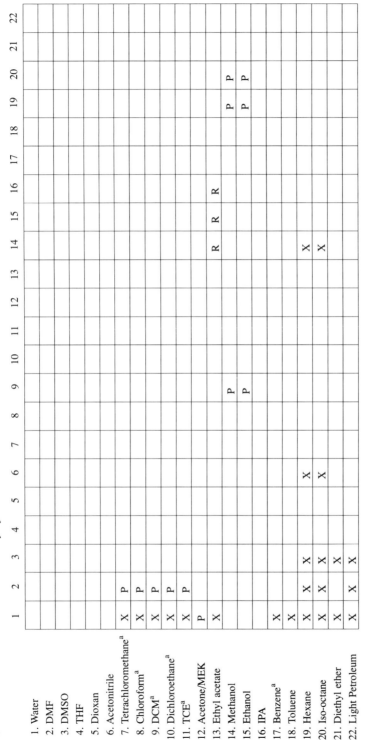

	1	2	3	4	5	6	7	8	9	10	11	12	13	14	15	16	17	18	19	20	21	22
1. Water																						
2. DMF																						
3. DMSO																						
4. THF																						
5. Dioxan																						
6. Acetonitrile																						
7. Tetrachloromethane[a]	X	P																				
8. Chloroform[a]	X	P																				
9. DCM[a]	X	P																				
10. Dichloroethane[a]	X	P																				
11. TCE[a]	X	P																				
12. Acetone/MEK	P																					
13. Ethyl acetate	X																					
14. Methanol									P				R						P	P		
15. Ethanol									P				R						P	P		
16. IPA													R									
17. Benzene[a]	X																					
18. Toluene	X																					
19. Hexane	X	X	X			X								X								
20. Iso-octane	X	X	X			X								X								
21. Diethyl ether	X	X	X																			
22. Light Petroleum	X	X	X																			

[a] Solvents that are either banned from use or in the process of being phased out.
All solvent combinations are compatible except: X – immiscible solvents P – partially miscible solvents R – solvents react with each other
DMF dimethyl formamide DMSO dimethyl sulfoxide THF tetrahydrofuran DCM dichloromethane
TCE 1,1,1-trichloroethane MEK methyl ethyl ketone (butanone) IPA isopropyl alcohol (propan-2-ol)

Table 5.3.5 *Continuous liquid extractors*

	Applicability	Relative precision	Disadvantages	Advantages
Liquid/liquid extractor. DCM dripping through water	General	* * * *	Emulsions possible	
Steam co-distillation extractor. Condensation of steam/DCM	Thermally stable, volatile analytes	* * * *	Emulsions possible	Small solvent volumes (20 ml). Any solvent can be used. Good for concentrating partially soluble analytes. Good for GC clean-up
Steam distillation extractor. Steam distilled through lighter-than-water solvent	Thermally stable, volatile and non-polar analytes	* * * * *	Volatiles can escape through condenser. Emulsions probable	Good for GC clean-up. Small solvent volumes (20 ml)
Flow Under Extractor (FUE). Heavier-than-water stream	General	* *	Poor selectivity. Large solvent volumes (700 ml)	Emulsions rare
Flow Over Extractor. Lighter-than-water stream	Non-polar analytes	*	Low recoveries for more polar analytes. Large solvent volumes (700 ml)	Better selectivity than FUE

DCM dichloromethane
* – low
***** – high

synthetic pyrethroid pesticides using a silica column: if an aliphatic hydrocarbon, such as hexane is used as the solvent, the pyrethroids are retained on the column, whereas a second wash with a stronger solvent, such as dichloromethane, will elute the pyrethroids and will also elute some unwanted lipids. As in so many clean-up techniques, there is a trade-off to be made. Optimum analyte extraction must be judged against acceptable levels of interferents.

This situation contrasts greatly with Gel-Permeation Chromatography (GPC) where the elution order is only dependent upon molecular size; it is independent of functional groupings or polarity. In cases where a range of unrelated compounds are to be collected, therefore, GPC is usually the technique of choice.

Solid-phase Extraction

A recent technique which provides a quick alternative to classical adsorption columns is the use of disposable solid-phase extraction (SPE) cartridges.[3] They provide a cheap and rapid extraction method, without the tedious column packing of adsorption chromatography or emulsion problems of liquid extraction.

The approach to SPE is essentially the opposite of that used for adsorption

Table 5.3.6 *Adsorbents for gravity-fed columns*

Adsorbent	Examples of use	Available forms	Precautions and problems
Alumina	Non-polar analytes with polar interferents, *e.g.* pesticides in food and wildlife, hydro-carbons, esters	Neutral, acid, alkaline (see Table 5.3.7). Variety of moisture content. Modified, e.g. silver nitrate	Do not allow the column to run dry. Do not expose to atmosphere for longer than necessary. Top column with drying agent. Width of column and elution speed affects retention. Not effective if interferents and analyte are of similar polarity
Silica gel	Alcohols, organic acids, strongly polar species. First choice for polar, H-bonding analytes. Not suitable for strongly basic analytes	Usually fully activated. 40 or 60 µm particles (40 give stronger retention). Can be deactivated	Do not let the column run dry. Do not expose to atmosphere for longer than necessary. Does not resolve structurally similar compounds well. May irrevocably bind some strongly basic analytes
Florisil	Steroid sex hormones. Aromatics from aliphatic interferents. Particularly pesticide residues. Vitamins	Usually fully activated	Do not let the column run dry. Do not expose to atmosphere for longer than necessary. Early problems with batch-to-batch variations (now largely overcome). May irreversibly bind some basic analytes
Activated carbon	Non-polar analytes from aqueous samples. Good for structurally similar compounds	Fully activated	Pre-wash before use
Diatomaceous (Fuller's) earth	Used more as a stationary support than adsorbent	Fully activated	Pre-wash before use

Table 5.3.7 *Types of alumina*

	Preparation	Applications
Neutral pH 6.9 – 7.1	Gamma alumina, plus small amounts of alpha alumina and sodium carbonate	Weak organic acids and bases. Hydrocarbons, esters, aldehydes, lactones, quinones, alcohols
Basic pH 10.0 – 10.5	Basic washing of neutral alumina	Acid labile compounds. Has strong cation-exchange properties in aqueous solution
Acidic pH 3.5 – 4.5	Acid washing of neutral alumina	Organic acids. Has strong anion-exchange properties in aqueous solution

chromatography. In the latter, the unwanted material is usually retained on the column while the analytes are eluted. In SPE, the analytes are usually retained on the cartridge as the sample or sample extract is washed through. They can then be washed off at a later stage using a different solvent. In some cases it is feasible to reuse cartridges, although with those made from disposable plastic the risks of cross-contamination between samples usually outweigh the small financial saving.

If the analyte adsorbs strongly to the cartridge (*i.e.* is adsorbed in a tight band at the head of the packing material) it may be fruitful to reverse the cartridge before washing it off. This technique is known as back-flushing.

A wide variety of adsorbents are available for many applications (Table 5.3.8). The most common use is for the analysis of aqueous samples: collecting non-polar analytes on a reversed phase cartridge, and then eluting with an organic solvent. This provides excellent sample concentration, as the extract from litre quantities of sample can be washed off with only millilitres of solvent.

Table 5.3.8 *Common solid-phase extraction sorbents*[a]

Sorbent	Applications	Examples of use
Octadecyl	Reversed phase extraction of non-polar compounds	Drugs, oils, food preservatives, vitamins, plasticisers, pesticides, hydrocarbons in aqueous extracts
Octyl	Reversed phase extraction of moderately polar compounds and compounds too strongly adsorbed by octadecyl	Priority pollutants, pesticides in aqueous extracts
Phenyl	Reversed phase extraction of non-polar compounds. Provides less retention of hydrophobic compounds	Not widely used
Cyanopropyl	Normal phase extraction of polar compounds	Amines, alcohols, dyes, vitamins, phenols. From aqueous or organic extracts
Silica	Adsorption of polar compounds	Drugs, alkaloids, mycotoxins, amino acids, flavenoids, heterocycles, lipids, steroids, organic acids, terpenes, vitamins. From organic extracts
Diol	Normal phase extraction of polar compounds (similar to silica gel)	Proteins, peptides, surfactants from organic extracts
Aminopropyl	Weak anion exchange extraction	Carbohydrates, peptides, nucleotides, steroids, vitamins. From aqueous or organic extracts
Dimethylaminopropyl	Weak anion exchange extraction	Amino acids. Aqueous extracts
Aromatic sulfonic	Strong cation exchange and reversed phase extraction (eliminates ion-pairing when used in place of ODS)	Amino acids, catecholamines, nucleosides, nucleic acid bases. Aqueous extracts
Quarternary amines	Strong anion-exchange extraction	Antibiotics, nucleotides, nucleic acids for aqueous or organic extracts

[a] S. K. Poole, T. A. Dean, J. W. Oudsema and C. F. Poole, 'Sample Preparation for Chromatographic Separations: An Overview', *Anal. Chim. Acta*, 1990, **236**, 3.

In addition to extraction of the analyte, some group separation clean-up is possible by careful solvent selection. The cartridge can be eluted with a liquid of low affinity for the analyte (Table 5.3.9 lists suitable solvents for some of the

Table 5.3.9 *Elution solvents suitable for some adsorbent phases*

Stationary phase	Applications	Incompatible solvents	Elution solvents (in order of eluting power)
C_{18}	Non-polar species in water	Strong acids and bases	Hexane, DCM, acetone, MeCN, MeOH, water
C_8	Non-polar species in water (less adsorbent than C_{18})	Strong acids and bases	Hexane, DCM, acetone, MeCN, MeOH, water
Cyclohexyl	Non-polar species in water	Strong acids and bases	Hexane, DCM, acetone, MeCN, MeOH, water
Phenyl	Non-polar species in water (less retention of hydrophobic compounds)	Strong acids and bases	Hexane, DCM, acetone, MeCN, MeOH, water
Cyano	Non-polar, polar and cationic species in water		MeOH, acetone, DCM, $CHCl_3$, hexane. acids, bases, and buffers
Amino	Polar/anionic species in water and organic solvents		MeOH, acetone, DCM, $CHCl_3$, hexane acids, and buffers
Diol	Polar species in organic solvents	Water	MeOH, acetone, DCM, $CHCl_3$, hexane
Florisil	Moderately polar species in organic solvents	Water	MeOH, ethyl acetate, DCM, $CHCl_3$, hexane
Alumina	Moderately polar species in organic media	Water	MeOH, ethyl acetate, DCM, $CHCl_3$, hexane
Silica	Moderately polar species in organic media	Water	MeOH, ethyl acetate, DCM, $CHCl_3$, hexane
Anion exchange	Anionic species in water		Acids and buffers
Cation exchange	Cationic species in water		Acids, bases, and buffers

DCM dichloromethane

more common phases) to remove some interferents, followed by a more powerful solvent used to wash off the strongly bound analyte.

All SPE cartridges can take only a limited sample loading. If constructed of plastics then organic solvents may leach materials from the plastic, this is a problem if electron capture detection is to be used. In effect, this means SPE is far more widely used when HPLC, rather than GC, is the intended end-determination technique, although this situation is changing with the increased use of GC–MS. This may be why SPE tends to be applied more to extraction from aqueous samples than to the clean-up of organic extracts.

Gel-permeation Chromatography (GPC)

Gel-permeation chromatography is used for removing interferents of large molecular size, *e.g.* fats. It also has applications in the biological field, in the separation of proteins, nucleic acids, carbohydrates, and other large molecular species. For biochemical applications a bed of polymeric packing, up to 1 m wide, is used to effect the separation, but for most analyses a column (analogous to those used for adsorption chromatography) is the preferred method.

The recent introduction of automated GPC equipment has caused a revival of interest in the technique for the clean-up of sample extracts. It is particularly useful in applications where adsorption chromatography cannot be employed, such as multi-analyte assays with determinands covering a range of polarities. The attractive feature of the technique for routine batch work is that it is easily automated and the columns are reusable. Samples can be loaded onto an auto-injector, fractions collected automatically, and the next sample injected after flushing the column to remove residual sample. The use of HPLC pumps also makes the analysis of batches of samples using GPC much less labour intensive, and thus more attractive for routine analysis.

One disadvantage of GPC is that it requires large volumes of solvent. In addition, when used alone, it does not provide a sufficiently rigorous clean-up for trace analysis (depending upon the detection system used). It is frequently used in conjunction with liquid/liquid partition, SPE, or adsorption chromatography. To be effective, it requires a 25% molecular size difference between the compounds to be separated.

Other Liquid Extraction/Clean-up Techniques

There are a plethora of other methods that are either rarely used or have specialist applications. *Distillation* used to be the preferred method for volatile analytes such as *N*-nitrosamines, but is seldom employed now as there are alternative techniques which are less time consuming. *Sweep co-distillation* (assisted distillation) is used occasionally, particularly in Australia, for the extraction of pesticide residues from lipid matrices. It requires specialised equipment, but once in routine use is ideal for processing batches of samples. An almost infinite variety of *chemical reactions* can be used, either for destroying the sample matrix or derivatising the analyte so that it can be extracted by a specified solvent. Treatment of the sample matrix with either an acid or a base are probably the most frequently used.[4] The use of these obviously depends upon the individual sample and determinand. *Preparative thin layer chromatography* is occasionally used to separate the components of small volumes of a solution. *On-line chromatography* is another technique that is occasionally employed. *Liquid chromatography-liquid chromatography* (LC-LC) is probably the most popular, often incorporating some kind of recycling and concentrating mechanism such as box-car switching. GC-GC and LC-GC combinations are also known.[5] These tend to find more use in research applications, as there are many parameters to optimise making the system and the conditions quite complex.

Purge and Trap Analysis

The sampling of volatile analytes from liquids is an exceptional case. In these situations, the obvious method of separating the determinands from the matrix is by heating, then collecting them from the gaseous phase. In the case of purge and trap analyses, the volatilisation is assisted by purging the liquid with an inert gas. The collection is usually on-line at the head of a GC column. The determinand can either be trapped on a small plug of adsorbent, or on the column itself.

Interference problems with this type of analysis are generally minimal, as only other volatile components pose a problem. The main difficulty from a quantitative point of view is the effect of band broadening as the trapped material is desorbed. Adsorbents tend to be of the same type of materials as those used for SPE (Table 5.3.8) and they are cooled to avoid leaching of the trapped material. However, the desorption of the trapped materials can seldom be controlled as precisely as would be desired, leading to characteristic chromatographic band broadening.

Samples or extracts for volatile analyses are usually screened first by headspace analysis. This gives poor quantitative results, but an idea of the analyte concentration is required before commencing a purge and trap assay in order to avoid overloading the adsorbent.

Applications of the technique include the standard US Environmental Protection Agency methods for trace organics in water, the analysis of food extracts for taints and flavourings, and the identification of scents in perfume samples.

5.3.3 Critical Aspects and General Precautions

Optimising Multi-analyte Clean-ups

The objective of all group separation techniques is to obtain an extract solution that contains all of the analyte originally sub-sampled, which is sufficiently free of interferents for the determination technique to accurately measure the analyte. The best method of checking that these two criteria are met, is the analysis of blank materials (to correct for interferents) and reference materials (to allow for analyte losses). To correct for interferents from the matrix, reagent blanks are generally not adequate; using a 'matrix blank', *i.e.* a sample which is known to be free of analyte, is preferred. Similarly, passing solutions of reference materials through the clean-up process may be insufficient as it does not allow for sample effects (*e.g.* lipids in the sample decreasing retention on polar adsorbents). Spikes, internal calibrants, or, ideally, matrix reference materials should be used.

When performing multi-analyte determinations a particularly careful check should be made for analyte losses. It is often impractical to check spike recoveries for each analyte each time a batch of samples is analysed. The analysis of pesticide residues, for example, can involve screening for upwards of 50 different compounds. In such cases, where the analyses are performed regularly, the only approach is to obtain recovery data for a different selection of analytes with each batch of samples. Each selected set should cover analytes with a wide range of polarities. This approach ensures that all the analytes are checked on a regular basis.

Unless all the desired determinands are of the same polarity, the optimum clean-up technique for one will not be optimum for another. Invariably a compromise must be reached; some losses or interferents may be unavoidable.

Solvents

The choice of solvent used for a partition or to elute from a chromatographic column is critical, as was the case for extraction from solid samples. A discussion of solvent suitability can be found in Section 5.2.

The choice of extraction solvent for liquid/liquid partition depends upon the analyte, its speciation, and the matrix. The partitioning solvent must be immiscible with the solvent used for extracting the sample. Two other factors to consider are the difference in density of the two solvents and the relative solubility of the analyte(s) and interferents in each.

Partition coefficients can be calculated from the ratio of theoretical solubilities in the two liquid phases, although it is important to bear in mind any possible speciation effects. The greater the difference in densities between the two phases, the greater the chance of a clean phase separation with no emulsion problems. It is usually more convenient to choose an extraction solvent that is denser than the sample phase, so that the lower phase contains the analyte, and can be easily removed through a tap. This density difference is the main reason why chlorinated solvents, particularly carbon tetrachloride, have always been popular for extraction from aqueous media. However, under the terms of the Montreal Protocol, they are now being withdrawn from use. The alternatives, such as ethyl acetate, tend to have a density closer to that of water, and so are more susceptible to emulsion formation.

In the case of emulsions, prevention is almost always better than cure (see Table 5.3.10). There is always a balance to be drawn between the desirability of shaking the phases vigorously to ensure complete extraction and the need for minimal mixing to avoid emulsion formation. With samples prone to foaming, gentle rotation or swirling for an extended period of time is probably the best compromise. Repeated rapid inversion of the flask interspersed with careful swirling can often give comparable partitions to the traditional approach of vigorously shaking the phases and dealing with any emulsions that may arise.

Table 5.3.10 *Methods of breaking down emulsions*

Action	Comment
Add an inert salt, *e.g.* sodium chloride	Risk of introducing contamination
Add a higher boiling alcohol, *e.g.* octanol	Risk of introducing contamination
Gentle warming	No contamination risks. Be alert to pressure build-up
Steaming	Risk of contamination
Gentle swirling	No contamination risk. Limited effectiveness
Centrifugation	No contamination risk. Requires expensive equipment
Increase volume of solvent	Remember to increase reagent blanks by corresponding amount
Remove as much clear solvent as possible. Repeat the extraction	Remember to do likewise to reagent blanks
Run emulsion through drying agent, *e.g.* anhydrous sodium sulfate	Can coagulate and block drying agent
Cool	No external contaminants introduced

If emulsions are formed and cannot be destroyed, the analyst must ensure that the entire emulsion is included in the final extract. A surprising amount of analyte can be entrapped in a small amount of emulsion. If the unwanted emulsion is aqueous, it can usually be dried by using a desiccant such as anhydrous sodium sulfate. The only danger with this approach is the possibility of the desiccant coagulating and entrapping a small proportion of the analyte.

pH and Speciation Effects

The pH of the system is also an important parameter to control, and is often actively used to adjust the selectivity of an extraction. Compounds that can be protonated respond to this treatment. Phenols and other weak acids show increasing polar characteristics as conditions become more alkaline. Amines and weak bases, conversely, become polar in acidic conditions, *i.e.* the pH must be above neutral if they are to be extracted into non-polar solvents. As well as working in the analyst's favour to improve the selectivity of extractions, the analyst should also be aware that pH effects can cause low extraction efficiencies.

Controlling solubilities by pH speciation can also provide an effective pre-extraction clean-up. For example, under alkaline conditions a solution containing pentachlorophenol can be safely washed with a non-polar solvent to remove non-polar interferents. If the sample is then acidified, the pentachlorophenol can be extracted with a second portion of the same solvent. A classic example of the use of this effect is in the sequential extraction of basic drugs from body fluids.

SPE and Adsorbent Columns

Solid-phase cartridges are notoriously prone to batch-to-batch variations. Nominally identical stationary phases from different manufacturers often show wildly different retention properties. Even different batches from the same supplier can vary. This is of particular concern when following methods published in the literature: the discrepancies between results cited in journals and those found in practice can often be traced to the erratic behaviour of SPE cartridges. It is essential practice to check the retention of reference materials on a representative cartridge from each new batch received.

The critical nature of solvent choice applies equally to SPE as to liquid partition. Apart from this, and flow rate control, another factor that can affect extraction efficiency is possible contamination of the sample. Aqueous samples, especially surface waters, frequently contain other organics, oils, and surfactants, which can either overload the cartridge or otherwise adversely affect the analyte adsorption.

Critical aspects of alumina, silica, and florisil are listed in Table 5.3.11. As with any form of chromatography, they relate to the physical path of the analyte through the column, its speed, and the proportion of active sites that are occupied, either by water or by polar components of the sample. Adsorbent columns must never be allowed to run dry, and the packing material must at all times be protected from atmospheric moisture.

5.3.4 Hints and Tips

Liquid Partition

This technique is extremely simple to use, but there are potential problems (see Table 5.3.12). The most serious is the formation of emulsions between the two immiscible phases. Table 5.3.10 details some of the methods by which emulsion problems can be tackled.

Apart from emulsion formation, the major risk from a quality assurance point of view is the possibility of analyte speciation changes. This is usually due to the pH of the two phases. The equilibria between species has a marked effect on the partition coefficient. The chemical structure of the analyte (and interferents, if known) should always be borne in mind, and the analyst should, at all stages of the analysis, be alert to the possibility of speciation effects.

Extraction efficiencies from aqueous media can be improved upon by altering the partition equilibria by the addition of water-soluble ionic species, a technique known as 'salting out'. For example, sodium chloride or sodium sulfate is added to the aqueous phase to aid the extraction of the more polar analytes. For most purposes a concentration of salt in the aqueous phase of about 3% is usually sufficient; this is normally added as a saturated solution.

The partition of analytes and interferents between two phases can be predicted theoretically, but with multi-analyte systems this can never be optimised for each analyte. Some losses must be accepted. For example, when removing fats using

Table 5.3.11 *Critical aspects of adsorbents*

Critical factor	Alumina	Silica	Florisil
Moisture content	Highly critical. Usually controlled at 9–11%. Less active the greater the water content	Highly critical. Must be kept free of adsorbed water	Highly critical. Must be kept free of adsorbed water
Fat loading	Highly critical. The more fat adsorbed the less retention of analytes. Different types of fat act differently *e.g.* vegetable oils compete for active sites more than animal fats (triglycerides)	Highly critical. The more fat adsorbed the less retention of analytes. Different types of fat act differently: compete with pesticides for active sites	Highly critical. The more fat adsorbed the less retention of analytes. Different types of fat act differently
Eluent speed	Speeding up the eluent flow increases retention of triglyceride fats relative to analyte	Critical	Critical
Column dimensions	Inside diameter affects retention: 12–15 mm optimal for 10% neutral alumina	Inside diameter affects retention	Inside diameter affects retention
Solvent	As with any chromatography, the eluent used is critical	As with any chromatography, the eluent used is critical	As with any chromatography, the eluent used is critical

Table 5.3.12 *Problems with liquid/liquid extraction*

Problem	Solution
Emulsions	Can be broken down (see Table 5.3.10)
Pressure venting: loss by nebulisation	Release pressure regularly through tap
Leaking taps	Always check the tap whilst pre-washing the glassware with solvent
Speciation problems: *e.g.* benzoic acid dimerises in benzene, amines become polar under alkaline conditions	Be aware of chemical structure of analyte

hexane/acetonitrile partition, non-polar analytes will also partition into the hexane layer.

The addition of water to the polar solvent in a two-phase system can be employed to adjust the partition selectivity. For example, the selectivity of dimethyl sulfoxide or acetonitrile extractions can be controlled by adding water. Solvation of π-electron functional groups is reduced as more water is added, whereas the hydrogen-bonding attractions remain largely unaffected.[6]

Some manufacturers add ethanol, as a preservative, to chloroform, which can affect liquid partitioning. However, ethanol can be fairly easily removed by washing with water. Other manufacturers use a variety of other preservatives and antioxidants in solvents, and it is best to check details in individual cases.

Adsorption Columns

In order to ensure reliable and reproducible adsorption, great care is necessary in the storage and use of sorbents. Commercial preparations generally have a distribution of active sites, *i.e.* a small proportion of the sites are extremely active. It is these sites that control the chromatographic reproducibility. If available, they will strongly or even irrevocably bind some of the analyte, causing tailing effects. It is therefore essential to precisely control the availability of these sites. If polar sorbents are left open to atmospheric moisture, these sites will gradually become hydrated. After activation and/or controlled deactivation, sorbents must therefore be stored either in a desiccator or in some other sealed vessel. On no account should they be exposed to the atmosphere.

It is essential that sorbents are activated and deactivated correctly and in a repeatable manner. Table 5.3.13 specifies the procedures for the activation of silica, alumina, and florisil, the three most common materials used, and also includes the conditioning of polymeric GPC beads.

The preparation of alumina can include the controlled deactivation of sites by the addition of water. The amount of added water is critical. The alumina can be deactivated in-house or purchased in a deactivated form. There are five grades of deactivated alumina available:

Activity Grade	I	II	III	IV	V
Weight % water added	0	3	6	10	15

To deactivate alumina in-house, a measured amount of water is usually added to alumina of Activity Grade I.

Other sorbents can also be partially deactivated by the addition of water, although this is less common.

Once deactivated/activated to the required degree, the sorbent must be protected from atmospheric moisture. The susceptibility of the material to hydration is dependent upon the degree of activation: the more active, the more hygroscopic. The activity also has a bearing on the choice of desiccant over which to store the material. It is an easy mistake to store a highly active sorbent over a less active desiccant so that, in fact, the sorbent desiccates the desiccant.

In order to exclude atmospheric moisture, extreme care must also be taken when packing columns. All operations should be carried out as swiftly as possible. The packing material is usually pre-mixed with the eluting solvent as a slurry, then poured into the column with gentle tapping to allow it to settle. For adsorbents of larger particle size (*i.e.* not silica) an alternative procedure is to pour the dry powder into a column already containing the solvent. In either case, the column should be firmly packed and settled, then topped up with a desiccant (such as anhydrous sodium sulfate) in order to minimise atmospheric exposure. The inside of the column should be carefully washed with solvent to ensure that there are no fine sorbent particles out of solution. Particular care should be taken rinsing ground-glass joints, PTFE stoppers, and taps.

Table 5.3.13 *Activation and controlled deactivation of adsorbents*

Packing material	Method of activation/preparation	Storage and precautions
Silica	Heat at 130–300°C for >2 h. Can be deactivated	Exact temperature and heating time not critical if within range. Higher temperatures denature silica and decrease capacity. Should store for 60 days in an airtight container (desiccator)
Alumina	Heat at ~500°C for >4 h. Cool in desiccator. Achieve activity grade by mixing with water with automatic shaking	Overheating to 800°C removes active hydroxyl groups. Can store in airtight flask for several months. Do not overshake or particle sizes are reduced
Florisil	Heat at 130°C for >5 h. Store at 130°C, or store in desiccator and reheat before use	Should store in airtight container for 2 days
Polymeric GPC Beads	Soak overnight in intended eluent to allow to swell	Ensure column never dries once packed

The column should be conditioned by allowing the eluting solvent to drip through and equilibrate. Typically 50 ml of solvent is used to condition a 5 g column. The extract is then placed on the head of the packing material in a tight plug, and the solvent allowed to drip through at a slow, steady rate. On no account should the column be allowed to run dry; *all* of the packing material should be immersed at all times. However, the meniscus of the eluting solvent should be kept as close as possible to the top of the packing material to avoid diffusion of the analyte through the solvent.

Once a batch of columns have been set up, the loading of polar contaminants (usually fats) is the factor that tends to vary most from sample to sample. The more fat on the column the less well retained the analyte (and the interferents), until a critical loading is reached at which no more material is retained. The critical loading depends upon the sorbent, its activity, and the type of fat. For example, it is accepted that 22 g of neutral 10% deactivated alumina will retain up to 400 mg of animal fat, but for vegetable fats and oils the maximum loading is nearer to 300 mg.

Due to the critical nature of activation and controlled deactivation of sorbents, activity is likely to vary slightly from batch to batch. It is therefore not recommended that more than one sorbent batch is mixed for the analyses of one batch of samples. It is also advisable to test each new sorbent batch against a fat retention specification, *e.g.* by measuring the weight of corn oil eluted through a column.

If there are still molecular interferents, *e.g.* sulfur compounds, aldehydes, and ketones, in the sample extract after alumina clean-up, the alumina can be modified by the addition of silver nitrate.[7] This has been found to remove such compounds.

Before adding the sample extract to a column, it is advisable to run a few drops of eluent onto a clean piece of glass, to check that no sorbent is passing through the cotton wool plug at the bottom of the column.

Solid-phase Extraction (SPE)

If used correctly SPE cartridges produce clean extracts rapidly and easily. However, if, for example, flow rates are too high, results tend to be somewhat erratic.

To obtain optimum results from solid-phase cartridges, care and patience are required. Rapid application of the sample extract followed by the eluting solvent, as if emptying a pipette, will reduce the effectiveness of the adsorbent. As with classical columns, the extract and eluent must be allowed to drip through at 1–2 ml min^{-1}, and the cartridge should not be allowed to run dry.

The same care with elution solvents and flow rates must be taken as with conventional gravity-fed adsorption columns; SPE is, after all, just miniaturised chromatography. Most manufacturers recommend pre-washing the cartridge with a series of solvents ending with the intended elution solvent. Sequential solvents should be miscible, the exception being when non-polar solvents are used after the adsorption of aqueous samples. This is the only case where the cartridge can be dried between solvents. When changing between all other solvents, it is important not to allow the cartridge to dry. Flow rates should be slow, leaving plenty of time for equilibration, with the solvent gradually dripping from the cartridge at typically 2 ml min^{-1}. Automated vacuum pumps greatly assist in processing multiple samples simultaneously, but many basic models are difficult to control. The suction is either 'off' or 'on'. Many analysts still use a manual syringe and rely on judgment: a tedious process if there are a number of samples to extract.

Even when used correctly, there are two major pitfalls to avoid: overloading the cartridge, and contamination from the cartridge construction materials. A 1 g cartridge typically has a capacity of 10–20 mg of adsorbed material. Therefore samples with a high level of interferents will soon tie up all the available adsorption sites.

Contamination from plastic construction materials is a particular problem when using element-selective sensitive detectors. Phthalates and other plasticisers, for instance, readily leach into non-polar solvents and give a strong signal on electron capture detectors (ECD). This generally makes disposable solid-phase cartridges unsuitable for methods that rely upon ECD.

Particulates in water samples may also clog SPE cartridges. This may be a sample homogeneity problem, as some analyte may be tied up in the sediment. It must be decided whether to sample the water, the particulates, or both, but the sediment should not be simply filtered out without investigation as to whether it contains analyte.

Pre-elution often provides an effective and simple method of additional clean-up. For example, an analyte of intermediate polarity in an aqueous extract can be retained on a reversed phase cartridge. Additional, retained, non-polar species can then be washed off using a solvent such as hexane. The analyte is subsequently eluted with a more polar solvent, such as dichloromethane. For a list of suitable solvents to use with different packing materials, see Table 5.3.9.

If a reversed phase C$_{18}$ cartridge is found to retain the measurand more strongly than desired, C$_8$ is an alternative, slightly less active, packing material.

Gel-permeation Chromatography (GPC)

One precaution, which is unique to GPC, is the need to allow the polymeric packing beads sufficient time to equilibrate with the eluting solvent. This is in order for the beads to swell to their full size. If the beads are not fully swollen, separations will not be reproducible because the size of the molecular gaps will vary from batch to batch. The usual practice is to allow the packing material to soak overnight. Once prepared, a column should never be allowed to run dry. If any part of the column should accidentally dry, then the entire elution profile must be re-established using reference materials.

Although GPC is a universal technique, it is most frequently used for the analysis of foodstuffs and similar samples. Its major use is the removal of high molecular weight fats. Animal fats contain a higher proportion of high molecular weight species than vegetable fats, and so are more amenable to GPC. Extracts containing animal fats usually only require the eluate to be split into two fractions, the first (containing fatty interferents) to be discarded and the second (containing a range of analytes) collected. Vegetable sample extracts tend to involve more detailed fraction collection, as there are smaller molecular interferents, which have similar retention properties to the analytes.

A lower pressure is required to force the eluent through GPC columns than that used for LC. Generally, standard or preparative HPLC pumps are used. The older type of gravity-fed columns must be supplemented with some type of pump to achieve the desired flow rates (1–4 ml min^{-1}). In order to facilitate the solvent flow, the cotton wool plug in the bottom of the column should be tamped a lot less firmly than for adsorption columns so that the column is not blocked. Once packed, the top of a gravity-fed GPC column can be gently tamped with a glass rod in order to settle the packing material evenly.

Coupled Chromatography

There are many critical parameters to consider with coupled chromatographic techniques, as they are generally complex in terms of the number of variables to be optimised. The most vital is probably the sample loading, both onto the preparative column and the fraction of sample passed onto the analytical column. These must be fully investigated for each individual sample type to ensure that neither column is overloaded.

Direct on-column injection is the most popular method of sample introduction for LC-GC. However, a very small amount of involatile material carried over to the determination gas chromatograph, is sufficient to cause unacceptable band broadening. Grob has estimated that if one-ten thousandth of the injected sample load carries onto the analytical column, it can be sufficient to ruin a GC system.[5] Also, any mobile phase used for LC-GC must be compatible with the GC column (in effect limiting the technique to normal phase systems). Possible alternatives to direct on-column injection are the use of temperature programmable injectors, allowing the splitless injection of large sample volumes.

The preparative column can also be easily overloaded and contaminated. This risk is lower for preparative and semi-preparative LC columns than for analytical columns, provided the manufacturers' specified loading is not exceeded. The column should have the same shelf life as an analytical column. Regular back-flushing should further prolong the column life. Back-flushing incorporated as part of the switching cycle also reduces the risk of cross-contamination between samples.

It must also be established that the collected fraction contains the analyte of interest. Chromatographic retention can vary with temperature; mobile phase composition and flow rate; and even sample type. All these must be carefully controlled once the required fraction has been established.

5.3.5 Associated Uncertainty

There is less uncertainty associated with extraction from liquids than the analogous extractions from solids. Matrix effects are generally minimal, particularly in aqueous samples. Apart from the possibility of emulsion formation or problems arising from speciation effects, extraction from a system with a high theoretical partition coefficient should approximate closely to 100%. Due to minimal matrix and binding effects, spiking recoveries are generally a more valid guide to losses than with solid samples. Exceptions to this are samples such as blood and other body fluids that contain proteins and other binding agents. The presence in an aqueous sample of oils, alcohols (*e.g.* extracting from wine), surfactants, and detergents can all affect partitioning.

Recoveries from SPE cartridges are more of a problem than liquid/liquid extractions, as there is always the possibility of a portion of the determinand being washed through when the remainder is adsorbed or *vice versa*. Recovery determinations using solutions of known concentration, and exhaustive washings with strong solvents, can provide an approximation to this. Reproducibility is greatly affected by the flow rate, it needs to be precisely controlled. Precision tends to improve with experience, and this should be remembered when training new staff.

Although not on the same scale as the uncertainty of analyte extraction, there is considerable uncertainty associated with extract clean-up in organic analyses. This can be reduced to a minimum by good analytical practices. The foremost of these is to take regular spiked extracts through the clean-up procedure to ascertain losses: ideally, at least one spike should be run alongside each batch of samples.

Extrapolating analyte losses from spiking recoveries assumes reproducibility of conditions. From this point of view, liquid partition is the technique with the least associated uncertainty, providing there are no problems with emulsions or speciation. It is less valid to assume that recoveries from adsorbent columns will remain precise, due to the slight batch-to-batch variations of sorbent activity, and the critical nature of, for example, the sample fat loading. Regular checks should be made on the exhaustive elution and fraction collection of both spiked samples and analytical standards.

Fraction collection from solutions of known concentrations generally gives a better model of sample retention for GPC than for adsorption chromatography, because standard solutions behave more similarly to those containing co-extractives. The uncertainty in predicting the retention characteristics of sample components from standard recoveries is therefore reduced. If column dimensions and solvent flow are controlled, there should be less batch-to-batch variation than with sorbents, as there is less variation in activity due to atmospheric moisture.

If parameters such as column temperatures, flow rates, mobile phase compositions, and switching-valve timing are precisely controlled, coupled chromatographic methods should give very good reproducibility. It is when these aspects are allowed to vary that the uncertainty increases.

5.3.6 References

1. D. E. Wells, 'Extraction, Clean-up and Group Separation Techniques in Organochlorine Trace Analysis', *Pure Appl. Chem.*, 1988, **60**, 1437.
2. T. L. Peters, 'Comparison of Continuous Extractors for the Extraction and Concentration of Trace Organics in Water', *Anal. Chem.*, 1982, **54**, 1913.

3. M. Moors, D. L. Massart, and R. D. McDowell, 'Analyte Isolation by Solid Phase Extraction (SPE) on Silica Bonded Phases – Classification and Recommended Practices', *Pure Appl. Chem.*, 1994, **66**, 277.

4. J. L. Bernal, M. J. Del Nozal, and J. J. Jiménez, 'Some Observations on Clean Up Procedures Using Sulphuric Acid and Florisil', *J. Chromatogr.*, 1992, **607**, 303.

5. K. Grob, 'Hyphenated High Performance Liquid Chromatography – Capillary Gas Chromatography', *J. Chromatogr.*, 1992, **626**, 25.

6. D. F. S. Natusch and B. A. Tomkins, 'Isolation of Polycyclic Organic Compounds by Solvent Extraction with Dimethyl Sulphoxide', *Anal. Chem.*, 1978, **50**, 1429.

7. D. C. Holmes and N. F. Wood, 'Removal of Interfering Substances from Vegetable Extracts Prior to Determination of Organochlorine Pesticide Residues', *J. Chromatogr.*, 1972, **67**, 173.

General Reading

8. S. K. Poole, T. A. Dean, J. W. Oudsema, and C. F. Poole, 'Sample Preparation for Chromatographic Separations: An Overview', *Anal. Chim. Acta*, 1990, **236**, 3.

9. K. Ballschmiter, 'Sample Treatment Techniques for Organic Trace Analysis', *Pure Appl. Chem.*, 1983, **55**, 1943.

10. S. M. Walters, 'Clean Up Techniques for Pesticides in Fatty Foods', *Anal. Chim. Acta*, 1990, **236**, 77.

5.4 Concentration of Organic Extracts

5.4.1 Overview

Due to the rigorous sensitivity requirements of trace analysis, it is extremely rare that a sample or sample extract can be measured directly. The determination techniques available are not sensitive enough to cope with the detection limits required without considerable concentration of the sample extract. Increasing the weight sub-sampled, reducing the volume of the extract presented to the measuring instrument, or taking larger aliquots of extract solutions at some stage in the procedure are all, in fact, the easiest and most common ways of improving the detection limit of a method.

A sample extract may need to be concentrated on a number of occasions during the course of an analysis. These solutions may be liquid partition phases; eluates or fractions collected from a chromatographic clean-up step; distillates; or extracts that need to be concentrated to a small plug for a chromatography column or SPE cartridge. Ultimately, the extract must be concentrated for the end-determination. Sufficient of the analyte has to be injected for the instrument to detect and quantitate.

In an ideal world, concentration would be unnecessary. Each time the extract solution has to be reduced another stage is added to the analytical method, a stage which, if not done with care, can cause significant analyte loss, with a subsequent increase in the overall uncertainty. Many volatile compounds are extremely easy to lose during concentration, *N*-nitrosamines being an excellent example. There is also the added problem that any interferents, either present in solvents, from glassware and apparatus, or from the sample itself, will be concentrated along with the analyte. The relative effect of solvent impurities, in particular, is markedly increased the more concentrated the extract.

In addition to these problems, it is undesirable to have a small final extract volume. Volume measurements are less accurate at low levels and the effects of evaporation are proportionally greater. If the analytical method leads to an end volume of only, for example, 1 ml, this is not much use when you have to fill three autosampler vials for different determinations, plus keep enough solution in reserve to fill vials for any confirmatory measurements. Nobody likes to be in the situation of relying on a one-shot injection.

The way to avoid these problems is to design experiments so that extracts are never concentrated to very small volumes. Obviously the degree of concentration required depends upon the analyte, the detection system used, and the specified reporting limit. With modern instrumentation and techniques, a final extract concentration equivalent to 1 g ml^{-1} is usually adequate to detect analyte concentrations below mg kg^{-1} levels. This means that, even allowing for extracts being split into aliquots during analysis, ideally, samples should not be concentrated to volumes of less than 2 ml.

If concentration to very low volumes is unavoidable, there are two general precautions to observe: never let solutions evaporate to dryness, and always keep solutions in an inert atmosphere. Provided such care is taken, losses of volatile analytes can be minimised.

5.4.2 Comparison of Techniques

Solutions that require concentration may be either aqueous or in organic solvents. Generally, different techniques are used for each of these.

Concentration of Solutions in Organic Solvents

Sample extracts in organic solvents are almost invariably concentrated by evaporation. A comparison of evaporation methods is given in Table 5.4.1. These can be sub-divided into two categories: techniques for reducing a large bulk of solvent to 5–10 ml [rotary evaporation, Kuderna–Danish apparatus, and Automated Evaporative Concentrator (EVACS)] and methods of further evaporating the 5–10 ml extracts produced by these techniques to the final desired volume (gas blow-down and Snyder tubes).[1,2] In the majority of cases a combination of two techniques is used: one from the former category to remove the bulk of the solvent, then switching to a less vigorous method from the latter group when the solvent volume becomes so low that analyte losses using a fiercer technique would be significant. If solutions are evaporated to low volume, a high boiling 'keeper' (*e.g. n*-octanol) is usually added to guard against accidental drying.

Of the methods of reducing bulk solvents, rotary evaporation is the technique of choice when analytes are involatile at the set temperature/pressure and are thermally stable. The distilled solvent is automatically collected, control of

Table 5.4.1 *Evaporative methods of extract concentration*

Apparatus	Advantages	Disadvantages	Examples of Applications
Rotary evaporator	Copes with high boiling solvents. Suitable for thermally unstable analytes	Volatile compounds generally lost. Low recoveries of some compounds due to entrapment in solvent vapour	Removal of alcohols, toluene, water, *etc.* from analytes of low volatility
Kuderna–Danish (K–D) Apparatus	Higher recoveries of volatile analytes. Can run multiple batches	Slow compared with rotary evaporator. Limited to reduction to ~1 ml. Unsuitable for thermally unstable compounds. Limited by solvent volatility	Removal of hexane, acetone, dichloromethane, *etc.* from analytes of high volatility
Snyder tube	Extends K–D evaporation to lower volumes	Labour intensive	As K–D
Gas blow-down	Extends K–D evaporation to lower volumes	Losses by nebulisation. Gas supply may contaminate sample. Limited to volumes > ~ 25 ml	Lower boiling solvents
Automated Evaporative Concentrator	Automated. Liquid level monitors for unattended operation	Requires 50°C difference in boiling points of solvent and analyte. Limited to reduction to ~ 1ml	General

external contamination is relatively easy, and a high degree of control can be exercised over the vigour of boiling. Table 5.4.2 compares the losses incurred when a solution of vanillin was reduced by rotary evaporation as compared with other heat-assisted techniques, at an identical temperature. Vanillin is a relatively volatile analyte, with a boiling point of 285°C. As can be seen, when the solution reaches low volumes (approaching dryness) there are significant losses when using techniques with no water-cooled condenser.

Many assays, unfortunately, are prone to sample losses by vaporisation of volatile analytes (*e.g.* polyaromatic hydrocarbons), thermal degradation, nebulisation, and splashing of low boiling point solvents. It can be difficult to transfer small volumes from the large flasks used without significantly increasing the volume again. The equipment is relatively expensive, particularly if it is necessary to reduce several batches of extracts at the same time.

In these cases the usual alternative is to use the Kuderna–Danish (K–D) apparatus, Figure 5.4.1. As described in catalogues, the tube is attached with a plastic clip, but these have a tendency to lose their elasticity after a time, and so a moulded glass wing and spring arrangement is preferred. The K–D is an extremely low-tech technique compared with rotary evaporation, but none the less effective. The entire apparatus can either be heated in a steam chest or in a low temperature water bath, *e.g.* if the analytes are thermally labile such as *N*-nitrosamines. The

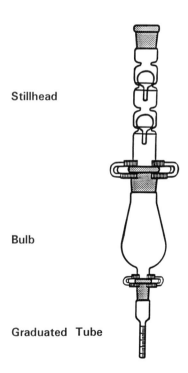

Stillhead

Bulb

Graduated Tube

Figure 5.4.1 *Kuderna–Danish apparatus*

Table 5.4.2 *Percentage losses of a 3 μg ml⁻¹ aqueous vanillin solution using different evaporation techniques*

Final volume sample	Method of evaporation	Recovery %	Comments
Dryness	Rotary evaporation	86	These samples were evaporated at 80°C. Maximum evaporation time 15 min
< 10 ml	Rotary evaporation	92	
10 ml	Rotary evaporation	94	
25 ml	Rotary evaporation	94	
Dryness	Air condenser	58	Evaporation time between 0.5 and 2 h
< 10 ml	Air condenser	76	
10 ml	Air condenser	88	
25 ml	Air condenser	95	
Dryness	Hot plate	17	Charring occurred when the sample reached dryness
< 10 ml	Hot plate	82	
10 ml	Hot plate	86	
25 ml	Hot plate	102	
Dryness	Heating block	72	At 80°C evaporation of a 10 ml aliquot took between 1 and 2.5 h to evaporate
1 ml	Heating block	88	
2 ml	Heating block	92	
5 ml	Heating block	86	
Dryness	Steam bath	71	Samples were evaporated in open 100 ml beakers. Evaporation took between 1 and 4 h
< 10 ml	Steam bath	85	
10 ml	Steam bath	92	
25 ml	Steam bath	98	

great advantage of the K–D is that batches of samples can be concentrated simultaneously at a very low capital cost. It is also a far better technique for volatile analytes as the continuous condensation of the solvent from the stillhead washes vaporised analyte back into solution.

Evaporation in a K–D apparatus normally takes place in three stages. First the volume of liquid in the bulb is reduced with the stillhead in place. Then, if the apparatus is in a low-temperature water bath, the stillhead can be removed when the solvent volume is reduced to about half the capacity of the bulb (if the solvent is boiling in a steam chest, the stillhead should be left on throughout). Once the volume of solvent is further reduced to about half the volume of the tube, the apparatus is removed and allowed to cool. Only when cool is the tube removed from the bulb, and the solution evaporated further (usually by gas blow-down).

Automated Evaporated Concentrator (EVACS) systems are a refinement of the water bath methods, in which evaporation is assisted by a stream of gas. The solvent level is automatically sensored and the gas flow stopped when it reaches a pre-set level. This allows for unattended operation, circumventing one of the main disadvantages of K–D, the need for unbroken vigil. The EVACS, of course, are much more expensive than K–D evaporators.

Of the two methods of further reducing extracts to low volume, gas blow-down usually takes preference. Snyder tubes can only be used one at a time, and there are the same problems with vigorous boiling and superheating as with any other heat-assisted technique. Gas blow-down can be performed very slowly and gently, with multiple samples applied to one split gas line. If the evaporation is too slow at ambient temperature, it is always possible to immerse the extract tube in a bath of warm water.

Concentration of Aqueous Solutions

Although aqueous solutions can be concentrated by vacuum-assisted evaporation, this is a slow and laborious process. It is usually much easier to effect concentration by extracting the analyte into a smaller volume of organic solvent. Methods such as solvent partition and adsorption onto a SPE cartridge or liquid partition column before eluting with solvent are common. These techniques are covered in detail in Section 5.3.

5.4.3 Critical Aspects and General Precautions

The greatest risk of unpredictable analyte loss during most analyses is at the evaporation stage. Losses tend to result from three main causes: those from over vigorous boiling, splashing and nebulisation, those from evaporation of volatile analytes, and those from oxidation or thermal degradation of unstable compounds. Great care must be taken to avoid all of these hazards.

Vigorous Boiling and Frothing

The obvious way to avoid splashing and nebulisation is to keep conditions as gentle as possible during the evaporation. For example, for most organic solvents a rotary evaporator water bath should never need to be heated to much above 40°C, and the bulb of the evaporator should just touch the surface of the bath rather than be immersed. In many cases even this may be too hot. It is particularly difficult to evaporate low boiling solvents such as ethers under vacuum without bumping. One solution is to leave the vacuum valve partially open so that the vacuum can be finely adjusted.

With the K–D apparatus, the problem is perhaps even more acute. It should be standard practice to add a few antibumping granules to the bottom of each tube, but these must be put in immediately before evaporation; bumping chips which have been standing in solution coagulate and lose their effectiveness. The antibumping granules must be completely inert (*e.g.* fused alumina) and should be washed with solvent, ideally in a Soxhlet apparatus, before use to avoid introducing contaminants. If an extract in a K–D does bump, the stillhead is liable to blow out of the top of the bulb and a large proportion of the solution lost.

All gas streams used for gas blow-down should be as gentle as possible. Under no circumstances should the liquid be allowed to splash. To avoid splashing and cross-contamination, the gas tube should not contact the liquid.

The other main cause of bumping is superheating of liquids in water baths. This does not necessarily require a high temperature, especially when evaporating volatile solvents. It can occur whenever the solvent level falls below the level of the water in the heating bath. As already mentioned, rotary evaporator bulbs should not be immersed, but should skim the surface of the water. Kuderna–Danish or gas blow-down tubes that are heated in a water bath should be continually adjusted so that as the solvent level falls the tube is raised, keeping the solvent meniscus above the water level.

Evaporation of Volatile Analytes

Some analytes are only slightly less volatile than the solvent that is being evaporated, and in these cases very strict temperature control is required. An extreme example is the *N*-nitrosamines, but evaporative losses of higher boiling compounds can also occur. With analytes such as *N*-nitrosamines, the only reliable method of concentrating extracts is to use a K–D (the solvent refluxing action, which is absent in rotary evaporation, is necessary to wash down volatilised analyte), which is slowly warmed in a water bath held at a precisely controlled temperature.

Thermal Degradation and Oxidation

The other major cause of potential losses is analyte degradation (usually oxidation) on reducing the volume. This becomes a severe problem when working with very low volumes. Even higher boiling and thermally stable compounds can be lost at very low volumes, and losses are almost inevitable if solutions are allowed to dry. Table 5.4.3 shows that the losses of organochlorine pesticides become significant when the use of a K–D apparatus is extended to volumes of less than about 5 ml. Burke *et al.* reached the same conclusion.[3] Similar results are

Table 5.4.3 Percentage losses of some organochlorine pesticides using different evaporation techniques[a]

Compound	To 5 ml on K–D then 1 ml with Snyder tube	Just to dryness on K–D	10 min beyond dryness with Snyder tube	To 1 ml with airline	To 0.5 ml with airline	Just to dryness with airline	10 min beyond dryness with airline
Hexachlorobenzene	17	100	83	1	13	42	100
α-Hexachlorocyclohexane	15	100	88	1	13	36	98
τ-Hexachlorocyclohexane	39	100	78	5	12	27	91
ß-Hexachlorocyclohexane	17	95	72	6	9	10	30
pp'-DDE	16	89	27	1	5	6	33
op'-DDT	13	80	15	9	6	1	26
Dieldrin	14	82	40	6	6	3	42
pp'-DDT	13	65	7	4	3	3	19
pp'-TDE	15	67	16	4	2	6	20

DDE = oxidised metabolite of DDT
DDT = dichlorodiphenyltrichloroethane
TDE = reduced metabolite of DDT
[a] Results obtained at the Laboratory of the Government Chemist.

found when gas blow-down solution volumes reach about 0.5 ml. Use of Snyder tubes reduces this problem. The two-ball version has been found to give the greatest recoveries. The EVACS incorporates a solvent sensor to avoid accidental drying. With all other techniques, the addition of a small amount of a high boiling solvent as a 'keeper' is recommended for evaporation to low volumes.

If it is essential to remove all traces of the original solvent (*e.g.* the removal of chlorinated solvents before GC-ECD) there are two approaches that may be used to avoid drying. It is possible to use a 'keeper', if a suitable one is available (*i.e.* of sufficient purity, and compatible with the determination system). Alternatively, small volumes of a suitable second solvent are added repeatedly and evaporated, so that the final aliquot is free of the first solvent. The second (final) solvent should have a higher boiling point than the original solvent.

For all techniques involving gaseous evaporation of low volumes, nitrogen is preferable to a stream of air. There is less risk of oxidation, and also piped or cylinder gases can generally be relied upon to be of higher purity than an air stream (oxygen-free nitrogen is available from most suppliers). When restoring pressure to a rotary evaporator bulb, nitrogen should also be used rather than opening to the atmosphere, especially if the analyte is susceptible to oxidation (*e.g.* vitamin E). The residue should then be redissolved in the required volume of the final solvent as swiftly as possible.

5.4.4 Associated Uncertainty

The uncertainty associated with concentration by evaporation is extremely high. This does not mean that recoveries are poor, or even that techniques are not reproducible and robust, but that there is always the *possibility* of analyte loss which cannot be predicted. All techniques except EVACS require the analyst to judge when the final volume has been reached and when to remove the extract. This results in varied final volumes, even within a batch of samples, and the ever-present risk of accidental drying.

5.4.5 References

1. E. A. Ibrahim, I. H. Suffet, and A. B. Sakla, 'Evaporative Concentration System for Trace Organic Analysis', *Anal. Chem.*, 1987, **59**, 2091.
2. S. K. Poole, T. A. Dean, J. W. Oudsema, and C. F. Poole, 'Sample Preparation for Chromatographic Separations: An Overview', *Anal. Chim. Acta*, 1990, **236**, 3.
3. J. A. Burke, P. A. Mills, and D. C. Bostwick, 'Experiments with Evaporation of Solutions of Chlorinated Pesticides', *J. Assoc. Off. Anal. Chem*, 1966, **49**, 999.

CHAPTER 6

Organic Analytes: Determination

6.1 Introduction to the Determination of Organic Analytes

Chapter 5 dealt with the methods available for obtaining a test solution. This solution should have sufficient concentration of the analyte(s) of interest and a low enough concentration of potentially interfering components to permit qualitative and (if necessary) quantitative determination at the detection limit with the level of uncertainty required.

6.1.1 Overview

The analysis of organic compounds at mg kg^{-1} levels and below requires determination techniques that are both sensitive and selective. Concentrations of analytes in sample extracts are typically in the range 0.001–10 µg ml^{-1}, and this places a heavy demand upon the measurement method. Additionally, there are invariably interfering compounds present in the extract due to the high concentration processes described in Chapter 5. This greatly restricts the measurement techniques available.

Before the 1950s, the range of determination techniques available to the trace organic analyst was fairly limited. Classical wet chemistry methods, relying on a colorimetric reaction to identify and quantify the analyte, were prevalent. Spectroscopic measurements were also used. All of these techniques suffer from one inherent disadvantage, the analyte is not separated from the other compounds in the extract solution. Consequently, the measurements are prone to interference effects.

The development of chromatographic separation techniques provided a breakthrough in analytical measurement. The advantages of being able to separate the compound of interest from the other substances in the sample extract before determination were obvious; retention times gave an increased degree of confidence in identification of compounds, and the detection sensitivity was generally increased by the reduction in the background signal as the analyte was separated from other interferents.

207

The development of chromatographic techniques was extremely rapid, from paper chromatography for the separation of amino acids through to early Gas–Liquid Chromatography (GC) and Thin Layer Chromatography (TLC).

Gas chromatography remains the workhorse technique for organic trace analysis, with the continual advancement and development of different types of detectors, columns, stationary phases, and injection techniques ensuring that the technology has kept pace with the increasing requirements of sensitivity and selectivity placed upon analytical measurement techniques. The introduction of capillary columns, particularly, has ensured that GC has kept a pre-eminent position amongst trace organic determination techniques. This has extended its application to complicated multi-analyte determinations, and although the sample loading on the column is lower than for packed columns, the improved signal-to-noise ratio resulting from the superior resolution often allows lower detection limits.

Thin-layer chromatography is still used for semi-quantitative screening applications because it is cheap and easy to use. In laboratories that do not have access to some of the more expensive modern techniques available, it is used more extensively.

Gas chromatography has two inherent disadvantages: it is not applicable to thermally labile compounds, and run times can be longer than one would ideally wish. Although liquid/liquid chromatographic techniques had been investigated since the early days of chromatography, it was not until the 1980s that High Performance Liquid Chromatography (also initially referred to as High Pressure Liquid Chromatography) came into widespread use. It provided another string to the bow of the trace organic analyst. It is more rapid than GC, and almost universally applicable. However, the detector with the greatest range of applications (the ultraviolet/visible absorbance detector) is generally not as sensitive as some GC detectors (*e.g.* the electron capture detector, ECD), and HPLC columns cannot match the resolution of capillary GC columns. Therefore HPLC tends to be used most for compounds that are thermally labile and so not amenable to GC, or for compounds that respond well to a particular detection system, *e.g.* analytes that absorb at an unusual wavelength or that fluoresce (or that can be derivatised to produce a fluorescent compound).

The most recent advancement in chromatographic techniques has been the coupling of capillary column GC with low-resolution benchtop mass spectrometers. Mass spectrometry was already well-established as a stand alone detection technique that provided the ultimate possible selectivity. However, without a method of separating the components of the test extract it suffered from interference effects and was inapplicable to multi-analyte determinations. Gas chromatography provided this separation method. Coupled GC–MS is very powerful technique offering both extreme selectivity, close to universal applicability, and, with more modern instruments, a sensitivity comparable to some of the more traditional element-selective detectors. Mass-selective detection has revolutionised GS analyses. It is probable that coupled LC–MS, which has recently become commercially available after some initial problems with interfaces, will have the same effect in the field of HPLC.

No identification based upon chromatographic retention alone is certain, the evidence is merely circumstantial. In order to increase the degree of confidence, it is standard practice to run a confirmatory analysis using a different system. With HPLC it is easiest to change the mobile phase and use the same column (*e.g.* aqueous acetonitrile rather than aqueous methanol), whilst for GC it is simpler to use a different column with a different stationary phase. As long as the two systems have different retention mechanisms, then matching retention times on two systems is usually taken as evidence of identity. Another great advantage of GC or HPLC with MS is that it negates this requirement for a second confirmatory determination.

Relying upon a chromatographic method of separating the analyte of interest is not the only approach to trace organic analyte determinations. In some cases it is possible to use a measurement technique that is so selective that co-extracted compounds in the sample extract do not interfere. Some examples of such techniques are given in Table 6.1.1. By their very nature these techniques are extremely specific, and so tend to be used only for selected applications. There is a great interest in developing such specific methods, particularly bioanalytical techniques such as enzyme linked immunosorbent assay, for as many analytes and sample matrices as possible. They have the advantage of being extremely rapid (and therefore cheap) and cutting out many of the laborious clean-up stages which are required for more traditional chromatographic analysis.

Table 6.1.1 *Determination techniques that do not require chromatographic separation*

Technique	Example of Application
Spectrophotometry	Urea, formaldehyde in effluents
Fluorescence spectroscopy	Morphine in blood samples
Polarography	Acrolein in water
Radioimmuno assay	Vitamin B_{12} in food
Probe mass spectrometry	Polyaromatic hydrocarbons in effluents
Enzymatic and biological reactions	Triazine herbicides in ground water
Chemiluminescent reactions	N-nitrosamines in cosmetics
Fluorine nuclear magnetic resonance	Fluorine-containing pesticide residues

6.1.2 Choice of Technique

The choice of technique for a particular analysis is largely governed by the resources and instrumentation available to the analyst, but there are still many other factors to take into consideration. A generous budget does not guarantee the success of a determination.

The decision as to which instrumental technique to use must always be taken with regard to the method as a whole. It cannot be taken in isolation. For example, many element-selective detectors will respond to certain solvents. Chlorinated solvents cannot be used during the early stages of a method if an ECD is to be used. On a more fundamental level, an analyst using GC with mass-selective detection may question whether a traditional multistep clean-up is really necessary. It may be that the crude sample extract can be examined directly, with minimal clean-up sufficient to provide a clean trace and avoid poisoning of the ion source. Another similar case is that of bioanalytical techniques (considered in Section 8.3) where there is seldom need for protracted sample preparation. The implications of the determination technique chosen must therefore be carefully considered before starting the analysis.

Table 6.1.2 lists some of the more common techniques available. When considering sensitivity and selectivity the entire method must again be considered: the cleaner the extract the greater the selectivity, and the more concentrated the extract the greater the sensitivity. However, there is a limit to how far a particular determination technique can be pushed by working on the sample preparation. The more clean-up stages required in a method the greater the risk of analyte loss and the higher the measurement uncertainty. It is also not advantageous to design methods that result in very small extract volumes. Small volumes are more difficult to measure precisely, there is the possibility of analyte loss on drying, and if there is only a limited volume of extract there will be limited scope for repeat analyses. For these reasons if the method results in an extract volume below about 2 ml, or if there are more than three separate clean-up stages, a better alternative is to re-examine the end-determination technique.

Factors that need to be taken into account when choosing the best determination technique for a particular purpose are now considered.

The Required Detection Limits

If the detection limits required are very low (*i.e.* of the order of µg kg^{-1} and below) this places great demands upon the determination technique. The signal-to-noise ratio, and therefore limit of detection, is obviously dependent upon the degree of clean-up obtained in the analytical method and the concentration of the sample in the final extract (the equivalent sample weight per injection), but some generalisations can be made. The detection techniques with the combination of sensitivity and selectivity that give the lowest detection limits are GC with ECD and HPLC with fluorescence detection, and these are usually used where possible. Combined GC with MS is often precluded because of lack of sensitivity, although some top-of-the-range models can now rival traditional GC detectors. Enzyme and biological kits are rarely useful for these levels of detection; they are more appropriate for screening at legal limits *etc.* in the mg kg^{-1} range.

The Degree of Quantification Needed

If only a qualitative or semi-quantitative result is required then a quick, crude but cheap technique is usually used, such as TLC or a bioanalytical kit. If a higher degree of quantification is needed then a technique must be chosen that gives a strong signal (ideally 10 times the background noise or greater) which is

Table 6.1.2 Some commonly used end-determination techniques for organic trace analysis

Technique	Relative sensitivity	Relative selectivity	Relative rapidity	Key applications	Inapplicable Areas	Incidence of use	Ease of automation	Relative complexity
GC	*****[a] (for ECD)	*****[a]	**	Compounds suitable for ECD, e.g. halogenated compounds. Non-UV absorbers	Thermally labile compounds Involatile compounds Very polar compounds.	*****	*****	**[a]
HPLC	***** (fluor.) *** (UV)	***	****	Fluorescent compounds Strong UV absorbers	Non-UV absorbers	*****	*****	**
Supercritical Fluid Chromatography	*****[a]	*****[a]	****	Not in routine use. Any chromatographic method as for GC/LC	Yet to be validated for most applications	*	***	***
TLC	**	**	*	Semi-quantitative detection of impurities, e.g. in pharmaceuticals	Non-UV absorbers	**	***	**
Ultraviolet/visible spectroscopy (UV/VIS)	**	*	*****	Compounds that absorb at unusually high wavelengths, e.g. >350 nm	Non-UV/VIS absorbers	***	*	*
Fluorescence spectroscopy	****	****	******	Fluorescent compounds	Non-fluorescent compounds	***	*	*
MS	****	*****	****	Compounds in complex mixtures	Mixtures of analytes (except GC–MS)	** (Except GC–MS)	* (Except GC–MS)	*****
Bioanalytical techniques	**	*****	****	Large molecules. Biologically active compounds, e.g. drugs. Pesticides in water	Extracts in organic solvents	**	**	**

[a] - dependent upon the detector used.
* – low
***** – high

unaffected by any interferents. Generally more reproducible results are obtained with HPLC than GC, mainly due to the more precise method of injection. For the same reason, packed-column GC generally gives better quantification than capillary GC. If a high degree of confidence is needed in a quantitative result from capillary GC, it is advisable to use an internal standard to compensate for any injection variation.

To be certain there are no interferents or other instrumental bias it may be necessary to determine the analyte by more than one technique.

The Structure of the Analyte(s)

The analyte to be determined may obviously lend itself to a specific detection method. The first choice for determining fluorophores would, for example, be HPLC with fluorescence detection. Apart from fluorimetry, most other selective detectors tend to be used in gas chromatography. Electron capture detection is ideal for chlorinated environmental contaminants, nitrogen–phosphorus detectors respond well to amines and azoles, organophosphorus pesticides lend themselves to flame photometric detection: these types of compound would therefore tend to be determined by GC.

If an analyte produces a specific biological response that can be analytically measured then a bioanalytical technique is suggested.

The Number and Range of Analytes

One major difference between organic and most inorganic trace analyses is the problem of multi-analyte determinations. Quite often the organic analyst is required to determine simultaneously a wide range of compounds with a variety of properties. In these cases the detection system used must be selective and sensitive enough to determine the entire range of analytes. This may mean changing the detection conditions as compounds elute. One option is HPLC with programmable UV/visible (UV/VIS) detection, although the resolution provided by HPLC is usually only sufficient to separate a dozen or so analytes. The best solution, providing that all of the analytes are volatile, is GC–MS. Not only is the detector virtually universal, it is also highly selective, and can be programmed to scan for different ions at different times. For example, some US Environmental Protection Agency (EPA) recommended methods describe the analysis of up to a hundred pollutants by capillary GC–MS.

If it is impractical to determine all of the analytes simultaneously in this manner, the extract can usually be split into fractions for multiple determinations on different instruments. This is a further reason why it is desirable to design methods that do not result in a small final extract volume.

The Acceptable Time to be Taken

For routine sample analysis, it is preferable for the end-determination to be automated. Most chromatographic instruments are now automated, and so the actual run time for each sample may not be important.

Capillary GC run times are in the region of 1 h each; this is an obvious disadvantage if there is a high turnover of samples. From this point of view HPLC, at ~15 min per run, is far preferable.

The Operator Skill Required

Many modern instrumental techniques, such as isotope dilution MS, need highly skilled operators. Even benchtop mass-selective detectors should only be operated and maintained by trained personnel. When deciding whether to use high-technology methods the need for staff training or recruitment must be considered alongside the capital cost of the instrument.

Which of these factors is given the highest consideration is dependent upon the purpose of the analysis. For example, analysis for a banned steroid in cattle would require a great certainty in identification, of the order provided by biological reactions or MS. A high quantitative certainty is necessary when testing pharmaceuticals for toxicological impurities. In many cases the only way of obtaining the required degree of confidence in a result may be to use more than one technique. The purpose of the analysis must be considered carefully before developing an analytical method.

The following sections detail the more frequently used techniques, concentrating mainly on chromatography and MS as these are by far the most popular. Ion chromatography, which does have some organic applications, is dealt with in Section 4.4 on inorganic analysis as this is its main area of application. The analysis of speciated compounds is also dealt with separately, in Chapter 7. Electrochemistry and spectrophotometry apply to both organic and inorganic species and are dealt with in Chapter 8.

6.2 Gas Chromatography

6.2.1 Overview

Gas chromatography (GC) was one of the first chromatographic separation techniques to be developed, and still has an eminent position today. As can be seen from Table 6.1.2, it compares extremely favourably with other determination methods in terms of both sensitivity and of selectivity.

Since its inception, advances in GC techniques and instrumentation have kept pace with the demands placed upon them. In particular, the development of a wide range of selective detectors has improved both its sensitivity and its selectivity for many analytes. Gas chromatography with ECD, for example, is still one of the most sensitive determination techniques available for halogenated compounds. The introduction of capillary columns was also a significant advancement, giving resolution unparalleled by any other chromatographic technique. More recently, interest has focused upon improving the method of sample injection, with temperature programmable injectors, both increasing the amount of solvent that can be injected and narrowing the bandwidth of the analyte plug once on the column.

The development that ensured GC a prominent position in analytical chemistry was the coupling of capillary columns to mass selective detectors (GC–MS). This gave an extremely powerful analytical technique: a method of both separating the analyte(s) from other compounds in the sample extract and positive identification. Gas chromatography–MS is widely used, both for screening samples for the presence of a vast variety of analytes, and for the positive confirmation of the presence of compounds that have been tentatively identified using another technique.

Gas chromatography cannot be regarded as a universal technique. There are some compounds for which it is not applicable. Examples of these are thermally labile materials, analytes that do not volatilise at column operating temperatures, and strongly polar analytes. However, polar analytes can be derivatised (*e.g.* the methylation of alcohols), thereby increasing the range of applicability.

6.2.2 Comparison of Techniques

The main factors to consider when designing a system are the type of column and separation mechanism required, and the method of detection that will be the most suitable for the required application. Once these have been chosen, the system can be fine tuned to obtain the selectivity and sensitivity required, by adjusting, for example, the oven temperature program, the carrier and make-up gas flow rates, and/or the injection method.

Column Stationary Phases and Types

The choice of column is fundamental to obtaining chromatographic separations. There are an extremely large variety of pre-prepared stationary phases available from manufacturers and suppliers. If no suitable phase is available it is possible to coat a stationary support, in-house, with a customised phase. These can give better performance than materials bought off the peg, although care must be taken to obtain an even loading. They are certainly cheaper.

Generally, solutes will separate better on a stationary phase of similar polarity. Thus, a mixture of hydrocarbons would be best resolved on a non-polar column. However, solutes will also have longer retention times on phases of a similar polarity: separating a mixture of polar compounds on a polar phase takes an inordinately long time. A compromise must be reached between resolution and retention.

Table 6.2.1 lists some of the common stationary phases and their key applications, based on resolution only. If retention times are too long, a phase should be used which is of slightly different polarity to that of the solutes. Manufacturers are continually launching new materials, and will often advise on suitable phases for novel or unusual compounds. An area of particular interest at the moment is the separation of optical isomers using cyclodextrin-based stationary phases, although as yet such phases do not have the thermal stability to be generally useful.

Apart from the stationary phase material, the column dimensions, degree of

Table 6.2.1 *GC stationary phases for different applications*

Analyte group	Examples of suitable phases for resolution of mixtures
Acids	OV-351, CPWAX-58 CB, SP-1000, BP-21, NB-351, FFAP
Alcohols, aldehydes, ketones, ethers, xylenes	Polyethylene glycol-based phases, CARBOWAX- 20M, CPWAX-52 CB, Supelcowax-10, BP-20M, NB-20M, HP-20M, Superox, DB-Wax
Alkaloids, carbohydrate derivatives	Phases of slightly lower polarity, methylphenylsilicones, OV-73, CPSIL-8 CB, SE-52, SE-54, BP-5, HP-2, RSL-200, DB-5, SPB-5
Amines, hydrocarbons $>C_6$	Non-Polar phases, separations based on boiling point differences, methylsilicones, OV-1, OV-101, CPSIL-5 CB, DB-1, SPB-1, SP-2100, HP-1, HP-101, BP-1, SE-30, DC-200, UCW-98
Hydrocarbons $<C_6$	Very non-polar phases, squalane
Halogenated compounds, oxygenated compounds, mid-polarity compounds	Methylphenylcyanopropylsilicones, DEGS, OV-210, OV-17, OV-1701, BP-10, CPSIL-19 CB, CPSIL-13 CB

This list is not intended to be exhaustive. Phases cited are examples only, and there is no implied recommendation of particular manufacturers.

FFAP – Free fatty acid phase
DEGS – Diethylene glycol succinate
Manufacturers: OV = Ohio Valley, CP = Chrompack, SP = Supelco, RSL = Alltech,
DB = J + W Scientific, NB = Nordion, HP = Hewlett Packard.

support coating and cross-linking, and many other factors all affect the chromatographic resolution.

Gas chromatography columns fall into two distinct types: packed and capillary. Both have very different properties and uses. Table 6.2.2 gives a summary of the major applications and limitations of each. Generally, capillary columns give better resolution but poorer quantitation than traditional packed columns. There is no theoretical reason for this poorer quantitation, but in practice factors such as the increased complexity of the injection mechanism leads to poorer precision for capillary systems. It is necessary to use sensitive detectors when using capillary columns for trace analysis, this is due to the smaller volume of gas eluted. The ideal applications of capillary columns are for the resolution of complex mixtures, multi-analyte assays, and the resolution of impurities in 'dirty' extracts. Capillary columns are essential for GC–MS, which relies heavily on complete chromatographic resolution and a minimal flow of gas through the detector.

Injection Methods

The method of introducing the sample extract into a gas chromatograph is dependent upon the type of column, the nature and thermal stability of the analyte, and the amount of sample injected. The most fundamental difference is between injection techniques for packed columns and those for capillary. With capillary

Table 6.2.2 *Capillary vs packed column GC*

	Capillary	*Packed*
Resolution	Approximately 10 times better than packed due to improved theoretical plate count	Poorer than capillary
Detection limits	Usually higher than packed. Sample loading is lower, but improved signal-to-noise ratio usually compensates	Usually lower than capillary
Quantitative precision	Poorer than packed. Analyte can be lost in the injection port. Poorer baseline limits, use of automatic integration	Better than capillary
Sample volumes	Limited to <500 ng, can be increased by solvent venting[1]	Larger than capillary
Uses	Fingerprinting and identifying complex mixtures such as polychlorinated biphenyls, oils and paints. Separation of complex mixtures	Accurate quantitation of mixtures of less than about 12 compounds
Limitations	Precision can be poor: internal standards are necessary to compensate for variability	Complex mixtures. Separation of closely eluting compounds

GC, the sample solution is injected into a glass-lined injection port, where it volatilises before passing onto the column. It is possible to control a variety of parameters, *e.g.* the temperature of the injection port (or it can be heated unevenly or at a programmed rate), the proportion of the volatilised sample that passes onto the column, and the manner and time at which the port is vented. In contrast, packed-column injection and on-column injection onto megabore columns is much simpler. The solution is injected directly either onto the silanised glass wool at the head of the column, or through the wool and straight onto the packing material.

Table 6.2.3 compares the various injection techniques available. For trace analytical work, the standard methods tend to be splitless injection for capillary columns (to achieve maximum sensitivity) and injection into the glass wool for packed columns.

GC Detectors

Detectors for trace analytical applications must be extremely sensitive. Some of the more common detectors are described in Table 6.2.4.

The best detectors for trace analysis are element selective, as the major factor in reducing sensitivity is the background signal from interfering co-extractives. The more selective the detector, the greater this is reduced. The chemical structure of the analyte will immediately suggest if it is amenable to selective detectors: *e.g.*

Table 6.2.3 *GC injection methods*

	Methods for capillary GC			Methods for packed GC	
	On-column injection	*Split injection*	*Splitless injection*	*Standard injection*	*Directly onto packing material*
Applications	High boiling compounds. Compounds likely to thermally degrade. Volatiles sampled in traps	High sample weights	Standard for trace analysis	Standard for trace analysis. Easy to automate	Analytes that would thermally degrade or adsorb to silanised wool
Reasons for use	Avoids condensation in injection port. Reduces sample decomposition	Reduces sample weight onto column	All of the sample onto column	Column not contaminated. Needle not blocked	Less possibility of loss. Can inject larger volumes
Occasions when inapplicable	Thick stationary phases, films, and shorter columns can give distortion of late eluting peaks.[2] Large injection volumes.[3] Polar solvents[3]	Rarely applied to trace analyses due to sensitivity loss	Sample sizes >10 μl. Solvents with boiling points within 25°C of solute. Determination of solvent impurities	Compounds that degrade or adsorb to silanised glass wool	Volatile analytes. Difficult to automate. Irreproducible

Table 6.2.4 *GC detectors suitable for trace analysis*[4]

Detector	Uses	Approximate detection limit	Selectivity	Linear range	Temperature limit (°C)	Relative complexity
Flame Ionisation Detector	Universal for hydrocarbons	2×10^{-12} g s^{-1}. Rather high for trace analysis. Poor S/N ratio due to low selectivity	Not applicable	$>10^7$	420	*
Nitrogen Phosphorus Detector	Selective for nitrogen and phosphorus	N: $\times 10^{-13}$ g s^{-1} P: $\times 10^{-14}$ g s^{-1} Sensitive enough for most trace analyses due to high selectivity	N/P 1:5 N/C 5×10^4:1 P/C 10^5:1	10^5	420	**
Flame Photometric Detector	Selective for sulfur or phosphorus	S: $<1 \times 10^{-11}$ g s^{-1} P: $<1 \times 10^{-12}$ g s^{-1}	S/C 10^3–10^6:1 P/C 10^5:1	S: $>10^3$ (square law) P: $>10^4$	420	**
Electron Capture Detector	Halogenated compounds	Response varies by compound down to 10^{-15} g. Workhorse detector for trace analysis	Not applicable	10^4	420	**

Detector	Selectivity	Sensitivity	Selectivity ratio	Dynamic range	Max. temp. (°C)	Rating
Thermal Conductivity Detector	Permanent gases, *e.g.* water	4×10^{-10} g ml^{-1} propane	Not applicable	$>10^5$	400	*
Mass spectrometer	Universal or highly selective	EI : 10^{-10} to 10^{-11} g; CI : as low as 2.5×10^{-14} g	Variable. Can be utterly selective	10^5	350	**** (Low res.) / ***** (High res.)
Hall Conductivity Detector	Selective for halogens, nitrogen, sulfur, or esters	Cl: 5×10^{-13} g s^{-1}; N: $2\text{-}4 \times 10^{-12}$ g s^{-1}; S: $2\text{-}4 \times 10^{-12}$ g s^{-1}. Not sensitive enough for routine use, but useful for confirmation of high positive results.	Cl/C $>10^6$:1; N/C $>10^6$:1; S/C $>5\times10^4$:1; NO/N $>10^2$:1	Cl: 10^6; N: 10^4; S: 10^4	400	**
Photoionisation Detector	Universal	2×10^{-13} g s^{-1}	Depends on ionisation energy	$>10^7$	350	**
Thermal Energy Analyser	Nitrosamines	10^{-10} g (dimethyl nitrosamine)	N–NO/C ∞; N–NO/NO 1-4:1	10^6	Above GC column limit	***

S/N = signal-to-noise; EI = electron ionisation mode; CI = chemical ionisation mode; res. = resolution; NO = nitrous oxides; N–NO = nitrosamines
* – low
***** – high

sulfur mode flame photometric detection would be the natural and obvious choice for any sulfur-containing compound.

Of the element-specific detectors, the ECD is the most sensitive, although not as selective as some of the others. It is extremely prone to interferents, both from the sample matrix itself, and from solvents, reagents, and apparatus used in the extraction. Despite this, it is the most widely used detector in low-level trace analysis.

The flame ionisation detector, although popular in other analytical fields, is rarely used for trace analysis due to its lack of selectivity and lower sensitivity.

The current trend is towards the introduction of GC–MS as a universal detector. Benchtop instruments with electron ionisation lack the sensitivity of some of the other techniques, but are improving. They are also ideal for the confirmation of results where the determinand has been found at levels exceeding the normal detection limits.

6.2.3 Gas Chromatographic Injection

The introduction of the sample into the gas chromatograph is the stage where there is most scope for problems to occur. This is particularly true of capillary column chromatography, where there are many parameters to optimise. If the split injection mode is used, the injection procedure is even more problematical. Table 6.2.5 summarises some of the potential pitfalls of capillary GC injection in both split and splitless modes.

Regular maintenance is essential for the correct operation of GC injection, both for capillary and packed-column instruments. If not performed, it is likely that spurious peaks will appear in chromatograms from contamination on septa, injection liners, glass wool, and other components of the injection mechanism.

Table 6.2.5 *Critical aspects of capillary GC injections and precautions to ensure valid results*

Split injection	Splitless injection
Difficult to operate quantitatively[5,7]	Risk of flooding injection port.[7]
Use of liner packing material aids split reproducibility[6]	The following recommendations have been made by Yang *et al.*[9]:
The 'stop flow' injection technique prevents aerosol splitting[8]	Injection volumes should be between 0.1 and 10 µl.
	A slow injection speed (roughly 1 µl s^{-1}) gives better chromatography
	The sampling time should be greater than 20 s
	The injection port should not be purged before 40 s have elapsed
	The solvent boiling point must be at least 25°C below that of the solute
	The initial oven temperature should be 15–30°C below the solvent boiling point

When using manual injections, the injection septum should be changed daily to prevent both contamination from particles of rubber passing onto the column, and gas leaks which cause drifts in retention times. When changing packed-column septa, it must be ensured that the carrier gas is switched off and that the head pressure has been allowed to fall. If this precaution is not observed, the pressure difference as the septum is unscrewed will cause the packing material to blow backwards. With all columns, but particularly capillary, septa should not be changed whilst the oven is hot, otherwise oxygen may diffuse into the system.

For auto-injection longer intervals can be left between septum changes, as the needle always punctures in the same place, but they should be changed after about 50 injections. It must also be ensured that the auto-injector needle has the correct type of bevelled end, as some of the wider needles designed for puncturing the septa of injection vials will shred the thicker injection septum.

The glass lining of capillary injection ports is prone to contamination, and should also be changed or cleaned regularly. Cleaning can be by the action of high purity solvents, or if the facilities are available, by heating the liner to a high temperature and volatilising any adsorbed material. Fats and other high boiling matrix components in particular will adsorb onto the liner, and the injection port should be examined for any traces of these whenever sample extracts containing such compounds have been injected. Once the liner has been cleaned it may need to be resilanised. If silanised glass wool packing is used in the glass liner, this should also be regularly changed.

The choice of temperature at which to run a capillary column injection port is often difficult. There are many analytes that thermally degrade at temperatures of 200–300 °C, and so require a lower injection temperature. Against this, at temperatures below about 200 °C there is an increasing risk of condensation of high boiling impurities and interferents. Often, a compromise must be reached. One solution is to use an injector with a programmable temperature control, and ramp the temperature once the analyte has left the injection port; this volatilises and purges any condensed material. The other option is to use on-column injection. However, even here the temperature must be chosen with care. The initial temperature should not exceed the boiling point of the solvent.

Hints and Tips for Injection

Manual injections onto packed columns should be as rapid as possible to minimise volatilisation of liquid at the end of the needle as it passes through the hot injection area. A technique that reduces this risk and which is often employed is to draw a plug of air into the syringe after it has been filled in order to provide a buffer between the heated area and the sample extract. The 'solvent flush' technique may also be used to improve quantitative results from manual injections onto packed columns: solvent is drawn into the syringe, followed by a small air plug, then the sample solution, then another air plug, and finally a second portion of solvent.

There is some contention about the merits of fast injection into splitless capillary liners. Yang *et al.*, have reported that a slow injection, under 1 μl s^{-1}, gives better chromatography: automation is obviously necessary to control such injections in a reproducible manner.[9] However, slow injections can lead to peak splitting due to solvent condensation. The optimum injection speed for a particular assay seems to be dependent upon individual factors such as the volatility of both the solvent and the solute, the temperature of the injection port, and the purging program. These should be tested empirically for each new analysis as part of the method development process. If on-column injection is being used, then a slower injection speed should be used.

One method of increasing the injection volume is to vent the solvent from the injection port. This requires a temperature programmable injector. The sample is injected with the port at a temperature slightly below the boiling point of the solvent with the injector valves open, and held at this temperature for about 3 min. The valves are then closed and the injection port heated rapidly, transferring the analyte onto the column. During this time the column is also held at a low temperature, so that the analyte forms a tight plug at the head of the column. The column temperature can then be ramped. If using this technique, it is essential to first compare chromatograms with those produced using conventional splitless injection to ensure that no analyte is volatilised and vented along with the solvent.

When using any type of splitless injection, it is important to allow the injection port time to clean between samples. This is achieved by opening the split valves once the run is underway; this ensures the liner is purged.

Autosamplers have greatly simplified the validation of GC injection (see Section 6.2.6), but they do cause some problems of their own. The possibility of solvent evaporation from crimp-topped vials should be guarded against, particularly on long, unattended runs. For most autosampler models, the vial caps should be crimped as tightly as possible. Once punctured, all further injections from an autosampler vial must be regarded as suspect. If replicate injections are required, and if the extract volume allows, they should be taken from separate vials.

Autosampler needles are vulnerable to bending, particularly when puncturing a new septum. Once the injector is assembled it is a good idea to puncture a new septum by hand before use.

6.2.4 Columns and Stationary Phases

All columns, whether capillary or packed, bought-in or home-made, will last provided that they are well cared for. If misused, resolution and repeatability will soon deteriorate.

One rule, which must be carefully observed, is that the recommended maximum operating temperature for the stationary phase is *never* exceeded. This is usually in the range 200–400°C, but can be lower for some specialised phases such as cyclodextrins. It is important to remember that, in addition to the oven temperature, one end of the column is also heated by the detector. Therefore, if it is necessary to heat the detector to high temperatures the column should first be removed. In many cases it is convenient to attach a length of deactivated tubing to each end of the column so that it, and not the stationary phase, is directly attached to the hot detector/injector.

Most stationary phases are susceptible to the action of moisture, oxygen, and other gas or solvent impurities. Even polyesters, which have good chemical stability, are hydrolysed by water at temperatures above 200°C. Methylsiloxanes, one of the most common classes of stationary phase, deteriorate in the presence of oxygen or solvent impurities which catalyse cross-linking. The only method of reducing these risks is to use extremely pure gas supplies, and to be careful which solvents are injected on to the column. Gases are usually scrubbed of oxygen, water, and hydrocarbons using chemical filters on the supply line. These are self-indicating, and it is important to change them when their usefulness has expired. Alternatively, gas generators are available for the production of pure nitrogen or hydrogen.

After prolonged use, columns invariably become contaminated, usually at the injector end, with high boiling and polar interferents from sample extracts. If not dealt with promptly these will cause carry-over into subsequent analyses.

For packed columns, the remedy is to repack either the whole column or just the contaminated injector end. The packing procedure should ensure that the stationary phase is as firmly packed as possible, and is unlikely to settle further. Procedures such as applying a vacuum to the detector end of the column or vibrating it at a high frequency are usually used (the vibration should be minimal, otherwise the support material will fracture, exposing 'active' silanol sites). Once packed to satisfaction, the head of the column is sealed with a plug of silanised glass wool. This should be handled as little as possible, as each tear or cut in the wool breaks the silanised surfaces.

Packing material should be kept in a desiccator, and protected from contamination or handling.

With capillary columns the procedure is even simpler. If contamination of the column is suspected, approximately 10 cm can be cut from the injector end. Special tools, supplied by most column manufacturers, should be used for this, as the cut must be sharp, clean, and perpendicular to the column length. It is a good idea to examine the cut using a microscope or magnifying glass to ensure that it is true. When reconnecting the column, consult the instrument manufacturer's handbook as to the length that should protrude into the injector (or detector). The correct length can be marked before reconnection (liquid paper is good for this) so that the mark is just on the visible side of the connecting nut. A new ferrule should be fitted, with the outside of the column carefully wiped to remove all traces of graphite, particularly if it is the end that is to be connected to the detector.

Cleaning of bonded-phase capillary columns with solvents is also possible. They can be removed from the GC and back-flushed with an organic solvent such as dichloromethane.

Capillary columns can be slightly protected by the insertion of a guard (pre-)column; usually coated with deactivated support material.

During storage, packed columns are usually capped with seals to exclude air. Capillaries must be sealed using a flame.

Hints and Tips with Columns and Stationary Phases

If a column is giving spurious peaks, *i.e.* it is contaminated with adsorbed material, it is seldom necessary to replace it or completely repack it. As the contamination is likely to be at the injector end, it is usually sufficient either to cut off a short length (capillary columns) or repack the top few cm (packed columns). The performance of non-polar packed columns may also be improved by the occasional injection of silanating reagents specifically prepared for regeneration purposes.

New columns must be conditioned before use. This entails heating the column overnight, at the normal operating temperature. The carrier gas must be flowing whilst this is done, but the column should not be connected to the detector due to the risk of contamination from eluting substances. Once conditioned, the column can be stored for a period of months before use.

Long narrow-bore columns made of metal often give better resolution than conventional packed columns, and metal capillary columns can usually be taken to a higher temperature. However, metal columns are liable to snap if cooled or heated too rapidly. The rate of cooling produced as an oven automatically resets to its starting temperature is sometimes enough to cause metal fatigue.

The retention of highly retained compounds can be reduced by the use of wall-coated open tubular capillary columns with thick film coatings, provided the solutes are also highly volatile.[10]

6.2.5 Gas Chromatographic Detectors

The component of the GC system that is most sensitive to contamination and interference is the detector. As with any other part of the system, they require regular maintenance and careful operation, in order to function to their full potential.

Detectors with a solid surface, such as ECD or Nitrogen Phosphorus Detectors (NPD) are susceptible to the condensation of high boiling compounds. To guard against this, it is important to keep a flow of gas passing through the detector at all times when the heaters are on. If the column is disconnected for any reason, an alternative gas supply (*e.g.* the make-up gas) should be left on. Air and moisture must be excluded from the hot detector. In cases where it is suspected that condensation has occurred the detector can be cleaned by leaving it at a high temperature (*e.g.* 400°C for ECD). The column should be removed beforehand to avoid overheating the packing material, and the detector sealed with a blanking nut to aid the flow of make-up gas. If necessary, the anode of most types of ECD can also be removed, and cleaned using solvents.

The flow of make-up gas (if any) is an important parameter to control with all types of detectors. Make-up gas is necessary with capillary columns in order to provide a reasonable working flow of gas for the detector, and is used occasionally for packed columns: mainly with ECD, as other, less sensitive, detectors can suffer from poor detection limits as the column eluent is diluted.

The physical configuration of spectrometric detectors, such as the Flame Photometric Detector (FPD), is very important. Capillary columns must be positioned so that the burning column exhaust is directly in the optical path of the detector. The manufacturer's handbook should state what length of column is required to protrude into the detector. The FPD fuel and oxidant gas ratios are also critical to the height of the flame and therefore the optimisation of the optical path. Experimental work is usually required to obtain the best ratios, with the gas flow rates subsequently carefully controlled.[4]

All element-specific detectors can suffer from interferents, which can either be misidentified as a positive result or mask positive results. Some possible sources of interferents for each detector are listed in Table 6.2.6. The method of analysis should be carefully chosen with these in mind, particularly in the use of solvents which may produce a response from the intended detector.

Table 6.2.6 *Common sources of interferents for element-specific GC detectors*

Detector type	Possible interferents
Electron Capture Detector (ECD)	Chlorinated solvents. If chlorinated solvents are used in the method, they must be completely removed by evaporation before injecting on the GC.
	Plastic or rubber materials. These can originate from tubing, vacuum seals, solid-phase extraction cartridges, gloves and many other sources. Phthalate plasticizers readily leach into fats or organic solvents.
	Multiple injections should not be made from auto-injector vials unless PTFE 'sandwich' septa are used.
	Glass, rather than plastic, stoppers should be used for all flasks containing organic solvents.
	Oxygenated compounds. The gas supply must be free from oxygen, and rigorous clean-up procedures are often required to remove such compounds from the sample extract.
	Any other halogenated compounds.
Nitrogen Phosphorus Detector (NPD)	Nitrogenated solvents such as acetonitrile, these are more difficult to remove by evaporation, so it is recommended that they are not used with NPD.
	Phosphate dishwasher detergents.
Flame Photometric Detector (FPD) (phosphorus mode)	Phosphate dishwasher detergents.
	Phosphate impurities in some solvents (even those marketed as high purity).
	Sulfur compounds, *e.g.* extracts from brassica vegetables cannot be examined by P mode FPD.
Flame Photometric Detector (FPD) (sulfur mode)	Rubber and plastics, if they come into contact with organic solvents.

The only common detector that does not respond linearly to analyte concentration is the sulfur mode FPD. This responds to the ionised S_2 molecule, and so the peak area is directly proportional to the square root of the analyte concentration (*i.e.* if the analyte concentration doubles, the peak area quadruples). Unfortunately, this is a simplification, as other sulfur species also give a signal. Therefore the response is not perfectly quadratic, particularly at low concentrations, where the signal from other sulfur species becomes significant.[11] For this type of detector it is essential that a calibration graph, covering the analyte concentration range, rather than a single-point calibration is used. The calibration graph must not be extrapolated to pass through the origin, as there are significant deviations from linearity at low concentrations. The phosphorus mode FPD is also non-linear in some situations when determining oxygen analogues of organophosphorus compounds. This is due to the degradation of the P=O bond. Degradation can be minimised by keeping metal transfer lines as short as possible.

Hints and Tips for Detectors

The sensitivity of GC detectors can usually be improved by optimising the operating variables of the instrument. This is particularly true of flame-based detectors, such as the FPD, and those dependent on a chemical reaction, such as the NPD. Adjustment is usually by trial and error. Computer-based optimisation packages are now available. These may be worth considering, especially if the instrument is already linked to a data processing unit.

The response of ECD to fluorinated compounds is poor.[12] If derivatisation is required, despite conventional practice, it is better to use a chlorinated acid anhydride than a fluorinated analogue.

Many benchtop GC–MS machines are supplied with the facility for chemical ionisation. This is particularly useful for the determination of large molecular species such as dioxins and some organophosphates. This increases sensitivity and selectivity and gives a large, distinctive molecular ion. To assist with the identifi- cation of complex isomeric compounds, for example dioxins or polynuclear aromatic hydrocarbons, mass profile monitoring can also be used.[13] This decreases the sensitivity, but gives a 10-fold increase in confidence of identification.

One danger with GC–MS is that, although interferents may not be apparent on the chromatogram, they can still affect the result. Compounds passing through the source reduce the incidence of ions or electrons, and therefore reduce the analyte signal. Many analysts believe that because GC–MS is so selective, crude and dirty extracts can safely be injected, with no clean up. This is often not true.

6.2.6 Associated Uncertainty

The uncertainty to be expected from a GC determination is one of the easiest stages to estimate in an analysis. It can be approximated as precision: expressed as the coefficient of variation (relative standard deviation) of replicate injections. This provides a total figure, inclusive of uncertainties in measuring small volumes of liquids, detector linearity, integration reproducibility, and all of the other factors that contribute to the final calculated result.

Some experimentally determined coefficient of variation (CV) values are given in Table 6.2.7. Obviously, they depend upon the analyte, concentration, detector type, and most of all (for manual injections) the experience of the analyst. Because of this inherent variation, the use of internal standards is strongly recommended, whenever possible.

Table 6.2.7 *Experimentally determined precision of GC injections*[a]

Injection type	Column type	Detector type	Number of replications	CV (%)
Manual	Packed	ECD	10	3.8
Manual	Packed	ECD	5	6.1
Manual	Packed	ECD	10	5.6
Automatic	Packed	ECD	9	1.5
Manual	Capillary	ECD	6	12
Manual	Capillary	NPD	6	10
Manual	Capillary	NPD	8	4.0

a Results obtained during a training program for inexperienced analysts at the Laboratory of the Government Chemist (without an internal standard).

It is immediately apparent that the use of an auto-injector considerably reduces the uncertainty associated with GC. The only point to remember is that precision measurements such as these do not consider the effect of solvent evaporation from vials, during long auto-injected sequences. Even allowing for this, the use of auto-injectors is strongly recommended where possible. The worst possible combination from an uncertainty point of view is manual injection into a capillary liner.

Another source of uncertainty, which precision measurements do not include, is that caused by differences in integration between standards and samples. Sample chromatograms are frequently 'noisy', which causes automatically set baselines to become unreliable. In these cases it is usually better to manually measure the peak heights, in spite of the uncertainty in the accuracy with which a ruler can be read. In any event, automatically set baselines should be clear and visible so that they can be checked by eye.

The solvent used for the final extract also contributes to the uncertainty. Glassware is usually calibrated gravimetrically using water. Wilkinson and McCaffery have measured the volumes of various solvents delivered by small volume syringes.[14] They found that, at volumes of 10 µl, the results from water cannot be extrapolated to more volatile solvents. Acetone (propanone), for example, gave a 15% deviation compared with 4% for water. Higher uncertainties therefore arise when the final extract is in a volatile solvent with a high coefficient of expansion.

6.2.7 References

1. M. Novotny, 'Contemporary Capillary Gas Chromatography', *Anal. Chem.*, 1978, **50**, 16A.
2. K. Grob, 'Band Broadening in Space and the Retention Gap in Capillary Gas Chromatography', *J. Chromatogr.*, 1982, **237**, 15.
3. L. Ghaoui, F.-S. Wang, H. Shanfield, and A. Zlatkis, 'Elimination of Peak Splitting and Distortion Associated with Liquid Sample On-Column Injection: Solvent Polarity Effects'. *J. High Resolut. Chromatogr.*, 1983, **6**, 497.
4. D. G. Westmoreland and G. R. Rhodes, 'Analytical Techniques for Trace Organic Compounds-II. Detectors for Gas Chromatography', *Pure Appl. Chem.*, 1989, **61**, 1147.
5. J. V. Hinshaw, 'GC Troubleshooting: Setting Up an Inlet Splitter', *LC.GC Int*, 1989, **2**, 24.
6. J. Volmut, E. Matisová, and P. T. Ha, 'Influence of Injector Liner Packing on the Analysis of Antiepileptic Drugs by Capillary Gas Chromatography', *J. High Resolut. Chromatogr.*, 1989, **12**, 760.
7. K. Grob, 'Injection Techniques in Capillary GC', *Anal. Chem.*, 1994, **66**, 1009A.
8. G. Liu and Z. Xin, 'The Glass Insert in Stop-Flow Split Injection', *Chromatographia*, 1990, **29**, 385.
9. F. J. Yang, A. C. Brown, and S. P. Cram, 'Splitless Sampling for Capillary Column Gas Chromatography', *J. Chromatogr.*, 1978, **158**, 91.
10. P. Sandra, I. Temmerman, and M. Verstappe, 'On the Efficiency of Thick Film Capillary Columns', *J. High Resolut. Chromatogr. Chromatogr. Commun.*, 1983, **6**, 501.
11. S. O. Farwell and C. J. Barinaga, 'Sulfur-Selective Detection with the FPD: Current Enigmas, Practical Usage, and Future Directions', *J. Chromatogr. Sci.*, 1986, **24**, 483.
12. S. M. Lee and P. L. Wylie, 'Comparison of the Atomic Emission Detector to Other Element-Selective Detectors for the Gas Chromatographic Analysis of Pesticide Residues', *J. Agric. Food Chem.*, 1991, **39**, 2192.
13. H. Y. Tong, D. E. Giblin, R. L. Lapp, S. J. Monson, and M. L. Gross, 'Mass Profile Monitoring in Trace Analysis by Gas Chromatography/Mass Spectrometry', *Anal. Chem.*, 1991, **63**, 1772.
14. I. J. Wilkinson and A. R. McCaffery, 'Volumetric Errors in the Delivery of Small Volumes of Liquids', *Lab. Pract.*, 1990, **39**, 67

General Reading

Large Volume Injections

15. J. Staniewski, H.-G. Janssen, C. A. Cramers, and J. A. Rijks, 'Programmed-Temperature Injector for Large-Volume Sample Introduction in Capillary Gas Chromatography and for Liquid Chromatography–Gas Chromatography Interfacing', *J. Microcol. Sep.*, 1992, **4**, 331.
16. J. V. Hinshaw, 'Very-Large-Volume Injection', *LC.GC Int.*, 1994, **7**, 560.
17. K. Grob, S. Brem, and D. Frohlich, 'Splitless Injection of up to Hundreds of Microliters of Liquid Samples in Capillary GC: Part 1, Concept', *J. High Resolut. Chromatogr.*, 1992, **15**, 659.

Cold Injections

18. C. Watanabe, H. Tomita, K. Sato, Y. Masada, and K. Hashimoto, 'Accuracy and Reproducibility in Splitless and Packed and Open Tubular Cool On-Column Injections', *J. High Resolut. Chromatogr.*, 1982, **5**, 630.

Autosamplers

19. J. V. Hinshaw, 'Autosamplers: Design, Operation and Troubleshooting', *LC.GC Int.*, 1991, **4**, 18.

6.3 High Performance Liquid Chromatography

6.3.1 Overview

High Performance Liquid Chromatography (HPLC) complements gas chromatography (GC) in the types of compounds that can be separated. In addition it has the advantage of shorter retention times and so more analyses per unit time.

Unfortunately, most HPLC detectors cannot rival the sensitivity of element-specific GC detectors. Standard UV detectors are approximately 10-fold less sensitive than NPD or FPD (depending upon the instrument type and analyte response) and anything up to a 100-fold less sensitive than ECD. Fluorescence detectors or the rarer chemiluminescence detectors have greater sensitivity but are limited to luminescent species (or derivatives).

The introduction of capillary and microbore columns, as in GC, provides a higher theoretical plate count, and therefore better resolution, than standard columns. This is often at the expense of the detection limit, as the reduced eluent flow restricts the amount of analyte that can pass through the detector at a given instant. However, capillary HPLC has other advantages: less mobile phase is used permitting the use of rare, toxic, or expensive solvents, and expensive stationary phases.

High performance liquid chromatography can be divided into two broad categories, normal phase and reversed phase. For normal phase chromatography a polar stationary phase (usually silica) is used to retain polar analytes. Reversed phase separations are based on the attractive forces between non-polar solutes and a non-polar functional group, which is bonded to a silica support. Most separations are now performed by reversed phase because of its wider field of application than normal phase. There are few compounds that are permanently retained on the column. Columns also tend to be more robust.

The separation of chiral enantiomers is increasing in importance, especially in the areas of pharmaceutical and agrochemical residue analysis where it is necessary to isolate and determine the biologically active enantiomer of isomeric compounds. This provides much more reliable information on the potential toxicity (or efficacy) of the determinand. Chiral HPLC is a well-established technique with many published applications. 'Pirkle'-type columns are much more robust than their GC cyclodextrin counterparts, although still very delicate by HPLC standards. They are available commercially.

6.3.2 Comparison of Techniques

Choice of Stationary Phase

The choice of stationary phase to use for a separation is governed by the chemical nature of the compounds to be resolved. Indications of likely retention behaviour can be gained from published results, previous work with similar compounds, or knowledge of chemical structure and functional groups; but the only real test is trial and error.

The main choice is between reversed phase and normal phase chromatography. This has implications for both the mobile and stationary phases and would be dependent upon the analyte and the possible interferents likely to be in the sample. The main differences between reversed and normal phase are summarised in Table 6.3.1.

Table 6.3.1 *Comparison of normal phase and reversed phase HPLC*

Reversed phase	*Normal phase*
80% of analyses	20% of analyses
Wide range of polar analytes	Only non-polar analytes
Uses viscous solvents	Uses non-viscous solvents
Higher back pressures	Lower back pressures
Highly reproducible	Lower reproducibility
Highly robust	Lower robustness

The more common stationary phases are listed in Table 6.3.2, along with typical applications and compatible solvents. The first choice for most applications is OctaDecaSilyl (ODS), because of its versatility and robustness. The presence of distinctive functional groups, such as amines or acids, in the analyte might suggest the use of a stationary phase with similar groups: most commonly a cyanopropyl (CN) or aminopropyl (NH_2) column. Silica columns are used mostly for the separation of extremely non-polar analytes, and are never used for basic compounds due to the risk of irreversible adsorption.

Choice of Column Dimensions

All stationary phases are available in columns of many shapes and sizes. The most common is the standard 25 cm column with a 0.46 cm internal diameter, which can be fitted with either externally or internally threaded nuts for connection to standard 1/16" bore pressure tubing. Table 6.3.3 shows some of the other choices available to the chromatographer. Cartridge columns, in particular, have gained in popularity over recent years: they are extremely convenient to use once a system is set up and running; columns can be interchanged in a matter of minutes. It is important to take care when installing columns to prevent slow leaks and dead-volumes arising; cartridge columns are particularly prone to such problems.

Table 6.3.2 LC stationary phases

Stationary phase	Description	Typical applications	Typical mobile phases	Incompatible mobile phases
Silica	Silica	General purpose normal phase. Non-polar compounds. Lower back-pressures than reversed phase columns	Hexane, iso-octane, ethers, alcohols	Water, bases
OctaDecaSilyl (ODS)	C_{18} hydrocarbon chain bonded to a silica support	Compounds of a variety of polarities. General purpose – reversed phase	Water, methanol, acetonitrile, tetrahydrofuran, aqueous buffers (pH > 2)	Hydrocarbons, pH >10
ODS-2	As ODS but free sites on silica either end capped with methyl groups or cross-linked	Gives narrower spread of adsorption sites than ODS, therefore less diffuse peaks. Higher back-pressure than ODS	As ODS	As ODS
ODS-3	As ODS-2, but greater degree of cross-linking	Gives less diffuse peaks than ODS-2. Higher back-pressure than ODS-2.	As ODS	As ODS
C_8	C_8 hydrocarbon chain bonded to a silica support	Less polar compounds than for ODS. Useful for analytes too strongly retained on ODS	As ODS	As ODS
Cyanopropyl (CN)	Cyanopropyl group bonded to a silica support	Reversed (RP) or normal phase (NP) use. Fatty acids, general utility	RP – water, alcohol NP – hexane, iso-octane, alcohol, ether	pH >10, pH < 2
Aminopropyl (NH₂)	Aminopropyl group bonded to a silica support	Usually normal phase. Polar compounds (NP). Carbohydrates (RP, weak retention)	RP – Water, alcohol NP – Hexane, iso-octane, alcohol, ether	pH >10, pH < 2
Pirkle	Phenylglycine enantiomer bonded to a silica support. Can be covalently or ionically bonded	Chiral separations. Ionic generally more effective than covalent but less robust	Hexane, iso-octane, small amounts of modifier	Solvents more polar than 20% propanol in hexane

Another alternative is to use capillary or microbore columns. For economic and environmental reasons many laboratories are replacing traditional columns with small bore versions. As yet they have found few applications in trace analysis due to the loss in sensitivity which can result from low flow rates. They are only suitable for coupling with the most sensitive of detection systems, invariably luminescence-based detectors.

Table 6.3.3 *Comparison of LC columns*

Column type	Advantages	Disadvantages
Traditional 25 cm × 0.46 cm packed column	Compatible with most fittings. Few back pressure problems. Better plate count than shorter columns	High solvent throughput. Can be problems changing columns ensuring the correct length of tubing protrudes past a previously used ferrule using adaptors and marrying nuts
Cartridge columns	Very easy to change columns once cartridge connected	Cartridge often prone to leaks and dead-volume problems
10 cm columns of either type	Shorter run times. Cheaper	Need smaller particles to obtain the same resolution. This can lead to back-pressure problems
Microbore and capillary columns	Increased plate count. Economical on solvents. Can be coupled to mass-selective or flame-based detectors	Sensitivity seriously impaired. Longer analysis times. Susceptible to small temperature variations. Unsuitable for buffered solvents

HPLC Pumps

An important consideration when optimising an HPLC system is the choice of pump. A pump that produces fluctuations in pressure as the pistons drive will lead to a noisy baseline, with a corresponding deterioration in detection limit. As with most analytical hardware, price is a rough indication of performance, so the choice is governed both by specification and budget.

The standard choice is a reciprocating piston pump. The pressure drop on the filling stroke of each piston is compensated by the upstroke of the other. A similar performance can be obtained from some top-range single piston pumps, as the filling stroke is almost instantaneous. A variety of pressure damping devices are available. A third option, which ensures smooth pressure, but is not often used due to its limited capacity, is the large syringe pump which empties itself slowly over one continuous stroke. These are available in sizes of up to 1 l of mobile phase, but once the reservoir is exhausted the analysis must be stopped and the pump refilled.

If gradient elution is required, then the pump must also be equipped with multiple reservoirs and a mixing valve.

HPLC Injection Techniques

The standard method of introducing a sample into a flowing solvent stream, for both manual and most autosampler injections, is to use a fixed volume loop connected to a six port injection valve.[1] Filling the loop (*i.e.* flushing through approximately three times the loop volume) ensures that injection volumes are reproducible. Loop calibrations are not particularly accurate. In most cases reproducibility is far more important than nominal accuracy, as analyses entail comparing a sample to a standard of known concentration. In the rare cases when an absolute measurement is required, uncertainty can be reduced by partially filling the loop with a volume measured from a calibrated syringe. The accuracy of this technique depends on the accuracy and precision of the syringe.

Most auto-injectors merely involve mechanisation of the manual injection steps. Some have additional features such as variable syringe speeds to deal with viscous liquids, or incorporating the needle as part of the injection loop so that it is continually flushed with mobile phase. In contrast with GC, because manual injection in HPLC is so reproducible, introducing automation does not produce a marked improvement in precision. Similarly, internal standards are less critical for LC than for GC, as there is not the same variation in injection volumes. The use of HPLC auto-injectors does not markedly reduce the uncertainty of a determination, although they obviously retain the advantages of economy and increased sample throughput.

An alternative to the six port injection valve is to use septum and syringe injections, analogous to GC injection. This older technique, which is now rarely used, gives less band broadening than injection valves, as the sample is introduced in a tight plug directly onto the head of the column. Without stopping the flow of the mobile phase, there is the risk of septum deterioration and leakage of packing material. Stopped flow injection requires skilled operation and the precision is largely dependent upon the experience of the analyst, but it does give better efficiency than valve injection.[2]

HPLC Detectors

Some of the more common HPLC detectors are compared in Table 6.3.4. The sensitivity of an HPLC assay is primarily dependent upon the detection system. The workhorse detector for HPLC is the UV/VIS absorbance detector. This has nearly universal applicability in the detection of organic solutes, as there are very few species that do not absorb radiation in the wavelength range 190–650 nm. For trace analysis, however, its use is restricted to either extremely strongly absorbing compounds or those that absorb at the more selective, higher wavelengths. This is due to inherent background noise from both the mobile phase and co-eluting interferents. For aromatic compounds, at a standard wavelength such as 254 nm (the absorbance wavelength of π-bond electrons) the sensitivity of HPLC with UV detection is generally lower than for GC, as indicated in Section 6.3.1.

If applicable, the detector of choice is the fluorescence detector. This probably accounts for half of trace analytical HPLC determinations. Background noise is all but eliminated by detecting the light emitted at 90° to the incident path, with an

Table 6.3.4 *HPLC detectors for organic solutes*

Detector	Uses	Sensitivity	Selectivity	Linear range	Relative complexity	Cost
UV/VIS	Compounds that absorb in range 190–650 nm. Includes most organic molecules, particularly those with delocalised π-electrons	***	* at low λ **** at high λ	Obeys Beer's law at low concentrations	*	Very cheap
Diode array	As UV/VIS. When spectral identification is required	*	*****	As UV	***	Moderate
Fluorescence	Fluorescent compounds. Most have large, planar, delocalized π-electron structures	*****	****		**	Moderate
Chemiluminescence	Chemiluminescent reactants. Activators or inhibitors of chemiluminescent reactions	*****	*****		****	Most home-made
Bioluminescence	Bioluminescent reactants	*****	*****		****	Most home-made
Electrochemical	Strong acids or bases	***	***		**	Cheap
LC–MS	Universal	*	*****		*****	Expensive

* – low
***** – high
λ = wavelength of detector.

enormous improvement in the detection limit. This is obviously restricted to molecules that fluoresce, e.g. molecules containing multiple aromatic rings, with large planar clouds of delocalised electrons. The use of other types of luminescence detectors, such as chemiluminescence or bioluminescence, is so rare as to be a novelty. If the analyst was fortunate enough to be determining a compound amenable to such techniques they would be the obvious choice. The background noise is even lower than with fluorescence, with a correspondingly improved limit of detection, and both types have unparalleled selectivity; in the case of bioluminescence, it is almost completely specific.[3]

The scope of application of all luminescence detectors can be improved enormously by derivatisation.[4] In most cases this involves reaction of the analyte with a luminescent reagent to form a luminescent product. The technique is quite common for fluorescence detection, and standard fluorescent reagents are available to react with specific functional groups.

Derivatisation may be either pre- or post-column. Post-column derivatisation is the most widely used because, as analytes are separated in their original form, published methods can still be used. So long as the reaction conditions are reproducible and the products are stable, the reaction need not necessarily go to completion. Equipment is available for either pre- or post-column derivatisation, or can be adapted and improvised in-house using three-way union connections.

There are two potentially serious disadvantages of luminescence derivatisations: the need for an extra pump, which increases the risk of pressure fluctuations, and the considerable increase in dead-volume produced by the reaction chamber. Particularly with adapted equipment it can be extremely difficult to synchronise the two pumps to avoid sharp changes in pressure. Controlling the reaction chamber temperature can also pose a problem with home-made equipment; with some reactions even mild fluctuations can severely reduce precision. For these reasons derivatisation is only usually used as a last resort, when no other detection system can give the required detection limit.

The refractive index detector is unsuitable for trace analysis, except when working at concentrations at the top end of the trace analytical range (10–100 mg kg^{-1}). It is not sufficiently sensitive to be useful at lower concentrations. It is the only truly universally applicable detector of all those mentioned here, but is not treated in detail here due to its limited applications.

Electrochemical detection is used for a few specialist applications. It is extremely sensitive for aromatic amines and phenols, more so than UV detection, but the degree of baseline noise is very flow-dependent.[5] Fluctuations in pump pressure reduces detection limits considerably.

The ideal detector for HPLC will identify as well as quantify eluting peaks. There are two alternative approaches. The more established method is diode-array detection, which gives identification by spectral matching. This is not absolute, as many compounds, particularly those of the same class in a multi-analyte method, have very similar spectra. It has to be remembered that peaks will shift slightly depending on the mobile phase used. Identifications can usually be made with about 99% certainty. The newer technique which is rapidly gaining in popularity is LC–MS. Initially this had many technical problems, mainly connected with the column–MS transfer mechanism, but these have mostly been resolved and there are now laboratories that use the technique as a matter of routine. Most interfaces will not produce library-searchable spectra. At levels of analyte concentrations below mg kg^{-1} neither technique has the sensitivity for most routine screening applications. They are extremely useful, however, for analyses at higher analyte concentrations and for positive confirmation of analytes provisionally identified using other techniques.

Choice of Detection Wavelength

The most universal detection method for the LC analysis of organic analytes is UV absorption. As such, the first question which most analysts ask is: 'What wavelength shall I use?'. The choice of wavelength is critical. The limit of detection is a trade-off between the maximum absorbance of the analyte, and minimising the interference from other compounds present and the mobile phase. The absorbance of the majority of organic solutes only begins to increase significantly at wavelengths below about 230 nm. It is at just such wavelengths that background noise also increases. The overall detection limit begins to worsen below about 210 nm.

Examination of the solute spectrum will often show another region where, although absorption is not as strong, background noise will be greatly reduced. Spectra can be obtained from reference books or by measuring the absorption spectrum. Aromatic rings, for example, show strong absorption near 254 nm, and this is an extremely popular detection wavelength for compounds containing such groups. Generally, the higher the wavelength the less the interference. At high wavelengths, absorption need not be very strong to give better detection levels than detecting at 210–230 nm, where the signal is greater but so is the noise level.

6.3.3 Solvents and Mobile Phases

Liquid chromatographic separation is based on the equilibrium between the solute in the mobile and stationary phases. The choice of mobile phase is therefore as critical to the analysis as the choice of column. When developing methods, it is usually easier to pick a column, which may not be ideal, but is likely to at least be suitable for the intended separation, and then experiment with different eluents.

Choice of Mobile Phase

The choice of mobile phase is restricted by the type of stationary phase. The most fundamental distinction is between reversed phase and normal phase systems. As can be seen from Table 6.3.2, for normal phase systems generally hexane- or iso-octane-based eluents are used, whilst those for reversed phase chromatography are invariably water-based. It is important that, when a pump is being transferred from normal phase use to reversed phase use (or *vice versa*), it should always be washed out thoroughly with a solvent miscible with both (*e.g.* isopropyl alcohol) before the changeover.

Table 6.3.5 Solvents for HPLC mobile phases

Solvent	Polarity index (alumina index)	Proton acceptance x_e	Proton donation x_d	Dipole–dipole x_n	UV cut-off (nm)	Boiling point (°C)	Refractive index	Toxicity
Normal phase								
n-Hexane	0.1	–	–	–	210	69	1.378	Chronic neurotoxic
Iso-octane (2,2,4-trimethylpentane)	0.1	–	–	–	205	98	1.391	Low
Ethyl acetate	4.4	0.34	0.23	0.43	260	77	1.370	Low
Diethyl ether (ethoxyethane)	2.8	0.53	0.13	0.34	218	35	1.353	Low
Methyl t-butyl ether (2-methoxy-2-methylpropane)	2.1	0.44	0.18	0.38	210	55	1.369	Low
Dichloromethane	3.1	0.29	0.18	0.53	245	41	1.424	Chronic carcinogen
Chloroform	4.1	0.25	0.41	0.33	245	61	1.443	Chronic carcinogen
Isopropyl alcohol (propan-2-ol)	3.9	0.55	0.19	0.27	205	82	1.377	Low
1,4-Dioxan	4.8	0.36	0.24	0.40	220	101	1.422	Teratogen
Reversed phase								
Water	10.2	–	–	–	200	100	1.333	None
Methanol	5.1	0.48	0.22	0.31	210	65	1.329	Mildly toxic[a]
Acetonitrile	5.8	0.31	0.27	0.47	210	82	1.344	Toxic[a]
Tetrahydrofuran	4.0	0.38	0.20	0.42	280	65	1.407	Toxic[a]
Isopropyl alcohol (propan-2-ol)	3.9	0.55	0.19	0.27	205	82	1.377	Mildly toxic[a]
1,4-Dioxan	4.8	0.36	0.24	0.40	220	101	1.422	Toxic[a],Teratogen, chronic carcinogen
Dimethyl formamide	6.4	0.39	0.21	0.40	310	155	1.428	Teratogen, Chronic carcinogen

[a] Refers to toxicity by inhalation

A selection of some of the more commonly used solvents for HPLC is given in Table 6.3.5. If water is regarded as the base mobile phase for reversed phase work and hexane (or iso-octane) as the base for normal phase systems, then the effect of the addition of modifying solvents can be roughly predicted from their position in the eluotropic series, and their specific intermolecular interactions. These can be expressed in terms of the molecular interactions, *i.e.* proton donation (x_d), proton acceptance (x_e) and dipole–dipole interaction (x_n).[6,7]

The eluotropic series (alumina index, Table 6.3.5) gives an indication of the ability of the solvent to elute a compound from the column. Generally the lower polarity solvents such as hexane have very low numbers with the higher polarity solvents such as water having very high numbers. For normal phase systems, the higher the series number the more powerful the solvent. Therefore, chloroform (trichloromethane) is a 'faster' solvent than diethyl ether (ethoxyethane) for normal phase separations. For reversed phase systems the lower the series number the more powerful the eluting solvent. Therefore, for reversed phase systems, acetonitrile is 'faster' than methanol. It is unlikely that the base solvent alone will be sufficient to elute the required solutes in either normal or reversed phase chromatography. It will be necessary to add at least one modifying solvent. Around 1 to 10% is usual for normal phase, whilst proportions of up to 80% and above can be required for reversed phase.

The solvent selectivity triangle (Figure 6.3.1) provides a representation of the molecular interactions of each solvent.[6] Each circle on the plot embraces a group of solvents with similar selectivity characteristics and generally the same

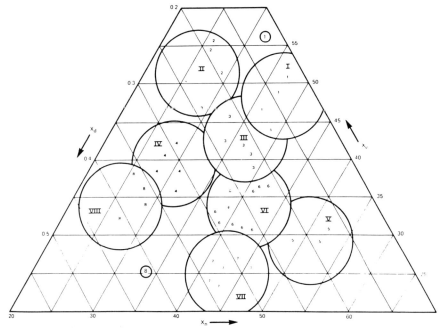

Figure 6.3.1 *Selectivity grouping of solvents*
(Reproduced from *J. Chromatogr. Sci.,* 1978, **16,** 223 by permission of Preston Publications, A Division of Preston Industries, Inc.)

functionality; this can be seen from Table 6.3.6. Solvents in the same region will tend to give similar resolution between compounds. Thus, if a water/dimethyl formamide mix failed to resolve two compounds, it is unlikely that water/methoxyethanol would fare any better. It is more advantageous to experiment with solvents with as differing interaction properties as practicable. Water/methanol and water/acetonitrile are the preferred choices for reversed phase (tetrahydrofuran is also used, though infrequently due to its high toxicity) with iso-octane/ether or iso-octane/isopropanol for normal phase (iso-octane with a chlorinated solvent is a further choice, but these are being phased out under the terms of the Montreal Protocol).

Table 6.3.6 *Classification of solvent selectivity in Figure 6.3.1*

Group	Solvents
I	Aliphatic ether, tetramethylguanidine, hexamethylphosphoramide, (trialkylamines)
II	Aliphatic alcohols
III	Pyridine derivatives, tetrahydrofuran, amides (except formamide), glycol ethers, sulfoxides
IV	Glycols, benzyl alcohol, acetic acid, formamide
V	Dichloromethane, dichloroethane
VI	(a) Tricresyl phosphate, aliphatic ketones and esters, polyethers, dioxane
	(b) Sulfones, nitriles, propylene carbonate
VII	Aromatic hydrocarbons, halogen-substituted aromatic hydrocarbons, nitro compounds, aromatic ethers
VIII	Fluoroalkanols, *m*-cresol, water, (chloroform)

Reproduced from *J. Chromatogr. Sci.*, 1978, **16**, 223 by permission of Preston Publications, A Division of Preston Industries, Inc.

The process of choosing a suitable solvent mix is usually one of trial and error. Automated software is available for attempting to model resolution for any combination of three or four solvents, but such systems have so far met with limited success. This is mainly due to the complexity of the equations that describe LC retention, and the unforeseen effect of other minor variables (*e.g.* column temperature). They can, however, be useful to give an idea of the type of eluents to try for complex or difficult separations.

Once a mobile phase has been found that resolves the compounds of interest, it is a fairly simple matter to adjust them to give a reasonable run-time. Adjusting the percentage of base solvent has little effect on the selectivity. For both normal and reversed-phase the lower the percentage of base solvent the faster the elution. Final adjustments to the retention time can then be made by varying the flow rate.

Analogous to ramping the temperature in GC, gradient elution is used in HPLC. Gradient elution means the mobile phase is altered during the separation. This is used when solutes of differing retention times are present and good resolution is required for all solutes. It is not used as a matter of course due to the reduction in reproducibility and the introduction of baseline variations as the mobile phase changes, but it is certainly an option worth considering in some cases.

If a positive result is found, it is standard practice to confirm the identity of the peak by using a different chromatographic system. This can be GC or LC. If LC is the choice, it is usually more convenient to use a different mobile phase with the same stationary phase, rather than trying to change the column. The same rationale applies to the choice of solvents for confirmatory systems as to method development. If it is required to confirm the identity of a peak, using the same column, then a mobile phase of *different* hydrogen bonding or dipole interactions from the first must be chosen.

Buffers and pH Modifiers

There are occasions when reversed phase systems produce tailing peaks whatever solvent modifier is used. This is generally because of an equilibrium between protonated and unprotonated forms of the compound. This will be dependent upon the pH of the mobile phase. For example, amines, amides, carbamates, and similar basic groups will be protonated in acidic media. Although the equilibrium between the protonated (polar) and unprotonated (non-polar) forms may still favour the unprotonated, there is still sufficient of the protonated form to cause peak tailing as it is retained on the column. The converse is true of weak acids, *e.g.* phenols and organic acids in alkaline media.

The remedy for this problem is to control the pH of the mobile phase. This can be by the use of a buffer or the addition of a small quantity of a modifier, such as ammonia, to ensure that there is either overall alkalinity or acidity. If adopting this latter approach, the risk of the modifier reacting with the analyte (*e.g.* ammonia replacing one of the N–R groups of an amide) must be remembered. Buffers must be washed out of the equipment thoroughly after use to prevent damage from crystallised salts.

Preparation of Mobile Phases

Only solvents of the highest purity should be used as mobile phases for trace analysis. Impurities give rise to noisy baselines and a corresponding worsening of detection limits. This is particularly the case when using UV detection at low wavelengths. Demineralised water should be regarded as a minimum requirement, with further purification used if available. Solvents should be filtered through small pore meshes, and degassed before use to remove air bubbles by either purging with helium or ultrasonicating. Some instruments are available with on-line degassing of the mobile phase reservoirs.

Instrumentation is now available that recycles waste mobile phase when there is no solute signal on the detector. Although attractive economically, such systems are not recommended for trace analysis. Mobile phase impurities become more concentrated with each pass through the system.

In separations dependent upon solute equilibrium between a stationary and mobile phase, the mobile phase itself is just as critical as the column. Generally, faults with the mobile phase are much easier to correct, if recognised, than those with columns. This is why the mobile phase tends to be the first part of the system to be checked if there are any problems.

The proportions of each component of a mobile phase are very important. The degree of accuracy to which these must be measured is dependent upon their effect on the properties of the total, which is usually inversely proportional to their volume. For example, a 60+40 mix of methanol/water can quite reasonably be measured with measuring cylinder accuracy, as an acceptable range of proportions from 58+42 to 62+38 would produce no significant changes in the chromatogram. However, if the mobile phase was to be modified to contain 0.6% ammonia, the volume added is absolutely critical. A concentration of 0.5% may give rise to tailing of basic peaks, whilst reactions with the solute could occur at over 0.7%. In this instance, accuracy of measurement less than that given by a Grade A pipette is unacceptable. Such critical amounts of small percentage modifiers are much more common in normal phase chromatography than reversed phase.

The most common problems with mobile phases are in the preparation. To avoid spikes in the chromatogram, as a result of air in the detector cell, it is necessary to thoroughly degas all solvents either before use or, if the equipment is available, continually during the analysis. This requirement can be incompatible with the precise measurement of mobile phase modifiers needed. Most modifiers, particularly for normal phase, are extremely volatile and will be lost during degassing. Under no circumstances should such phases be degassed continuously. Preferably, degassing should be done *before* the modifier is added.

Phases with small amounts of volatile modifiers require careful handling in other respects. If the LC pump has mixing valves, these should not be relied upon to mix the components. Apart from being insufficiently accurate, the valves will be warm enough to give significant losses by volatilisation; this may even occur if the solvents have been premixed. Unfortunately, the only remedy for this is to change the pump. Particularly with unattended, automated analyses, the utmost effort has to be made to avoid evaporation, which can involve anything from covering the reservoir with Parafilm to controlling the laboratory temperature.

To achieve the detection limits required for trace analysis, it is vital to minimise background noise. A major contribution to this is the mobile phase.[8] High purity solvents are a first requirement, but at low wavelengths the absorbance of the mobile phase must also be considered. The UV cut-off of some of the more common eluents is given in Table 6.3.5. It is unfortunate that, for many compounds, an adequate response is only achieved by using wavelengths below 220 nm, where background absorption starts to become significant. Eventually a limit is reached at which increases in sensitivity resulting from stronger solute absorbance are negated by the increase in background noise. Of the three common reversed phase solvents (water, acetonitrile, and methanol), acetonitrile has the highest UV cut-off, due mainly to solvated impurities. Ultra-pure Spectroscopic Grade acetonitrile is available (though expensive), and using this optimum wavelengths of around 200 nm can be achieved in reversed phase chromatography.

6.3.4 Practical Advice

Care of Columns

Unless a particularly rare or specialised stationary phase is required analysts are now seldom required to pack or re-top their own columns. It can usually be safely assumed that identical columns from the same manufacturer will give comparable chromatography, although there have been reported instances of suppliers introducing end-capping or other production alterations without prior warning. This assumption does not hold for equivalent phases manufactured by different companies. The size, shape, porosity, and surface coverage of the packing material particles have a marked effect on retention, and each new column must be fully validated for its intended purpose.

Most HPLC columns, particularly reversed phase columns, are fairly robust. However, there are a few simple precautions to be taken, which will extend the useful life of the column and prevent premature degradation.

All columns are received from the manufacturers stored in a suitable solvent and with the ends sealed with blanking nuts. This is how they should be kept. Leaving columns lying around with the ends open to the atmosphere only tends to shorten their lifetime, as does leaving them full of acids, bases, other corrosive solvents or buffers. Where such eluents have been used, they should be washed off with at least three column volumes of a water/methanol mixture as soon as the analysis is complete. Apart from protecting the stationary phase, buffer salts can crystallise within the pump or pipework, and once blocked or seized up can be difficult to clear.

Any type of stationary phase can be dissolved or denatured by using the wrong solvent. Documentation supplied with manufactured columns usually warn about solvents that should be avoided. This information should be kept for referral, but a rough guide is given in Table 6.3.2. Obviously any silica-based phase, which covers the great majority of columns, will dissolve in alkaline solution. The pH of the mobile phase should therefore not exceed 10. The other major risk is from inadvertently passing aqueous solvents through normal phase columns.

The other most frequent method of ruining columns is to allow the mobile phase reservoir to run dry, thereby pumping air through the system. With the increased use of auto-injectors and unattended analyses, such unfortunate accidents will happen to most analysts at least once. The only advice that can be offered is to take the precaution of calculating, from the run-time and the flow rate, the volume of solvent that will be required for a sequence of injections.

If a column has been adversely affected by allowing it to run dry, using an unsuitable solvent, or merely giving it such prolonged or intense use that the chromatography has deteriorated, it is always worthwhile attempting to regenerate it. For contaminated columns, regeneration could be as simple as reversing the direction of flow to back-flush the column head. If this does not suffice, most reversed phase columns can be cleaned by purging with isopropyl alcohol (propan-2-ol) for 30 min or so. At its most complex, regeneration can entail the sequential flushing with a series of solvents, one of which is designed to replenish or clean the active part of the stationary phase. Such procedures are specific to the stationary phase used. For example, a sequence of iso-octane, isopropyl alcohol, acetonitrile, dilute acetic acid, acetonitrile, isopropyl alcohol, and finally back to iso-octane can work wonders for the theoretical plate count of tired cyanopropyl columns.

As well as denaturing or overloading the stationary phase, the physical structure of the packing can also be damaged. All columns have a maximum recommended operating pressure. If this is exceeded the packing material will physically break down, giving rise to a small void at the top. This will result in a characteristic peak broadening. Some laboratories retain the equipment to pack columns, making it fairly easy to top up the stationary phase, but if this is not the case the column may have to be written off. It is tempting to reverse the direction of flow, as there may be better chromatography with the void at the base of the column, but results will soon deteriorate again as the packing resettles. It is more sensible not to exceed the pressure limitations in the first place. Most pumps are fitted with a pressure monitor, and this should be adjusted so that the system shuts down if the back pressure approaches the safe limit for the column (usually in the region of 200–400 bar).

Care and Maintenance of High Pressure Pumps

The pump is the one item of chromatographic hardware that most analysts are likely to take for granted. Operation is simple, there is only one parameter (the flow rate) to set, and maintenance requirements are fairly minimal. However, these requirements do have to be considered, or results may be adversely affected.

The most frequent problems with HPLC pumps are caused by air bubbles in the mechanism. The risk of these can be minimised by thoroughly degassing the mobile phase, see Section 6.3.3. Signs to look out for are a lower than expected back pressure, no solvent dripping from the detector waste pipe, and visible bubbles in the pump inlet tube which make no progress along the pipe with each piston stroke. If the pump does become disabled in this manner it must be purged. Most models have a column bypass valve, which means that high flow rates can be forced through the pump without any problems of back pressure. It is a good idea to do this each time the mobile phase reservoir is changed, as this is when there is the greatest risk of air bubbles forming. If this does not clear the problem, there is no alternative but to undo each connection on the pump, starting from the piston inlet and working systematically to the column head, and draw mobile phase through each individual component using a hand-held syringe.

Another routine precaution to avoid air bubbles is to store the pump inlet filter in a reservoir of solvent when not in use, so that it is not allowed to dry out. Buffers, salts, and corrosive solvents must be washed from the pump after use, as salt crystallisation can wreck a piston.

Even when a pump is working properly, it is often good practice to check regularly the actual flow rate. This is done by measuring the volume of solvent discharged over a period of time. There can sometimes be a significant discrepancy between the measured and the nominal flow rate, particularly with cheaper pumps working at low flow rates. It is a good idea to check the flow of any new pump before applying it to a routine method.

As described previously, the pressure cut-out should be set at below the maximum operating pressure of the column. This normally does not exceed 400 bar but is frequently lower.

The only other form of occasional maintenance which must be remembered is oiling of the pump heads and changing the piston seals. This should be done in accordance with manufacturers' instructions. It is only required very occasionally.

Detector Configuration and Maintenance

Like HPLC pumps, most detectors require little maintenance. If problems do arise, they usually require a service engineer's visit. However, there are some considerations, particularly on how the equipment is set up, that should not be neglected.

Most systems have a choice of two outputs, 1 V and 10 mV. The former is intended for connection to data systems and computers via analogue-to-digital converters, whilst the 10 mV lead is for standard X–Y plotters, chart recorders, and computing integrators. If in doubt, refer to the integrator or plotter reference manual. Inadvertently using the wrong connection can lead to inexplicable losses in sensitivity. The signal cable should be sited well away from any power leads, to avoid chromatographic noise caused by electrical induction.

Some instruments have the facility for the user to set the response time (occasionally referred to as the output filter). This can be anything from 0.5 to 6 s. The slower the response the smoother the baseline, but this is countered by a broadening of peaks with a corresponding reduction in efficiency. For trace analysis, the response time is best kept as low as the baseline noise allows.

Noise will also arise from dirty or cracked flow cells. With some instruments it is possible to clean and replace these fairly easily as the cell is self-aligning, but in other cases dismantling the flow cell will lead to more problems than it solves. Re-aligning the delicate optics of the detector can be very difficult. It is worth trying to back-flush solvent directly through the detector at low pressure: this will often clear dirt or air bubbles from the cell. If this fails, it is a job best left to the service engineer.

Most UV lamps are very expensive, and there is obviously economic pressure to get the maximum life possible from each. However, old lamps working at less than their optimum power lead to losses in sensitivity and, in extreme cases, poor run-to-run reproducibility. It is recommended that the strength of a lamp is checked regularly, and that it is replaced when it falls below the manufacturer's specification. Accumulated use of 200–400 hours can be obtained from most lamps. As the greatest strain is caused by the instrument being turned on, many analysts have found that leaving lamps on continuously (including overnight and over the weekend) significantly extends their life.

6.3.5 Hints and Tips

Flow Rates

Water (a common reversed phase solvent) is a much more viscous solvent than the common normal phase solvents (usually iso-octane or hexane), and so back-pressures are invariably higher when using reversed phase solvents. This can restrict the flow rate to less than the ideal, making an alternative normal phase system more suitable. Typically, flow rates of 0.5 to 1 ml min^{-1} are used for reversed phase chromatography, whilst 2–3 ml min^{-1} is more common for normal phase. High flow rates reduce the number of theoretical plates but this is only a problem when trying to resolve two closely eluting compounds. To decrease solute retention times, it is also not unusual to heat normal phase columns to 40 or 50°C.

Use of Guard Columns

A guard column of the same stationary phase as the analytical column is often advantageous in trace analysis, as the low detection levels required often mean that large sample weights are injected onto the system. Environmental and food samples, particularly, are likely to contain polar compounds that often bind to the head of the analytical column and significantly reduce its useful life. A disposable guard column, although increasing the dead volume of the system, protects the analytical column to some extent. It is particularly recommended when using unusual and expensive stationary phases.

Columns and Connections: Dead-volume

In order to maximise the resolution of an HPLC column, it is absolutely vital to restrict the connecting pipework volume to a minimum. If this is not done, diffusion will cause marked broadening of peaks, with corresponding loss of resolution. It is not uncommon for dead-volume to cause a peak with a baseline width of 0.5 min on a well set up system to broaden to 2 min.

The key areas in which to minimise dead-volume are between the injection port and the column head, and between the base of the column and the detector. These lengths of tubing should be as short as possible. This is sometimes difficult, particularly if using an auto-injector or detector with a long connection already attached. In general, most instruments are supplied with far too much pipework. It is good practice to replace them with shorter lengths where practicable or, if the long section is part of the instrument, cut the tubing as short as possible. If a longer connection (greater than about 5 cm) is unavoidable, then extremely narrow bore tubing, *e.g.* 0.12 mm internal diameter can be used, although this is prone to blocking, especially by crystallised buffer salts. As the band broadening of a solute is proportional to the 4th power of the internal diameter of the tubing, reducing the tubing diameter is much more fruitful than reducing its length by the corresponding volume.

Another common source of dead-volume is the swaged fittings themselves.

When connecting pipework to the injection port, the column, or the detector, it must be ensured that the end of the pipe is cut smoothly at a 90° angle, and fits flush against its connection. Steel tubing is best cut with dedicated tools and then filed by hand to obtain a smooth finish. The end of the pipe should then be pushed firmly against the column (for example) and the ferrule swaged on by tightening the locking nuts. Once swaged into position, the ferrule should only be regarded as fit for this one intended connection. When changing columns, the ferrule should be cut off and replaced with a new one, as a protruding pipe length suitable for one column is unlikely to abut perfectly against another. It is tempting to overlook this procedure, but the temptation should be resisted. As well as leading to band broadening, poorly fitting connections at the head of the column can give rise to a characteristic splitting of peaks.

To ease the changing of fittings, disposable Teflon ferrules are available. These can be pulled off and replaced by hand, rather than the necessity of cutting the tubing to replace steel ferrules. They cannot stand the same pressures as steel, but are perfectly adequate for low pressure systems (less than ~150 bar) or the low pressure detector end of the column. Swaging on Teflon ferrules requires some practice; too loose and they blow off, too tight and they block the system (especially if using Teflon tubing, which will squeeze closed). They can also be awkward to remove if stuck inside the thread of a column nut.

Trace Enrichment

The use of solvent gradients can be utilised to effect a concentration step at the determination stage of an analytical method. This is referred to as trace enrichment. A larger than usual sample volume is injected onto the column, with the initial mobile phase being of an insufficient strength to elute the analyte. This is pumped through for a few minutes to (hopefully) elute the bulk of the non-polar interferents, and then the strength of the mobile phase is increased to take off the analyte. This can be an extremely useful technique.

Column Overload

The majority of analysts would regard sample overload as purely a detection problem: as long as the detector is operating within its linear range, then there should be no problem. In fact, it is also possible to overload the column.

Overloading the column has the effect of reducing resolution due to band broadening. There are two distinct causes. The first, mass overload, *i.e.* where a solute is so concentrated that the mass of it cannot be retained in a tight band, is rarely applicable to trace analysis. The second, volume overload, is more pertinent. This occurs when the injection volume is so large that the sample is no longer injected as a tight plug, which is the ideal case. The temptation in trace analysis is always to increase the injection volume in order to increase sensitivity, but it must be remembered that there will be a price to pay in terms of resolution.

Choice of Detection Wavelength

Once a suitable region of the spectrum has been identified, a wavelength should be chosen where the absorbance is not changing rapidly with wavelength, *i.e.* the maximum of a peak. This is to guard against disproportionate variations in the absorption if there should be a minor variation in the set wavelength of the detector. If an internal standard is being used, the same applies to its spectrum. This should be a minor problem if it is borne in mind when actually choosing the internal standard; it makes it far easier if the internal standard has a similar spectrum to that of the analyte.

For fluorescence detection, suitable wavelengths can be obtained by taking spectra from a scanning fluorimeter if no literature references can be found. If this is unavailable, then information from the absorbance spectrum can be used. The absorbance maximum can be used as the excitation wavelength and the emission wavelength set at 30 nm higher.

Problem Recognition

There are many column-related faults that can be baffling when first encountered, but once identified are instantly recognisable if they recur. Some of the more common ones are listed in Table 6.3.7.

The most common symptom, which has already been described, is the peak splitting or tailing that is caused by voids and dead-volumes. Column overloading is another frequent problem that many analysts fail to spot; it is assumed that if the response is within the linear range of the detector then the response must, by definition, be linear. The risk of this is greater for shorter columns.

One problem which can occur, and is not described in Table 6.3.7, is the baffling discrepancy between chromatograms obtained using an 'old favourite' column and those from a new 'identical' one from the same manufacturer. The problem may not necessarily be an increase or decrease in retention time; there may be changes in selectivity. This could be due to a variety of causes, invariably due to unpublicised manufacturing changes. Suppliers have been known to change the sizes or shapes of the packing material, or even to start end-capping ODS material, without giving any indication to their customers. If significant chromatographic changes cannot be explained, it is always worthwhile contacting the supplier just to check that such changes have not been made.

6.3.6 Associated Uncertainty

The uncertainty associated with HPLC determinations is, in most cases, negligible when compared with the sample preparation stages of an analytical method.

With manual injections, overfilling the sample loop gives excellent reproducibility and precision for major components of a mixture. However, as can be seen from Table 6.3.8, there is a marked decrease in precision as the size of a chromatographic peak decreases. This emphasizes the need for 'clean' extracts for trace analytical work. Ideally, the analyte peak should be one of the largest on the trace, and well resolved from any other components. If the analyte can only be

Table 6.3.7 *Recognition of column-related problems*

Symptom	Possibe cause	Suggested remedy
Peak tailing or splitting	(i) Voids or dead-volume in column or pipework	Check fittings between column and injector. Use smaller detector cell. Retop column. Back-flush column.Replace column
	(ii) Blocked inlet frit	Replace inlet frit
Distortion of larger peaks (flattening of the tops)	(i) Column overload	Reduce sample size
Increasing peak tailing as capacity factor increases	Secondary retention effects: (i) Reversed phase	Basic compounds: add ammonia or triethylamine. Acidic compounds: add ethanoic acid or ethanoate salt. Ionic compounds: add salt or buffer. Try a different column.
	(ii) Normal phase	Basic compounds: add ammonia or triethylamine. Acidic compounds: add ethanoic acid or ethanoate. Polyfunctional compounds: add water. Try a different column.
	(iii) Ion-pair	Basic compounds:add ammonia or triethylamine.
Acidic or basic peaks tail	Inadequate buffering	Use 50–100 mM buffer concentration. Use buffer with pK_a the same as the pH of the mobile phase
Retention times drift	(i) Poor temperature control (particularly critical for normal phase)	Thermostat column
	(ii) Insufficient column equilibration	Give column longer to equilibrate
	(iii) Changes in mobile phase	Prevent evaporation of volatile components (see Section 6.3.3)

Table 6.3.8 *Coefficient of variation (CV) of trace components in an anthraquinone reference material compared with that of the major component when analysed by HPLC* [a]

Injection number	Anthraquinone	Impurity peak 1	Impurity peak 2	Impurity peak 3	Impurity peak 4	Impurity peak 5
1	99.795	0.054	0.043	0.020	0.020	0.068
2	99.805	0.052	0.042	0.017	0.027	0.057
3	99.802	0.049	0.044	0.021	0.030	0.054
4	99.800	0.051	0.043	0.022	0.026	0.058
5	99.808	0.052	0.044	0.017	0.020	0.054
6	99.782	0.052	0.048	0.023	0.031	0.064
7	99.787	0.053	0.045	0.021	0.030	0.064
8	99.795	0.053	0.046	0.023	0.024	0.059
9	99.800	0.052	0.047	0.018	0.021	0.058
10	99.790	0.057	0.045	0.023	0.026	0.064
CV (%)	0.008	3.9	4.2	12	16	7.9

Sample: anthraquinone dissolved in glacial ethanoic acid (ten replicate injections). Column: 15 cm × 2 mm Spherisorb S5 ODS2. Injection volume: 5 μl (Rheodyne 7410 full loop). Mobile Phase: 75:25 methanol/water at 0.2 ml min^{-1} Cell: 8 μl 10 mm pathlength. Detection:UV absorption at 254 nm. Peaks are quantified as the percentage of the total integrated area.

[a] Results obtained at the Laboratory of the Government Chemist.

determined as a shoulder on a large interfering peak or as one quill in a porcupine of peaks then the uncertainty of the integrated area can make a significant contribution to the overall uncertainty of the analysis. When developing a method, a balance must be drawn between the inevitable uncertainty of adding extra stages to the sample preparation and clean-up and the minimum clean-up requirements to obtain a chromatographic trace that can be integrated with a degree of precision.

Injectors and Injecting

As already noted, measurement uncertainty resulting from HPLC injections is so minimal as to be inconsequential when compared with the other stages of a trace analytical method. If performed correctly, precision and reproducibility is excellent.

The only real area where errors can occur is in sample carry-over.[9] Such cross-contamination between injections is extremely rare, but it is still a common sense precaution to thoroughly flush out both the syringe and the injection loop, if using manual injection, between samples. Injecting three times the nominal loop volume, so that most of the sample is used for cleaning, is the usual practice.

Carry-over in auto-injectors is virtually unheard of. Most have some kind of automatic purging and washing procedure. If this is user-controlled, it should be run at least each time the mobile phase is changed. Many models include a washing sequence between injections.

To minimise the solvent front on the chromatogram, sample extracts should ideally be made up in the mobile phase. This is more important for early eluting peaks, as these could co-elute with a large tailing solvent peak. Leaving the needle in the injection port after the valve has been switched to the load position also helps to prevent small baseline shifts as it is withdrawn.

6.3.7 References

1. J. Ruzicka and G. D. Christian, 'Flow Injection Analysis and Chromatography: Twins or Siblings?' *Analyst*, 1990, **115**, 475.
2. N. K. Vadukul and C. R. Loscombe, 'Examination of Simple Injection-valve Coupling Techniques in High-performance Liquid Chromatography', *Chromatographia*, 1981, **14**, 465.
3. A. Townshend, 'Solution Chemiluminescence – Some Recent Analytical Developments', *Analyst*, 1990, **115**, 495.
4. U. A. Th. Brinkman, 'A Review of Reaction Detection in HPLC', *Chromatographia*, 1987, **24**, 190.
5. G. D. Reed, 'An Evaluation of Electrochemical Detection in Reverse-phase HPLC', *J. High Resolut. Chromatogr. Chromatogr. Commun.*, 1988, **11**, 675.
6. L. R. Snyder, 'Classification of the Solvent Properties of Common Liquids', *J. Chromatogr. Sci.*, 1978, **16**, 223.
7. J. C. Berridge, 'Techniques for the Automated Optimization of HPLC Separations', John Wiley, Chichester, UK, 1985.
8. J. W. Dolan, 'Solvent Selection, Part 1– UV Absorption Characteristics', *LC,GC Int.*, 1994, **11**, 631.

9. M. Saha and R. W. Giese, 'Primary Contribution of the Injector to Carryover of a Trace Analyte in High-performance Liquid Chromatography', *J. Chromatogr.*, 1993, **631**, 161.

Capillary HPLC

10. K. Jinno, 'Advantages of Miniaturized Liquid Chromatographic Columns', *LC.GC Int.*, 1989, **2**, 30.
11. N. Vonk, W. P. Verstraeten, and J. W. Marinissen, 'Miniaturized Columns for the Routine HPLC Lab: High Speed and Minibore Performance', *J. Chromatogr. Sci.*, 1992, **30**, 296.
12. E. S. Yeung and W. G. Kuhr, 'Indirect Detection Methods for Capillary Separations', *Anal. Chem.*, 1991, **63**, 275A.

Alternative Retention Mechanisms

13. K. Jones, 'Affinity Chromatography – A Technique for the 1990s', *LC.GC Int.*, 1990, **4**, 32.
14. A. P. Foucault, 'Countercurrent Chromatography', *Anal. Chem.*, 1991, **63**, 569A.

6.4 Thin Layer Chromatography

6.4.1 Overview

Thin Layer Chromatography (TLC) was one of the first chromatographic techniques to be developed. For many years it was regarded as the poor relation of HPLC. As HPLC came into routine laboratory use, interest in TLC waned. It is only in more recent years, with the advent of High Performance Thin Layer Chromatography (HPTLC) and advances in automation and quantitation, that TLC has enjoyed something of a renaissance.

Thin layer chromatography has always had the advantage of being relatively cheap and easy to use. It can give much better separation than HPLC, being particularly useful for difficult separation problems such as mixtures of pesticide residues and low concentrations of analytes in the presence of a high concentration of co-extractives. When comparing TLC separation to HPLC, one difference that is also frequently overlooked is that TLC is universal: the entire plate is examined, so that even compounds that do not elute can be detected as such. The HPTLC technique gives yet a greater degree of resolution than classical TLC.

Although traditionally considered a qualitative or semi-quantitative technique, recent advances in detection systems have enabled the possibility of accurate quantitation of plates. However, such apparatus is expensive. A fully automated HPTLC system with densitometry quantitation can cost two or three times the price of a basic HPLC system. This can negate what is still seen as the major advantage of TLC, its low cost.

6.4.2 Comparison of Techniques, General Precautions, and Critical Aspects

As with all other chromatographic methods, TLC can be described in terms of the sample introduction method (spotting), the stationary phase (the plate), the mobile phase (the eluting solvent), and the detection/quantification technique used.

Spotting the Sample

The sample solution can be applied either as a spot or as a streak. In either event, it is critical to minimise the area of the solution on the plate in order to reduce band broadening.

The optimum concentration range for the sample solution is from 0.01 to 1.00% w/v. At higher concentrations the plate can become overloaded and a characteristic tailing effect is observed as the plate develops. Figure 6.4.1(a) shows a correct separation (✓) and the effect of overload (✗). The solvent chosen in which to apply the sample should be as low in the eluotropic series as possible, *i.e.* as non-polar as is possible for the particular solute. A polar solvent will spread out over the stationary phase before the plate begins to develop, leading to a ring effect as shown in Figure 6.4.1(b). The solvent chosen for the dissolution of the sample should also be as volatile as possible, so that it evaporates rapidly before the spot has an opportunity to diffuse. The plate should be dried between the application of each spot, or the same ring patterns appear as are shown in Figure 6.4.1(b).

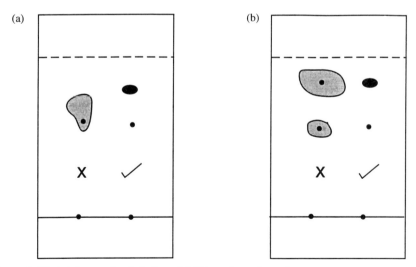

Figure 6.4.1 *(a) An overloaded plate (b) Effect of a polar solvent*

The traditional method of spotting the sample solution on to the plate has been to use graduated glass capillary tubes. It is important to take care when bringing these into contact with the plate, as it is very easy to scratch the surface of the adsorbent. Templates are available to fit over the plate, acting as a guide in order to increase the precision of application. It is even possible to automate the application with some modern apparatus.

Streaking is usually used as a method of application when the plate has a pre-sorbent zone of deactivated Kieselguhr. The solution is applied 5–10 mm below the interface between the deactivated zone and the analytical layer. Using streaking application on to a deactivated pre-sorbent zone gives less band broadening than spotting directly onto the adsorbent, and so improves resolution. When measuring relative retention (R_f) values on such a plate, the origin is taken as the bottom of the analytical layer, not the point of application.

Choice of Plate

As with any chromatographic technique, the choice of stationary phase is critical. Similarly the particle size, thickness of coating, and the manufacturer can have a pronounced effect upon any given separation. There can even be marked batch-to-batch variations between plates obtained from the same supplier.

Some of the more common adsorbents for TLC are listed in Table 6.4.1. In addition to these, most HPLC stationary phase (*e.g.* ion-exchange phases, cyanopropyl and aminopropyl bonded phases) have been applied to separations with varying degrees of success. Adsorbents can be modified by impregnation

Table 6.4.1 *Some common adsorbents for TLC*

Adsorbent	Description	Examples of Applications
Silica gel G	'G' refers to 13% calcium sulfate binder	Silica gel is the most widely used TLC adsorbent
Silica gel H	'H' contains no binder	Steroids, amino acids, alcohols,
Silica gel F254	'F' contains a fluorescent agent,	hydrocarbons, lipids, aflatoxins,
Silica gel F366	which is excited at 254/366 nm Any UV absorbing spot on the plate quenches the background fluorescence	vitamins, alcohols
Neutral alumina	All forms of alumina are	Sterols, dyestuffs, vitamins,
Acidic alumina	available with or without	alkaloids. Basic alumina finds
Basic alumina	binder	the most applications
Cellulose	As paper chromatography, but smaller pore sizes.	Hydrophylic compounds *e.g.* sugars, amino acids, nucleic acids
Reversed phase TLC	As HPLC stationary phases, usually C-18	Very polar compounds. Not widely used
Kieselguhr	Specially treated diatomaceous earth.	Usually used as a deactivated pre-sorbent zone for other phases
Cyclodextrins	Large complexing agents, enantiomer-specific	Chiral separations

with a wide range of compounds, including undecane, mineral oils, phenoxy-ethanol, oxalic acid (ethanedioic acid), and dimethyl sulfoxide.

Plates for TLC can be prepared within the laboratory or purchased pre-coated. The adsorbent can be bonded to aluminium, plastic, or glass (glass microscope slides are often used for small home-made plates). More and more laboratories are now opting to use manufactured plates, as the range available and the precision and robustness of plates has improved in recent years. Preparing plates in-house is an extremely messy and time-consuming business, even when using commercial plate spreaders. Manufactured plates are also easier to modify if required.

The adsorbent particle size of most TLC stationary phases lies between 5 and 50 µm. The technique of HPTLC merely refers to the use of the same adsorbents, but with much smaller particle sizes to give improved resolution. The film thickness is generally 0.25 mm, although plates with thicker films are available. These are generally only used for preparative TLC, where the sample loading is much higher.

One modification that is particularly useful for trace analysis is to coat a plate so that the film thickness varies from about 1.0 mm, where the sample is applied, to the standard 0.25 mm at the top. A higher sample loading is therefore possible on the lower half of the plate, allowing high concentrations of interferents to be removed, without compromising the resolution provided by a thinner film. Another worthwhile modification, already mentioned, is the use of a concentration zone (or pre-sorbent zone) of deactivated Kieselguhr. This prevents the applied solution diffusing and spreading before the plate development begins. It is usually used in conjunction with automated HPTLC systems with automatic sample applicators, but is equally relevant to other areas of TLC.

Care must be taken during the storage of TLC plates to ensure that the initial activity does not change due to uptake of water or oxidation. It may be necessary to reactivate plates by heating for 3 to 4 hours at 150°C before using them.

Developing the Plate

Solvents used for TLC plate development are analogous to those used for normal phase HPLC. Table 6.4.2 gives some examples of solvent systems used for particular applications. These are usually binary or ternary mixtures of solvents; again, analogous to HPLC mobile phases. Solvent properties are described by the eluotropic series, usually based upon either water solubility or alumina retention (the 'alumina index').[1]

The development of the plate is usually ascending in a saturated tank, but it can also be descending, horizontal, or by the use of thin layer electrophoresis. With any of these methods, the use of two-dimensional development (turning the plate through 90° and developing it again perpendicular to the original line of development) adds an extra degree of certainty to any identification based upon R_f values. Some automated HPTLC systems also use circular development, which gives a similar increased degree of resolution.

It is essential that the atmosphere of the development tank is saturated with solvent vapour before the plate is placed in it. If the tank has not reached

Table 6.4.2 *Some typical mixtures, adsorbents and suggested mobile phases*[2]

Analyte	Adsorbent	Suggested mobile phase
Aflatoxins	Silica gel	Toluene / ethyl ethanoate / propanone (3 : 2 : 1) with 1% ethanoic acid
Aliphatic hydrocarbons	Silica gel+ 20% AgNO$_3$	Hexane (twice developed)
Alkaloids	Alumina	Trichloromethane
Amino acids	Silica gel	Propan-1-ol / 34% aqueous ammonia (67 : 33)
Aromatic hydrocarbons	Acetylated cellulose	Propan-1-ol / propanone / water (2 : 1 : 1)
Barbiturates	Silica gel GF	Ethyl ethanoate / methanol / aqueous ammonia (82 : 14 : 4)
Carboxylic acids	Silica gel G	Hexane / ethoxyethane / ethanoic acid (80 : 20 : 1)
Carotene	Silica gel G	Propanone / light petroleum (10 : 90)
Gangliosides	Silica gel HPTLC	Trichloromethane / methanol / water (2 : 1 : 1)
Metal ions	Cellulose polygram cell 400	Butan-1-ol saturated with 3 M hydrochloric acid
Monoterpenes	Silica gel	Benzene
Nucleotides	Cellulose	Saturated aqueous ammonium sulfate / 1 M sodium ethanoate / propan-2-ol (80 : 18 : 2)
Organochlorine pesticides	Silica gel	Heptane / propanone (98 : 2)
Organophosphorus pesticides	Silica gel HPTLC	Hexane / propanone (5 : 1)
Serotonin	Silica gel	Ethyl ethanoate / trichloromethane (3 : 2)
Sterols	Alumina with AgNO$_3$	Hexane / ethyl ethanoate 20 : 1
Sulfanilamides	Silica gel	Two-dimensional: (1) Trichloromethane / methanol (95 : 1), (2) ethyl ethanoate / methanol / aqueous ammonia (30 : 10 : 1)
Vitamin K	Silica gel	Benzene

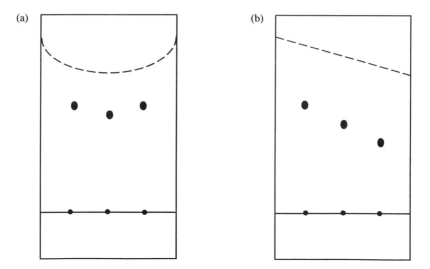

Figure 6.4.2 *(a) Effect of an unsaturated chamber (b) Effect of direct sunlight*

equilibrium then the solvent front will migrate unevenly, as shown in Figure 6.4.2(a). Care must also be taken to ensure that there is no temperature gradient across the chamber (*e.g.* such as that which would be caused by placing it in direct sunlight), or the solvent migration will again be uneven (Figure 6.4.2b).

Plate development generally takes around 20 min, although it may be as rapid as 5 min for HPTLC. It is obviously vital to stop the development before the solvent front reaches the top of the plate whilst leaving sufficient time to obtain the best possible resolution, so a close eye should be kept on the progress of the solvent.

Visualisation and/or Quantification

Thin layer chromatography traditionally has been regarded as, at best, a semi-quantitative technique, relying largely upon the visual comparison of spot intensities to give approximations of analyte concentrations. Whichever visualisation method is used, this involves the laborious spotting of many standard solutions, with a range of concentrations, on every plate. This situation has been improved in recent years by the development and introduction of densitometers (*e.g.* the Zeiss TLC scanner). These are very expensive, but provide an easy, routine, and reliable method of quantifying TLC results.

Prior to densitometers, the only method of quantification was to physically scrape off the spot, redissolve the analyte, and determine its concentration by an alternative technique. This procedure is extremely tedious, and suffers from two major potential problems: the visible spot which is scraped from the plate may not contain all of the analyte, and it is difficult to desorb the analyte which is bound to the plate. The latter is the more serious problem. One of the principal causes has

Table 6.4.3 *Some methods of TLC plate visualisation*

Visualisation method	Comments	Examples of applications
Part I: Universal methods		
Direct visual examination	Usually non-destructive	Dyestuffs, food colorants
Pre-coated fluorescent plate UV absorption 254 or 366 nm	Usually non-destructive gives UV-absorbing spot on fluorescent background. Non-UV absorbers can be sprayed with a reagent to give a UV-absorbing product	UV absorbers, *e.g.* compounds containing unsaturated rings
Iodine vapour	Usually destructive. Do not allow the plate to overdevelop in the iodine tank. Record the spots quickly (*e.g.* photograph) before the iodine sublimes	Most organic compounds, *e.g.* pharmaceuticals
Char with sulfuric acid at elevated temperature	Destructive	Most organic compounds *e.g.* pharmaceuticals, pesticides
Part II: Examples of analyte-specific methods		
Ninhydrin	Destructive. Gives distinctive purple spots	Amino acids
Water	Usually non-destructive.	Lipids, steroids
2,4-Dinitrophenylhydrazine	Destructive. Gives yellow or red spots	Aldehydes, ketones
Ammonium vanadate / anisidine	Destructive. Gives pink background	Phenols
Cerium(VI) ammonium nitrate	Heat plate and cool before spraying	Alcohols
Alizarin	Destructive. Gives violet spots	Amines
(i) Choline esterase (ii) Acetyl choline (iii) pH indicator	Produces ethanoic acid background with colour-less spots for choline esterase inhibitors (no conversion of acetyl choline to ethanoic acid)[3]	Nerve gases, insecticides

been shown to be the binder used in the production of the plate, and so the use of unbound plates can sometimes overcome the problem.

All measurements of R_f values rely upon knowing the migration distance of the solvent. The first thing to do when removing the plate from the tank is therefore to mark the position of the solvent front. This should be done rapidly, as the solvent soon dries. The method of visualisation should then be chosen to suit the analyte.

Some of the more common techniques are listed in Table 6.4.3, but the list of possible derivatisation reagents, in particular, is extensive and can only be touched upon here. More extensive lists of reagents can be found in the literature.[1,2]

Many quality assurance schemes require a record to be kept of developed plates. For this purpose, plates are usually photographed. As with visualisation for measurement purposes, the photograph should be taken as quickly as possible after development, particularly if the reagents used do not give a permanent stain, *e.g.* when iodine vapour is being used. The use of Polaroid film, which develops in a matter of minutes, is an advantage, as there may still be the opportunity to take a second photograph if there is a problem with the first. Sometimes, it is even easier to judge relative spot intensities on the photograph than looking at the plate itself.

6.4.3 References

1. J. C. Touchstone, 'Practice of Thin Layer Chromatography', John Wiley, New York, 3rd edn, 1992.
2. R. Hamilton and S. Hamilton, 'Thin Layer Chromatography', John Wiley, Chichester, UK, 1987.
3. P. J. Wales, C. E. Mendoza, H. A. McLeod, and W. P. McKinley, 'Procedure for Semi-Quantitative Confirmation of some Organophosphorus Pesticide Residues in Plant Extracts', *Analyst*, 1968, **93**, 691.

General Reading

4. 'Modern Thin Layer Chromatography', ed. N. Grinberg, Marcel Dekker, New York, 1990.
5. E. R. Schmidt, Chromatography and Mass Spectrometry – An Overview', *Chromatographia*, 1990, **30**, 573.
6. H. Jork, 'Advances in Thin Layer Chromatography', *Int. Lab.*, June 1993, 18.
7. C. F. Poole and S. K. Poole, 'Instrumental Thin Layer Chromatography – A Report', *Anal.Chem.*, 1994, **66**, 27A.

Applications

8. P. M. Scott, 'Recent Developments in Methods of Analysis for Mycotoxins in Foodstuffs', *Trends Anal.Chem.*, 1993, **12**, 373.
9. H. S. Rathore and T. Bagum, 'Thin Layer Chromatographic Methods for use in Pesticide Residue Analysis', *J. Chromatogr.*, 1993, **643**, 271 (Bibliographic Review).
10. A. Mohammed, N. Fatima, J. Ahmed, and M. A. Majid-Khan, 'Planar Layer Chromatography in the Analysis of Inorganic Pollutants', *J. Chromatogr.*, 1993, **642**, 445.
11. G. Szepesi. 'The Role of Thin Layer Chromatography in Steroid Analysis', *J. Planar Chromatogr.– Mod. TLC*, 1992, **5**, 396.
12. I. Ojanperu, 'Toxicological Drug Screening by Thin Layer Chromatography', *Trends Anal. Chem.*, 1992, **11**, 222.
13. J. Sherma, 'Modern Thin Layer Chromatographic Pesticide Analysis Using Multiple Development', *J. Assoc. Off. Anal. Chem. Int.*, 1992, **75**, 15.

6.5 Organic Mass Spectrometry

6.5.1 Overview

Mass spectrometry (MS) is a very powerful detection and identification tool for the analysis of trace organic compounds. The technique involves the collision of the organic compound with electrons or molecules and their subsequent breakdown to give a mass spectrum of the resulting fragments. In this spectrum there may be as few as one ion (often the molecular ion) or many more ions depending on the analyte and the conditions used to ionise the compound. The pattern and intensity of the spectrum gives both quantitative and qualitative information about the analyte. Examples of its use are the quantitative analysis of pesticides extracted from foodstuffs, the identification of trace amounts of controlled drugs, and the detection and quantitation of drug metabolites extracted from body fluids.

Its most common use, by far, in trace analysis is linked to a chromatographic technique, most commonly GC. This separates the components in a mixture and introduces them into the mass spectrometer. There are a range of options available to the analyst: which type of mass spectrometer to use, the ionisation technique, the chromatographic technique, and which data acquisition mode to use. In addition to these options, as with most analytical techniques, there are parameters that can be varied for each of the options, to gain optimum performance. All aspects concerned with the chromatography are covered in Sections 6.2 and 6.3. This section is concerned with the MS component.

For many years MS was regarded as an expensive technique and therefore only used when other techniques such as a chromatographic analysis, with a detector of low specificity, would not suffice either on the grounds of sensitivity or specificity. Although conventional magnetic sector instruments are still not cheap, the introduction of benchtop instruments (quadrupoles and ion traps) has made the technique financially accessible for routine use.

MS as a Confirmatory and Routine Technique

Where an analyte has been previously analysed by a chromatographic technique with a detector of limited specificity, a mass spectrometer will often be employed (possibly using the same or similar chromatographic conditions) to confirm the identity of the compound. With the advent of less expensive bench-top instruments, mass spectrometers are now also being used for routine trace analysis. Qualitatively specific and quantitatively sensitive information is gained in the same time as it takes to run a chromatogram.

MS as a Technique for Identifying Unknown Compounds

Chromatographic techniques generally, though not invariably, rely solely on retention time for a qualitative identification. The use of a mass spectrometer as a detector results in the production of a spectrum which, depending on the compound and ionisation mode, provides more precise qualitative information.

Since the type of spectrum obtained in mass spectrometry depends mostly on the ionisation mode, spectra are labelled with acronyms appropriate to the ionisation mode used, *i.e.* electron impact (EI), chemical ionisation (CI), and chemical exchange (CE). These names also convey the type of information obtainable from the spectrum. Patterns characteristic of a specific class of compound can be obtained from the lower mass ions in EI spectra. In other words, groups of related compounds will tend to have similar spectra in the lower mass range (in EI mode). The higher mass ions (obtained from CI and the higher mass range of EI) will give more specific information on the identity of the analyte. Obviously this is most useful when used in conjunction with either low mass ion information or chromatographic information.

In positive ion electron impact (EI+) MS the mass spectrum of a compound varies only marginally with the specific instrument or conditions used. Thus it is often possible to produce a spectrum which can be compared with entries in spectral libraries; these may be in the form of computerised databases or in reference books.[1-3] A spectrum can often be matched with an entry in a computerised database in a quantifiable way, *i.e.* with a 'score' indicating the closeness of the match. There are, however, some limitations.

(i)   Library entries are limited. No single database contains entries for even 10% of all known organic compounds.

(ii)   Some compounds do not have very characteristic spectra.

(iii)   The relative abundances of the ions in the spectrum can vary slightly depending upon the instrumental conditions. This may be sufficient to produce library mismatches. At or near the limit of detection spectral distortion may also become marked.

(iv)   Unlike Fourier transform infrared and nuclear magnetic resonance spectroscopy, MS does not usually differentiate between isomers.

MS as a Technique for Quantification of Components in a Mixture

Mass spectrometry can be used for the quantitative determination of organic compounds using the signal from a single ion, the combined signal from a group of ions, or the total ion current. It is therefore possible, from a single injection, to achieve unambiguous identification of an analyte and an accurate quantitation. The precision of the measurement will be dependent on the total system, not just the mass spectrometer. The reproducibility of the peak area for ten 2 ng injections of methyl stearate is shown in Table 6.5.1. For a number of injections, where the concentration of the analyte is close to the detection limit, the reproducibility appears to be improved if the integrated peak areas from the ion chromatograms is used rather than measuring subtracted spectra.

Gas chromatography–MS systems vary in specification and therefore in sensitivity. The cheapest models, working only in EI mode, have detection limits comparable to traditional element-selective detectors such as the FPD and NPD. Positive ion chemical ionization (CI+) can markedly increase sensitivity for relatively large electropositive compounds (relative molecular mass 250–800). For

Table 6.5.1 *Reproducibility in GC–MS*

Injection number	%$M^{+ a}$	Peak area from chromatogram		
		m/z 74	*m/z* 298	%M^+
1	10.16	1136	88	7.7
2	8.28	1011	81	8.0
3	7.40	1258	86	6.8
4	10.88	934	89	9.5
5	6.64	910	61	6.7
6	9.74	1259	90	7.1
7	7.75	1326	100	7.5
8	7.57	1952	139	7.1
9	7.52	1903	140	7.4
10	6.24	1948	124	6.4
$\bar{x}$	8.22			7.43
σ_{n-1}	1.5			0.9

[a] Obtained from the background subtracted mass spectrum, *i.e.* the ratio of the molecular ion (*m/z* 298) to the base ion (*m/z* 74).

electronegative compounds, top of the range instruments with negative ion chemical ionization (CI–) can now rival the sensitivity of the ECD. However, whatever the specification of the instrument, the response of each analyte is dependent upon the ionisation mode and the conditions used. For multi-analyte determinations, there will always be some analytes which respond poorly.

6.5.2 Comparison of Mass Spectrometers

The three most common methods of separating ions are the magnetic sector, the quadrupole mass filter, and the ion trap. These are compared in Table 6.5.2. As there are few commercially available instruments using time-of-flight separation, these are not considered.

The ability of a mass spectrometer to distinguish between ions of different masses is described in terms of its resolution (also called resolving power).

The resolution, R, of a mass spectrometer is given by Equation (1), where M is the mass to be measured (strictly the mass/charge ratio, m/z, but as most ions have a charge of 1 it is commonly referred to as the 'mass number', in either atomic mass units or Daltons), and ΔM is the smallest mass difference that can be separated from mass M.

$$R = \frac{M}{\Delta M} \tag{1}$$

For example, the minimum resolution required to separate a mass of 300 from a mass of 300.03, is 10 000, as calculated from Equation (1), *i.e.*

Table 6.5.2 *Comparison of mass spectrometers*

	Magnetic sector	*Quadrupole*	*Ion trap*
Cost	High	Low – medium	Low – medium
Resolution	High (*e.g.* 200 000) or low (*e.g.* 2000)	Low (*e.g.* 1000)	Low (*e.g.* 1000)
Running costs	High maintenance. Skilled operator	Low maintenance. 'Black box' operation	Low maintenance. 'Black box' operation
Sensitivity (scan mode)	Not suitable. Insensitive	Fast scan speed over full range (*e.g. m/z* 50 – 800). Medium sensitivity	Limited scan range (*e.g. m/z* range = 50) but sensitivity as high as SIM for quadrupole
Sensitivity (SIM mode)	Very sensitive, especially low resolution. Limited to jumps of ≈ 50	Good sensitivity. Can monitor any ion from whole range	No sensitivity advantages for SIM over scan mode
Library matching	Good	Good	Can be complicated by secondary ionisation
Applications	Complex mixtures of structurally similar compounds, *e.g.* dioxins in fly ash	Wide analytical use. Standard for routine GC–MS	Alternative to quadrupole. Not as widely used

SIM = selected ion monitoring.

$$\frac{300}{(300.03-300)} = 10\,000$$

There is, not surprisingly, a trade-off between resolution and sensitivity. As resolution is increased, fewer ions of any type are transmitted through the system, and consequently the sensitivity is reduced. However, for the same resolution, a sector instrument is considerably more sensitive than quadrupole and ion trap instruments.

Sector Instruments

Most modern magnetic sector instruments are double focusing. The main benefit of these instruments is the high resolution that can be achieved. They can provide resolution of up to 200 000. One of its most common uses is in the analysis of dioxins. Some sample types, such as complex environmental samples, require high mass-spectral resolution to separate interferences which can co-elute even when using capillary GC.

Sector instruments tend to be expensive and consequently tend to be used for high resolution specialist analysis, leaving the routine mass spectrometric work for the quadrupoles and ion traps. A great deal of maintenance is required to enable high resolution to be established and maintained. If possible such instruments should be reserved for high resolution work. Swapping between low resolution routine work and high resolution specialised tasks merely results in long down times.

Quadrupole Instruments

Quadrupole instruments cannot achieve the high resolution of sector machines and they have a limited mass range. Older machines were limited to ions with m/z lower than 1000, whereas newer machines have a mass limit of m/z 2000–4000. However, for the purposes of trace analysis it is likely that the majority of analytes will have a relative molecular mass below 1000. Analytes of higher mass are unlikely to be volatile, precluding their analysis by GC–MS.

Quadrupole instruments are more robust than sector machines and require less maintenance. A further advantage of quadrupoles over sector machines is their rapid scanning. Older sector machines require a full 1 s or more to scan from m/z 30 to 500. Even newer sector machines with laminated magnets can barely scan this range in 0.3 s. In contrast, a quadrupole analyser can easily scan this range in 0.1 s. Clearly, for mass spectrometers coupled to capillary chromatography columns from which components can elute very rapidly, rapid scanning is an asset because several scans can be acquired for each peak. This will improve both the reconstruction of the peak shape and the deconvolution of unresolved peaks.

As has already been mentioned quadrupoles will tend to be much cheaper than floor standing sector instruments. For routine trace analyses of medium to low difficulty the analyst's main choice is between the bench-top quadrupole and the ion trap.

Ion Trap Instruments

Ion trap instruments work by a different mechanism from quadrupoles but yield, more or less, the same information for the same price. There is no appreciable difference in the quality of the spectra obtained or the detection limits available. They are both better than most GC detectors but not as good as a sector instrument working in a low resolution mode.

Earlier ion traps often produced spectra that were quite different from conventional EI spectra; if there were too many ions in the chamber, secondary ion–molecule reactions produced CI-type spectra, thus preventing library searching. This problem has been virtually eliminated from newer ion traps by the incorporation of automatic gain control.

One advantage of ion traps is that, unlike sector and quadrupole instruments, there is little difference in sensitivity between scan mode and selected ion monitoring mode. However, ions can only be scanned over a limited range; for most instruments this is a range of between 30 and 50 m/z units.

6.5.3 Methods of Introducing the Sample into the Mass Spectrometer

Gas Chromatography–Mass Spectrometry (GC–MS)

This technique has been established as a routine analytical tool for many years. Until about 1980 all GC–MS applications used packed columns, reflecting GC practice at the time. Various interfaces, particularly the jet separator interface, were employed to remove most of the carrier gas from the eluent before entering the MS source, without removing an equal proportion of analyte at the same time. With the increased use of capillary columns with helium as the carrier gas flowing at ~1 ml min^{-1} the mass spectrometer pumping systems can easily deal with the entire flow from the GC. This results is more efficient chromatography and lower detection limits.

Direct Probe Insertion

The limitation of GC–MS is the volatility of the analyte. Until a few years ago, the only way to obtain spectra for relatively involatile molecules was to employ direct probe introduction. With programmed heating, probe introduction often allows good spectra to be obtained from simple mixtures, *i.e.* those with about three components. This technique is of limited use for trace analysis. Obviously there is no possibility of obtaining spectra for each component in more complex involatile mixtures. For such a task, LC–MS is necessary.

Liquid Chromatography–Mass Spectrometry (LC–MS)

This technique extends the scope of chromatographic MS to involatile and polar compounds. The relative merits of GC–MS and LC–MS are summarised in Table 6.5.3.

Table 6.5.3 *GC–MS and LC–MS properties*

GC–MS	LC–MS
Well-established technology	Evolving technology
Eluent easy to handle	Eluent more difficult to handle
Volatile compounds only	Volatile or involatile compounds
Library searchable spectra	Library spectra not normally searchable
Ultra-high chromatographic resolution (100 000 plates)	High resolution (20 000 plates)

Unlike GC–MS where there is generally no choice of interface, LC–MS offers a range of interfaces according to the needs of the analyst. The earliest commercial LC–MS interface was the moving belt. This had many problems, such as very poor detection limits and possible sample degradation. Moving belts were followed by the thermospray interface. This was an improvement on the moving belt in terms of sensitivity but proved very compound-specific. The tip temperature proved quite critical; good sensitivity for a compound could only be obtained within a 5°C range of tip temperature, which rendered the interface suitable for confirming suspected compounds, but of limited use for handling mixtures of unknowns.

Successors to thermospray include the particle beam interface, which gives searchable EI-type spectra, and the electrospray, which is a useful technique for compounds that do not ionise when subjected to other LC–MS techniques. Electrospray interfaces also have the advantage that, depending on the conditions under which they are operated, they can give solely relative molecular mass information or can cause pseudo molecular ion fragmentation. At present there are three LC–MS interfaces commonly available: electrospray, particle beam, and atmospheric pressure chemical ionisation.

Electrospray interfaces can give rise to multiply charged ions, allowing samples of high relative molecular mass to be determined. They normally produce only the unfragmented molecular ion, analogous to CI-type spectra, which identifies a compound solely on the basis of its relative molecular mass. This is currently the favoured method. It is reliable and, because the whole of the analyte is converted to a single ion, it offers good quantitation and sensitivity.

The major applications of LC–MS at the moment are the quantitative analysis of specified involatile compounds in a mixture and as a confirmatory technique for established LC–UV/VIS analyses. The electrospray interface is suited to these applications, and for the quantitation of compounds that have been identified spectrally using a different LC–MS interface. It is not suitable for the identification of unknowns, as it produces no characteristic fragmentation.

Particle beam interfaces do produce characteristic fragmentation spectra. This may grow in importance but at present there are doubts about its ability to cope with matrix interferents.

Atmospheric pressure chemical ionization (APCI) is similar to electrospray, but produces some fragmentation. The big advantage of APCI is that it is able to tolerate high LC-column flow rates (double that of electrospray). Usually electrospray LC–MS is marketed with a combined electrospray/APCI source.

Supercritical Fluid Chromatography–Mass Spectrometry (SFC–MS)

Supercritical fluid chromatography (SFC) was originally recognised as being a bridge between GC and LC, but the technique has yet to fulfil its early promise. However, SFC is viable when it is considered as an introduction technique for MS. Capillary SFC uses extremely narrow bore columns so that MS pumping systems can, in many cases, handle the entire eluent from a capillary SFC system. In addition, a moving belt LC–MS analogue using packed column SFC has also been tried. Capillary SFC–MS gives spectra that are similar to conventional EI spectra through charge exchange (CE).[4] The detection limit for capillary SFC–MS was found to be poor, making it unsuitable for trace analysis. However, packed column SFC–MS has the potential to offer enhanced detection limits even over LC–MS.

Capillary Zone Electrophoresis–Mass Spectrometry

A further introduction method, which is being developed for MS, is capillary zone electrophoresis. Advances in the technology and commercialisation of this technique are expected.[5]

6.5.4 Ionisation Techniques

Electron Impact (EI)

The commonest form of ionisation is by EI. Analyte molecules are bombarded with a stream of electrons, which produces a positive molecular ion, *i.e.* an ion having the same atomic composition as the neutral molecule. The molecular ion will possess excess energy, which generally causes fragmentation of the molecular ion. These smaller ions produce a characteristic spectrum. In some cases, however, the molecular ion is particularly unstable or a fragment ion is particularly stable so that no molecular ion is seen. This behaviour often occurs with aliphatic amines and alcohols. In other cases the molecular ion is so stable that little fragmentation occurs. This is usually the case for polynuclear aromatic hydrocarbons.

Electron impact is most suitable for compounds that easily lose electrons, *e.g.* where the charge can be delocalised. It is used when structural information is required from the fragmentation pattern.

Positive Ion Chemical Ionisation (CI+)

In many cases failure to observe a molecular ion is a disadvantage. The greater the degree of fragmentation the lower the sensitivity, as there is less of each fragment monitored, and so a larger uncertainty in the quantitation. If it is suspected that the molecular ion is absent from the spectrum, a softer form of ionisation is needed. The commonest alternative is CI. In CI+, a reactant gas such as ammonia, methane, or isobutane (2-methylpropanone) is first ionised by a stream of electrons. The resulting ions cause ionisation of analyte molecules, by proton exchange, to give protonated analyte molecular ions. The ions produced may be dependent on the reactant gas employed. For an analyte of mass M, ammonia tends to give not only $[M+H]^+$ ions but also $[M+NH_4]^+$ ions; the proportion formed under a given set of conditions depends on the nature of the analyte. To limit the tendency of the analyte to undergo EI-type ionisation, the concentration of reactant gas is kept at a much higher concentration than that of the analyte. Therefore, on statistical grounds, CI predominates.

Comparison of CI and EI is shown in Table 6.5.4. Chemical ionisation mass spectra generally contain protonated molecular ions and (sometimes) a few fragment ions. Clearly, these spectra are not searchable against database spectra, since stored spectra are recorded in the EI mode. However, CI MS can often be more sensitive than EI MS because the total charge is not distributed among a large number of ions and, as the ions monitored are generally larger, there is less background noise.

Table 6.5.4 *Comparison of ionisation techniques, EI and CI MS*

EI	CI
Very low pressure	Relatively high pressure
Molecular ion sometimes present	Molecular ion normally present
Extensive fragmentation	Limited fragmentation
Spectra searchable against databases	Spectra not searchable

Chemical ionisation is carried out at a higher pressure (~1 Torr) than EI (~10^{-6} Torr). Due to this the possibility of source contamination is increased, and so the filament lifetime is reduced.

Negative Chemical Ionisation (CI-)

Some electrophilic analytes, particularly halogenated compounds, do not readily form positive ions and therefore do not respond well to either EI or CI+. The technique of negative chemical ionisation can be used in these cases. This produces deprotonated negative molecular ions (of mass *M*–1), with the same sensitivity advantages as CI+. Since only electrophilic compounds respond well, it is particularly useful for analysing, *e.g.* halogenated compounds, in complex extracts. An example of the sensitivity of CI– for halogenated compounds is illustrated in Figure 6.5.1. This shows the response for an injection of 10 fg of the pesticide octafluoronaphthalene. In practice, for absolute identification, information from at least three ions arising from the analyte is required.

Fast Atom Bombardment (FAB)

This is a further method of ionisation which has proved increasingly popular. In FAB, the sample is held in a liquid matrix, usually glycerol, and is bombarded with a stream of fast neutral xenon or argon atoms. This is a useful ionisation technique for large and labile compounds, particularly those of biological interest. In common with CI spectra, FAB spectra are not searchable against MS databases.

6.5.5 Scanning Modes

Full Scan Mode

In full scan mode the whole mass range is scanned: this means, for a GC–MS system, all *m/z* values from about 35 to 800. The full spectrum produced gives unequivocal identification of the compound. However, sensitivity is poor as the greater the number of ions acquired the less time the detector dwells on each individual ion.

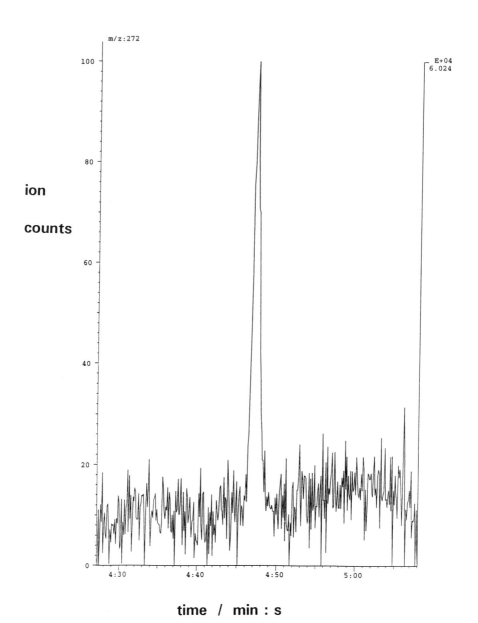

Figure 6.5.1 *Ion chromatogram of the molecular ion of octafluoronaphthalene* (10 fg *injection) by negative ion chemical ionisation (SIM)*

Selected Ion Monitoring (SIM)

In SIM the instrument is set to acquire only a few ions. Sensitivity is increased as time is limited to monitoring only the ions of interest. Clearly SIM is of little use in examining unknowns. However, it is a good confirmatory technique provided that a sufficient number of structurally significant, *i.e.* diagnostic, ions are monitored. The detection of these ions in the correct ratio is taken as confirmatory identification.

The ions chosen should be structurally significant, ideally the most intense in the spectrum (to maximise sensitivity), and unlikely to result from co-eluting interferents (to minimise specific interferents). However, these three objectives may be mutually exclusive, particularly when compounds fragment into many small ions. An example is the monitoring of m/z 182 and m/z 303 (M^+) as the diagnostic ions for cocaine. The base ion has an m/z of 82 but this is found in *all* tropenes not just cocaine.

By performing SIM, the detection limit can be improved approximately 10-fold for quadrupole instruments and up to 10 000-fold for magnetic sector instruments. Originally sector instruments could only alternate between monitoring ions with a 10% mass difference. Now, however, by use of the Hall Probe to monitor the magnetic field with a high degree of precision, there is scope for larger jumps between masses.

Quadrupole instruments are particularly suited to the use of SIM time programmes, *i.e.* acquiring different ions specific to different analytes during different regions of the chromatogram. They are therefore suited to screening samples for a large number of analytes. Ion traps are also used for multi-analyte screens, although they are usually operated in limited scanning mode (over an m/z range of approximately 30) as this offers the same sensitivity as SIM. Because it is less easy to switch acquisition between different ions on sector instruments, these are more suited to the analysis of single analytes or classes of analytes (*e.g.* dioxins), where high resolution is a necessity.

6.5.6 Tandem Techniques

In essence a tandem mass spectrometer consists of two mass spectrometers joined in series. A tandem mass spectrometer can be purchased as a dedicated instrument with the option of using it as a conventional mass spectrometer. This technique is useful for tackling particularly difficult samples, possibly with many interferents.

The first mass spectrometer acts as a clean-up stage either instead of or in addition to a conventional separation stage such as GC. The analyte is ionised as gently as possible, to form the unfragmented molecular ion (referred to as the 'parent' ion). This is then passed through the first mass spectrometer, separating it from any potential interferents. The parent ion is then fragmented by collision with a reactant gas. The resulting fragments (referred to as 'daughter' ions) are passed through the second mass spectrometer to produce a mass spectrum.

There are three main types of tandem techniques: hybrid tandem mass spectrometers (sector-quadrupole), triple stage quadrupole (quadrupole-quadrupole), and ion trap-quadrupole mass spectrometers. In the first two types the gas cell normally contains an additional set of quadrupole rods to collimate the ions (hence the term 'triple stage quadrupole'). These only operate magnetically using radiofrequency energy, as any applied voltage within the gas cell would produce arcing. On newer machines quadrupole rods in the collision cell are being replaced by hexapole rod assemblies.

Hybrid instruments offer a greater degree of selectivity; the initial sector instrument can pass an ion at high resolution before it is transmitted to a second stage. However, triple stage quadrupoles are more stable and robust.

There are three main ways of using a MS–MS system.

Daughter Ion Spectra

In concept this is the simplest MS–MS technique. The first MS is kept fixed to acquire the parent ion of the analyte sought. The second MS is scanned. Analytes are identified by their daughter ion spectra.

This type of MS–MS is very useful for confirming trace analytes in complex matrices, particularly as relatively little sample preparation is required. With a hybrid MS–MS the chances of detecting and quantifying an analyte in a complex mixture are increased. Unequivocal identification can be helped if collisional conditions are chosen such that not all of the parent ion is fragmented, *i.e.* the daughter ion spectrum gives an indication of relative molecular mass. Production of a daughter ion spectrum sometimes makes it possible to distinguish between isomers, *e.g.* 1,2-, 1,3-, and 1,4-disubstituted benzenes. This is normally impossible by conventional MS.

Parent Ion Spectra

The converse of daughter ion MS–MS is to scan the first mass spectrometer whilst the second remains fixed. The instrument is thus configured to pick out members of a related class of compound, which give similar daughter ion spectra. For example, if the second MS is set on m/z 149 any compound which gives a signal is likely to be a phthalate. As the mass is known of each of the parent ions which produces a 149 fragment, some idea can also be gained of which particular phthalates are present.[6]

Constant Neutral Loss Monitoring

The final option is to scan both the first and second MS simultaneously. They are both scanned at the same speed at a fixed range apart. For example, if the first MS always has an m/z value 35 units above that of the second MS, it is probable that any signal is the result of the parent ion fragmenting by the loss of chlorine. This configuration could thus be used for qualitatively screening samples for chlorinated compounds. This, again, is very useful for detecting trace quantities of an analyte in a mixture.

6.5.7 General Precautions

Sample Stability

Pyrolysis of samples can occur in several parts of the system, for example in the transfer line or the source. For this reason, careful temperature control has to be maintained. In general, no parts of the system should be heated unnecessarily; excessive temperatures can produce degradation and spectral distortion, with reduced molecular ion intensity. If parts of the instrument (*e.g.* the transfer line) have cold spots then ghosting can occur as a result of the deposition of trace amounts of analyte. Other problems that may occur are tailing of chromatographic peaks and low sensitivity.

Source temperature is an important factor in MS. Generally, a hot source will result in more extensive fragmentation. In EI, excessive fragmentation may lead to spectral distortion through the loss of high-mass ions. In CI, the aim is to achieve as little fragmentation as possible so as to obtain molecular mass information. This may not be achieved because of fragmentation brought about by an excessively hot source.

Instrumental Stability

Mass spectrometers are sensitive to their environmental conditions. Voltage stabilisation equipment is necessary in areas where the mains voltage is subject to variation and spikes. Mass spectrometers perform better at constant room temperature; in any case, these instruments produce a large amount of heat and some temperature control is normally essential. On sector machines, cooling water is employed to keep the magnet power supply at constant temperature. At high resolution, say 10 000, drift of the mass calibration can be quite pronounced if room and cooling water temperatures are not controlled to within 1–2°C.

Reproducibility

To achieve good reproducibility, all instrumental factors, such as source temperature, should be kept constant. Reproducibility will, of course, be affected by all parts of the system so that, in GC–MS, cool on-column injection will give better quantitative results than split injection. Similarly, in LC–MS, loop fill injection will give better results than partial fill injection. Reproducibility is also improved by allowing an instrument time to stabilise before use; it is not advisable to power up an instrument and begin work immediately.

Safety

Many of the safety aspects of MS are general to several types of analysis, for example, the use of toxic chemicals. However, some less common potential hazards exist. Several parts of the instrument operate at a very high voltage so there is a possible electrical hazard. For instruments that have non-armoured glass components under vacuum, safety precautions need to be in place. Although MS is a destructive technique, a toxic hazard still remains from sample residues left in various parts of the system. Vacuum pump oil is generally the most obvious source of contamination, and so discarded oil should normally be treated as toxic. Similarly, care should be taken while handling source parts during cleaning if toxic samples have been introduced.

6.5.8 Critical Aspects

Tuning

Most instruments have a software-controlled tuning procedure, usually with a visual representation of the reference peaks. Tuning, and calibration of the mass scale, should be done daily. The tuning parameters should be adjusted sequentially to obtain maximum sensitivity and mass resolution over the mass range of interest. However, the settings that give maximum sensitivity are not necessarily the best. Values should be chosen so that small changes up or down do not have a marked effect on the reference peaks; it is better to sacrifice a little sensitivity if it means the instrument is operating on a 'plateau'. Operating on a 'cliff' can adversely affect reproducibility.

In sector machines, which are normally tuned on an ion of perfluorokerosene (PFK), tuning affects sensitivity and resolution. It is not particularly difficult simply to achieve a high resolution. The demanding aspect is to achieve the maximum possible sensitivity at a given resolution. The tuning of a sector machine at high resolution, say 10 000, is very much an iterative process. All the source and other tuning controls are adjusted in turn several times to achieve gradual improvements in performance. The electron voltage and ion repeller voltage should be operated on a plateau. The optimum setting of these will, in fact, vary from instrument to instrument. The slit settings obviously play a major part in determining resolution. Very occasionally, to achieve an acceptable combination of resolution and sensitivity the flight tube will need baking (*i e.* heating overnight). Specific advice on this should be sought from the manufacturer.

Quadrupole instruments are often calibrated with perfluorotributylamine. This allows calibration to a relative molecular mass of over 600. Tuning is performed to give a spectrum resembling a corresponding sector mass spectrum, though the tuning can be adjusted to enhance or suppress the low end of the spectrum. Sensitivity can be increased by increasing the detector voltage, but with the high concentrations of co-extractives found in trace analytical samples there is the risk of overloading the detector at settings over 200–250 kV.

Tuning a quadrupole is simpler than performing high resolution tuning on a sector machine. Most modern quadrupole instruments incorporate software that enables automatic tuning.

Scan Speed

Scan speed is also a critical factor. It is widely appreciated that, if scanning is too slow, spectral distortion can occur, particularly in capillary column GC–MS.

Furthermore, accurate quantitative measurement from full scanning operation becomes difficult if the peaks are based on few data points. It is less widely realised that excessively fast scanning is also undesirable because weak ions can be missed. Although fast scanning gives excellent peak profiling, all mass spectrometers have a minimum interscan settling time. With a very fast scan, therefore, an instrument has a relatively high dead-time and a significant proportion of the analysis time is not spent on acquisition.

Solvent Delay

In GC–MS the filament of the ion source should not be turned on whilst the solvent front is passing through the detector or it is likely to burn out. For each chromatogram, most instruments allow the programming of a 'solvent delay' before the detector is activated. This may need to be up to 10 min, depending upon the solvent, the injection volume, and the solvent retention time.

6.5.9 Hints and Tips

Calibration

Many compounds have spectra containing abundant ions at low mass. In general, however, low mass ions are not very diagnostic. Although mass spectrometers operate at very low pressure, there is, in practice, always some air in the system. It is thus often, but not always, convenient to scan down to only m/z 33. This avoids recording the ions of oxygen, nitrogen, and water.

Calibration at Low Masses

When scanning at low masses the calibration can be unreliable resulting in some spectral mismatch. The mass range is only usually calibrated against reference gas ions, the lowest of which is rarely with m/z below 60–70. To improve calibration at low masses the reference table can be augmented with ions m/z 44 (carbon dioxide), 40 (argon), 32 (oxygen), 28 (nitrogen), 18 (water), or 17 (hydroxyl), any of which will be present from the air.

Spectral Contaminants

Any spectrum will include a contribution from system contaminants such as common sample impurities or column bleed. Hydrocarbons and adipate and phthalate plasticisers are often present in samples, and GC columns can bleed to produce siloxanes. Some of the more common contaminants are shown in Table 6.5.5.

Slow bleed is seen as a gradual temperature-dependent rise of the baseline. Bleed peaks occur because of column contamination: silicon compounds released by the septum can accumulate in the injection liner and on the front of a cold column. Column bleed and the appearance of bleed peaks are common to all GC

Table 6.5.5 *Values of m/z for some common system contaminant ions*

Compound type	Characteristic ions
Hydrocarbon	43, 57
Adipate	129
Phthalate	149
Siloxanes	73, 147, 221, 355
Cyclosiloxanes	207, 281

systems. However, these phenomena are best illustrated by GC–MS. Figure 6.5.2 shows the injection of 300 ng of cocaine onto a GC–MS system in which the GC column had been left for several days at 60°C with the usual helium carrier gas flow. Trace 6.5.2(d) shows the reconstructed (or total) ion chromatograph. Most of the peaks are due to siloxanes with a base ion of m/z 73; these appear in trace 6.5.2(a). Cocaine has a base ion with m/z 82 and is shown in trace 6.5.2(b). The ion with m/z 149 shown in the ion chromatogram 6.5.2(c) indicates the presence of a phthalate.

The best way of preventing column bleed peaks interfering with standard or analyte peaks is to inject some solvent onto the column, take the column through the programme to be used for the analysis (or one with a higher maximum temperature), and repeat the process if necessary. Two cycles of injection/ programming are occasionally beneficial for systems in poor condition. Contaminant peaks can be reduced by keeping the injection liner rigorously clean; this also prevents adsorption of trace components.

Low bleed columns designed specifically for GC–MS are being introduced. These offer the advantage of improved detection limits through reduction of chemical noise (reduction of signal-to-noise due to chemical effects); this is of particular significance in ultra-trace SIM work, typically at the picogram level. However, analyte selectivity can be affected and the analyst should not assume that the low bleed version of a column will perform in exactly the same manner as the conventional version.

Contaminants are often present at a much higher level than the compound(s) of interest. Background spectral subtraction can usually be employed to remove the spectrum of a component which partly overlaps an analyte. Background scans can be removed from before or after the chromatographic peak, or from both sides. Hence it is often possible to perform successful spectral subtraction for both early and late eluting peaks, providing that the background is not too intense relative to the signal, *e.g.* less than a 100-fold. An example is the analysis for cannabis in the presence of phthalates.[7]

Interference in Probe Analysis

In probe work m/z 44 (carbon dioxide) is sometimes seen as a result of analyte thermal decomposition before it reaches the source. For example, some antibiotics produce a CO_2 signal that is more intense than any of the other ions. To allow for

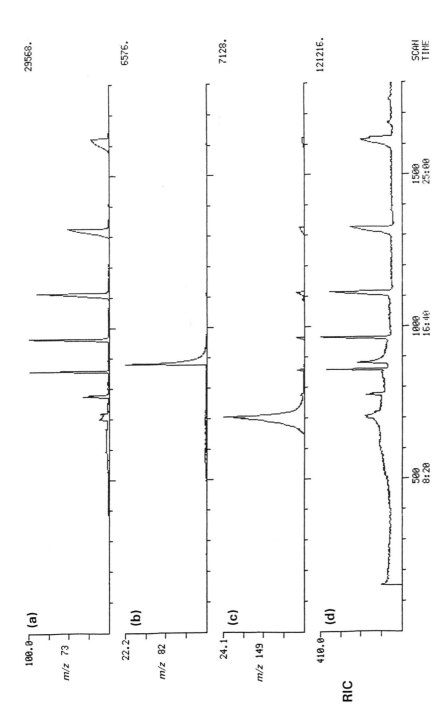

Figure 6.5.2 *Reconstructed ion chromatograph (RIC) + mass chromatograms from a 300 ng injection of cocaine onto a non-conditioned system*

this it is usual to only acquire ions above an m/z of 50, so that the 'false' ion at 44 is not apparent.

In probe–MS analysis, several compounds may come off the probe at the same temperature making normal background subtraction impossible. With some data handling systems it is possible to perform a different form of spectral subtraction. A spectral search is carried out on the total scan; the library will often give the best match for the component of highest concentration. The library spectrum of the best match can then be subtracted from the sample spectrum and the resultant spectrum searched against the library, generally producing a match for the component of next highest concentration. In some cases, it may be possible to show the presence of several components. It is most likely to be successful for components that have significant structural differences.

6.5.10 Uncertainty in Identification

Mixed Spectra

When sample introduction is by probe insertion, the analyst will often be aware that mixed spectra can occur. This situation is less likely to occur in GC–MS. If it is suspected that a spectrum is not due to a single component, ion chromatograms should be plotted for several of the ions. Even if two compounds co-elute and background subtraction is not possible, there will often be an indication of whether a spectrum is mixed, *e.g.* non-common ions from co-eluting compounds will often have different peak shapes.

Spectral Distortion

At or near the detection limit of an instrument significant spectral distortion can occur. This is manifested in the relatively low intensity of the heavier ions in the spectrum. Where a slow scanning speed relative to the peak width has been employed, the concentration of the compound varies during the scan. For example, the apex of a chromatographic peak may be passing through the detector as a scan starts, but by the time the scan is completed only the tail is left. As most scans work from high to low mass, this would result in overabundant higher-mass ions relative to the lower mass ions.

Another distortion effect is manifested in SIM. This occurs when the limit of determination is such that the major intense ion can be quantified, but the less intense qualifier ion is below the limit of accurate determination. This alters the apparent ratio of the two ions.

Random Ions

Random ion events occur as a result of very small instabilities in electronic circuitry and in electron multiplier operation. Random ions are not true ions but electrical interference. For example, if an ion of m/z 400 appears in a quadrupole mass spectrum where no such ion was really present, this denotes that the random event occurred somewhere in the system just as the rods were set to allow an *m/z* 400 ion to pass. Random ions are usually only significant when an instrument is working close to its detection limit. They appear as very narrow 'peaks', resembling a vertical line, or 'spike', rather than a two-dimensional peak.

Mass Discrimination

The relative abundances of ions in a mass spectrum, produced using a magnetic sector instrument under standard conditions, are generally accepted to be reproducible, independent of the instrument, and library searchable. Although instrument settings, sample concentration, and the actual instrument will have minor influences on the spectrum, distortion on sector machines is rarely particularly pronounced. On quadrupole instruments, marked spectral distortion can occur. Dirty quadrupole rods, in particular, cause high mass suppression. The time that the rods can operate between cleaning depends on the number and type of samples the instrument handles. A single injection of silicones can be sufficient to render high masses virtually absent in subsequent spectra. Rod cleaning can frequently be accomplished by simply baking the instrument housing. Some instruments are supplied with small 'pre-filter' rods, which accumulate the majority of the contamination and are easier to clean than the main rods.

6.5.11 Uncertainty in Quantitation

Suppression

Signal suppression can be a serious problem in SIM MS. It occurs as co-eluting compounds reduce the source sensitivity and therefore the signal response. The co-eluent will not usually be apparent, because it does not fragment into the ions that are being monitored, but it can commonly cause a signal to drop to 10–30% of its unsuppressed value. This has a profound effect when comparing sample extracts with standards prepared in solution that are free of interferents. Unless the problem is addressed, quantitative results cannot be assumed from SIM. There are a number of approaches to circumvent the effect, but none is guaranteed.

Correction Against a Constant Reference Signal. Most sector instruments have the option of bleeding a constant flow of PFK reference gas into the source during operation in order to calibrate the mass scale. It is also common practice in SIM to acquire and plot a PFK ion; any suppression at the time of analyte elution will cause a reduction in PFK response, somewhat like a negative peak. The PFK trace will show up any suppression and allow a correction to be made. For example, if suppression reduces system sensitivity by half, this indicates that the apparent result should be doubled. This is a semi-quantitative approach. The only absolute way to correct for suppression is to use Isotope Dilution MS (see Section 6.6).

Benchtop quadrupoles do not have the facility to introduce a reference gas during a run, and so an alternative compensation method must be used.

Table 6.5.6 *Examples of applications and techniques*

Analysis type	Technique	Reference
Tetra- to octa-chloro dioxins and furans	EI GC–MS, high resolution SIM	8
Large biomolecules	Electrospray–MS	9, 10
Bile acids	Thermospray LC–MS–MS	11,12
Drugs of abuse	EI/CI GC–MS and probe–MS	13
Polymer additives	SFC–MS	14,15
Peptides and proteins	FAB–MS	16
Clinical	FAB electrospray	17
Sulfazo dyes	Capillary electrophoresis–MS–MS	18–20

Use of Internal Standardisation. Internal standards should always be used for capillary GC analyses. However, they are of limited use for correcting suppression effects. As suppression is caused by substances co-eluting with the analyte, an internal standard must be chosen that elutes at the same time in order to mimic the signal reduction. Internal standards will not, therefore, be of use for multi-analyte determinations.

Use of Matrix Matched Standards. Matrix matched standards are recommended for capillary GC analyses. They may or may not help with signal suppression, depending upon the similarity between the matrix extract used to prepare the standards and the sample matrix extract. The effectiveness is likely to vary from sample type to sample type.

Varying the Tuning Parameters. Increasing the repeller voltage can sometimes reduce suppression, as ions are forced out of the source as soon as possible after they are formed. This has a minimal effect upon either sensitivity or filament life. Lowering the emission current can also help, but as this markedly reduces sensitivity it may be impractical.

Varying the Oven Temperature Programme. It may be that, by reducing the rate of change of the oven temperature, the co-extractives causing the suppression can be chromatographically separated from the analyte.

Enhancement

Mass spectrometry enhancement is much rarer than suppression and normally will occur only with a badly tuned source. If a source parameter, *e.g.* repeller setting, is not operating on a plateau, the small but significant change in source conditions that occurs when the analyte enters may cause increased sensitivity and thus signal enhancement.

Scan Mode at Low Signal Strength

Uncertainty in a result can occur simply from signal strength. In MS, very weak signals are difficult to measure accurately. This arises for statistical reasons, related to ion counting. Quantitative measurement of small amounts of analytes is thus more accurate when performed in SIM rather than scanning mode.

6.5.12 Applications

Table 6.5.6 gives a selection of examples of some of the techniques mentioned.

6.5.13 References

1. National Institute of Standards and Technology, 'NIST/EPA/NIH Mass Spectral Database', Gaithersburg, MD, USA, 1992.
2. Mass Spectrometry Data Centre, 'Eight Peak Index of Mass Spectra', The Royal Society of Chemistry, Cambridge, UK, 4th edn, 1991.
3. 'The Wiley/NBS Registry of Mass Spectral Data', John Wiley, Chichester, UK, 1989.
4. G. A. MacKay and G. D. Reed, 'Application of Capillary SFC, Packed Column SFC and Capillary SFC MS in the Analysis of Controlled Drugs', *J. High Resolut. Chromatogr.*, 1991, **14**, 537.
5. R. D. Smith, J. H. Wahl, D. R. Goodlett, and S. A. Hofstadler, 'Capillary Electrophoresis–Mass Spectrometry', *Anal. Chem.*, 1993, **65**, 574A.
6. K. L. Busch, G. L. Glish, and S. A. McLuckey, 'Mass Spectrometry/Mass Spectrometry – Techniques and Applications of Tandem Mass Spectrometry', VCH, New York, 1988.
7. H. S. Sachdev and G. D. Reed, 'Spectroscopic Analysis of Illicit Drugs and their Containment Materials', *Spectrosc. World*, 1990, **2**, 27.
8. United States Environmental Protection Agency, 'Method 1613: Tetra- through Octa-chlorinated Dioxins and Furans by Isotope Dilution HRGC/HRMS', Revision A, Office of Water, Woburn, MA, 1990.
9. R. D. Smith, J. A. Loo, R. R. O. Loo, M. Busman, and H. R. Udseth, 'Principles and Practice of Electrospray Ionization–Mass Spectrometry for Large Polypeptides and Proteins'. *Mass Spec. Rev.*, 1991, **10**, 359.
10. C. E. Edmonds, J. A. Loo, R. R. O. Loo, H. R. Udseth, C. J. Barinaga, and R. D. Smith. 'Applications of Electrospray Ionization (ESI) Mass Spectrometry and Tandem Mass Spectrometry in Combination with Capillary Electrophoresis for Biochemical Investigations', *Biochem. Soc. Trans.*, 1991, **19**, 943.
11. C. Eckers, N. J. Haskins, and T. Large, 'On Line Negative-ion Liquid Chromatography–Mass Spectrometry for the Analysis of Bile Acids using Low- and High-resolution Mass Spectrometry', *Biomed. Environ. Mass Spectrom.*, 1989, **18**, 702.
12. C. Eckers, A. P. New, P. B. East, and N. J. Haskins, 'The Use of Tandem Mass Spectrometry for the Differentiation of Bile Acid Isomers and for the Identification of Bile Acids in Biological Extracts', *Rapid Commun. Mass Spectrom.*, 1990, **4**, 449.
13. J. Yinon, 'Forensic Applications of Mass Spectrometry', *Mass Spectrom. Rev.*, 1991, **10**, 179.
14. G. A. MacKay and R. M. Smith, 'Supercritical-fluid Extraction and Chromatography with Mass Spectrometry of Labile Polymer Additives', *Anal. Proc.*, 1992, **29**, 463.
15. G. A. MacKay and R. M. Smith, 'Supercritical-fluid Extraction and Chromatography–Mass Spectrometry of Flame Retardants from Polyurethan Foams', *Analyst*, 1993, **118**, 741.

16. R. M. Caprioli, 'Continuous-flow Fast-Atom-Bombardment Mass Spectrometry', *Anal. Chem.*, 1990, **62**, 477A.

17. R. Caprioli and A. H. B. Wu, 'Mass Spectrometry', *Anal. Chem.*, 1993, **65**, 470R.

18. E. D. Lee, W. Mück, J. D. Henion, and T. R. Covey, 'Capillary Zone Electrophoresis–Tandem Mass Spectrometry for the Determination of Sulphonated Azo Dyes', *Biomed. Environ. Mass Spectrom.*, 1989, **18**, 253.

19. J. Yinon, T. L. Jones, and L. D. Betowski, 'High-sensitivity Thermospray–Ionization Mass Spectrometry of Dyes', *Biomed. Environ. Mass Spectrom.*, 1989, **18**, 445.

20. E. D. Lee, W. Mück, J. D. Henion, and T. R. Covey. 'Liquid Junction Coupling for Capillary Zone Electrophoresis–Ion Spray Mass Spectrometry', *Biomed. Environ. Mass Spectrom.*, 1989, **18**, 844.

6.6 Isotope Dilution Mass Spectrometry (IDMS)

6.6.1 Overview

The technique of IDMS is generally accepted as a definitive method for both organic and inorganic analytes due to its high precision and accuracy. It was initially developed during the 1950s for elemental analysis and extended into the field of organic compounds in the 1970s. Thus whilst elemental analysis by IDMS is fairly well developed, in the case of organic analysis it is still developing.

The basic principle of the technique is that a known amount of an enriched isotope of the element to be determined (or an isotopically labelled analogue of the analyte in the case of organic analytes) is added to the sample and the changed isotopic ratio is measured on a portion of the sample using a mass spectrometer, so enabling the analyte concentration to be calculated.[1] This technique is accurate, sensitive, and relatively free from interference effects and is generally classified as a definitive method. A major advantage is that once isotope dilution has taken place, loss of analyte during isolation has no influence on the analytical result, thus making this the best form of internal standardisation. The use of IDMS in the inorganic analytical field is discussed in Section 4.3.

With the increasing use of MS in the field of analytical chemistry, essentially due to the advent of new, compact and economic instrumentation, *e.g.* ones with the ion trap detector, IDMS is playing an increasingly important role in trace analysis. Notwithstanding this, there are several reasons why, in spite of the unrivalled accuracy and precision available with this technique, it is not more widely used. The method requires spiking with an isotopic analogue of the analyte and this should be of a known high purity. Such isotopic analogues are very expensive. In principle, the method is simple and allows for knowledge or control of all the variables that can lead to error. In practice, achieving accurate results requires careful design of the experiment and considerable attention to detail, and hence is quite time consuming. Such factors have led to a slow build-up of the technique, although in some analytical areas, *e.g.* clinical analyses, where matrices can lead to difficult analyses, it has become popular.

The type of mass spectrometer and the way in which data are collected can influence the analytical result in terms of accuracy and precision. Experimental work has shown that quadrupole spectrometers can give poorer precision when compared with magnetic sector instruments operating at the same mass resolution. Using a mass spectrometer in a scanning mode to acquire ion abundance data is much less sensitive than acquiring data in a SIM mode, leading to poorer precision. Care must also be taken in the choice of isotopic analogue in IDMS so as to ensure that the monitored analogue ion is as interference free as possible.

6.6.2 Comparison of Techniques

There are a number of alternatives available when carrying out IDMS, including choice of inlet system, ionisation mode, scanning mode, and method of calculation of results.

Inlet Systems

A range of inlet systems is normally available. These include direct, GC, HPLC, and SFC. The GC inlet is most widely used for IDMS, utilising capillary column GC–MS.

Ideally the analyte and labelled analogue should have the same retention time on the GC column in order that the relevant ions can be monitored simultaneously under the same MS conditions. The use of a capillary column may bring about a considerable separation between the analyte and the labelled analogue even if there is only a small difference in mass. Under these circumstances it may be preferable to use a packed column. It should be noted, however, that when using a packed column some form of separator is necessary to remove the carrier gas; this will introduce a mass discrimination effect into the measurement.

It may be necessary to use HPLC and SFC inlets for involatile compounds not amenable to GC–MS; however, care must be taken to avoid mass discrimination in the use of these inlets. Indeed it may be preferable to form a volatile derivative of the analyte and employ GC–MS. This could also confer the advantage of a higher relative molecular mass and so help to avoid interfering ions. The direct inlet (heated direct insertion probe) may be used for involatile analytes. Care must be taken to ensure that the analyte is evaporated from the probe quickly and evenly (a temperature programmable probe is useful). However, even using this method of analyte introduction some separation can occur between the analyte and the labelled analogue. A comparison between direct and GC introduction techniques for the same level of the same analyte shows that the GC inlet gives superior reproducibility. Table 6.6.1 shows the results obtained for naphthalene-d_8.

Ionisation Mode

The selection of the ions to be monitored is of the utmost importance since it will influence both the specificity and the sensitivity of the IDMS analysis. Normally, for the greatest specificity, the molecular ion will be monitored if it is sufficiently intense and in this context the optimal mode of ionisation should be used. Where molecular ions are weak under EI ionisation they may be strong under CI. The use of the HPLC-coupled ionisation techniques of thermospray, electrospray, ionspray,

Table 6.6.1 *Comparison of inlets for a sample of naphthalene-d₈*

	Direct probe	Inlet GC
Mean naphthalene-d_8 (mol%)	42.89	43.33
Sample standard deviation	1.19	0.23
Coefficient of variation (%)	2.77	0.53

and particle beam should also be considered. If fragment ions, as opposed to molecular ions, have to be monitored, they should be at high mass values so as to keep interferences from instrument background, column bleed, and co-eluting sample components to a minimum.

Scanning Mode

When acquiring spectra, mass spectrometers are normally scanned through the mass range of interest, *e.g.* by scanning the magnetic field of a magnetic sector instrument. When acquiring ion abundance data from a small number of ions, as is the case in IDMS, the technique of SIM can be utilised. Whilst either of these modes of operation may be used in IDMS, the SIM technique has significant advantages which render it the method of choice in virtually all cases.

In order to achieve high accuracy and precision the ion abundance measurement of each isotopic species must be made as near to simultaneously as possible, *i.e.* fast switching between the masses. The cycle time in SIM mode is far more rapid than that of the scanning mode, and the length of time spent measuring each ion of interest is longer in the SIM mode, leading to enhanced sensitivity. Indeed sensitivity can be up to 1000 times greater in SIM mode than in the scanning mode. One drawback of using the SIM mode is that a systematic error of mass discrimination is introduced when using this technique on magnetic sector instruments employing electrostatic analyser (ESA) voltage switching. This can easily be corrected for since calibration ratio solutions are equally affected.

Calculation of Results

There are two ways in which results may be calculated in IDMS analysis, either by a graphical method or by a bracketing procedure.[2-7] It is useful here to consider the basic IDMS equation. Consider the case of a mixture of N moles of analyte and M moles of isotopically labelled analogue internal standard; the inverse isotope ratio R is given by Equation (1).[8]

$$R = \frac{(N/M)P_A + q_A}{(N/M)P_B + q_B} \qquad (1)$$

where P_A is the molecular ion intensity of the analyte, P_B is the ion intensity due to the analyte at the mass corresponding to the molecular ion of the labelled analogue, q_B is the molecular ion intensity of the labelled analogue, and q_A is the ion intensity due to the labelled analogue at the mass corresponding to the molecular ion of the analyte.

The relationship between the MS measured isotope ratio R and the mole ratio is a hyperbolic function, not a linear one. There are four situations that can occur in IDMS analysis.

P_B and q_A are Zero

The first is the ideal case in which the analyte does not have an overlapping ion at the same mass as the molecular ion of the analogue and the analogue does not have an overlapping ion at the same mass as the molecular ion of the analyte. Thus a plot of R against N/M is a straight line with a y intercept of zero and a slope of P_A/q_B. This slope will be close to unity but deviation will occur if there is a significant mass discrimination between the two masses or if there is an isotope effect that influences the ionisation efficiency or fragmentation of the labelled analogue.

P_B is Zero

The second case occurs where the analyte does not have an overlapping ion at the same mass as the molecular ion of the analogue but the analogue does have an overlapping ion at the same mass as the molecular ion of the analyte. The plot of R against N/M will still be a straight line but with the y intercept at q_A/q_B.

q_A is Zero

The third case is the reverse of the second. The analogue does not have an overlapping ion at the same mass as the molecular ion of the analyte but the analyte has an overlapping ion at the same mass as the molecular ion of the analogue. The plot of R against N/M will now be a half-hyperbola with a y intercept at the origin.

P_B and $q_A \neq Zero$

The fourth case is the most complex but it is the situation that is most frequently encountered. Both the analyte and analogue have overlapping ions at the masses of interest. The plot of R against N/M will be a half-hyperbola with a y intercept of q_A/q_B.

Calibration samples are mixtures of different amounts of the analyte calibration material to a fixed amount of the labelled analogue material; the ion ratios for these different mole ratios are recorded. The calculated inverse isotope ratios are represented graphically and the unknown concentration of the analyte estimated from this graph. However, precision and accuracy are improved when the calibration data are fitted to an equation from which the concentration can be predicted. Linear regression analysis can only be applied to the first two situations.

Apart from the problem of non-linearity the calibration curve approach does have the further disadvantage in that measured ion abundance ratios can change with time since calibration and sample measurements cannot be made simultaneously.

The alternative is to use a bracketing procedure. This involves the measurement of each sample between measurements of calibration standards whose ion abundances most closely surround the ion abundance ratio of the sample. Measurement must be made to a strict protocol and the molar ratio of analyte to labelled analogue in the sample is calculated by linear interpolation between the bracketing mixtures. Excellent precision and accuracy (relative standard deviation better than 1%) can be achieved using this procedure.

6.6.3 General Precautions

In order to obtain high accuracy and precision when undertaking IDMS the following factors should be taken into account.

Purity of Isotopic Analogue

It is essential that the isotopic and chemical purity of the isotopically labelled analogue is known as accurately as possible. Mass spectrometric measurements can be made in an iterative manner to establish isotopic purity. A survey of manufacturers has shown that, in general, labelled analogues are available to a purity of 99% or better. A check of the isotopic purity of a labelled analogue can be made using a certified reference sample of the unlabelled analogue. A 50:50 mixture is measured and this is used to estimate the mass bias effect. The labelled analogue is then measured to determine its percentage purity, using the estimated mass bias correction to correct the signal. A correction is then applied to the 50:50 mixture, which modifies the mass bias effect, and the process is repeated until no further change in values is obtained. In such cases accurate isotopic purity measurements can be made with only one or two iterative cycles.

Spiking Level

For optimum accuracy and precision the ratio of analyte to labelled analogue should be that which produces equal ion signal responses in the MS measurement, *i.e.* there should be a similar amount of labelled analogue compared with the analyte. Care should be taken to ensure that this can be achieved. In some cases this may mean carrying out an initial, quick, semi-quantitative analysis to establish the analyte level. Experimental work has shown that the standard deviation of measurements on 1:1 mixtures are significantly better than for other ratios. Errors introduced by, for example, an inaccurately set electrical zero level of the mass spectrometer, are minimised if such a ratio can be used.

Carrier Effect

Notwithstanding the comments made in the section 'Spiking Level', in certain circumstances, particularly at the ultra-trace level, sensitivity can be improved by using a large excess of labelled analogue to reduce adsorptive losses of the analyte during GC–MS (the 'carrier effect').[9] This has the effect of decreasing the precision of the analysis and so whenever possible elimination or minimisation of adsorptive losses is preferable to the use of a 'carrier'.

Analyte Level

For highest accuracy and precision the analyte level should be high. In order to achieve this attention should be directed towards the extraction, clean-up, and extract concentration stage. This has been clearly demonstrated by experimental work where a dilution of the analyte/isotopic analogue mixture by a factor of 1000 (2.5 ppm to 2.5 ppb) resulted in a worsening of the standard deviation of the ratio measurement by a factor of 50. However, the concentration should not be so high as to overload the sample introduction system, *e.g.* the capillary column if using GC–MS.

Instrumental Stability

In the ideal situation the analyte and labelled analogue ion are monitored simultaneously. However, this is not possible unless the instrument is an isotope ratio mass spectrometer with multiple detectors which are set for a specific application, *e.g.* at mass 45 and 44 for labelled and unlabelled CO_2. Consequently measurements must be made sequentially and any instability in the instrument will cause errors. Normally repeated alternate measurements are made on the analyte and analogue ions to minimise the effects of any instability present.

Safety

Normally the use of isotopically labelled analogues does not introduce any additional hazards over and above those that may be present in the use of mass spectrometer systems, or are present in the analysis of the analyte itself. The isotopes used in IDMS are stable isotopes. There is, in general, no reason for using radioisotopes for this type of analysis; this would introduce a radiation hazard.

6.6.4 Critical Factors

There are a number of critical points in the analytical procedure for IDMS analysis that must be considered.

Homogeneity

For accurate sampling the material must be homogeneous. This is a particularly critical stage of the analysis since, together with the precise addition of the labelled analogue and the accurate measurement of calibration solutions, it is the only step where eventual losses are not compensated for by the presence of the labelled analogue. It is also important to ensure that full equilibration between analyte and labelled analogue is achieved in order to ensure that the analyte and analogue behave identically during the analytical procedure. It is, however, difficult to ensure that full equilibration has taken place.

Choice of Isotopic Label

The labelled analogue of the analyte will, in general, be labelled with a stable isotope such as ^2H, ^{18}O, ^{15}N, or ^{13}C. The relative molecular mass of the analogue should, if possible, be at least three mass units greater than that of the analyte. This is to avoid overlap of the natural isotopes of the analyte with the mass value of the labelled analogue.

Care must be taken in the choice of analogue. This is particularly true in the case of ^2H, where isotopic exchange is liable to take place during isolation of the analyte from the sample, giving rise to random and systematic errors. Again, in order to avoid discrimination during the various steps of the analytical procedure the label should not be in a position that affects solvation properties, pK_a values, or derivatisation kinetics (where applicable).

Interfering Ions

For an organic sample extract the ion current at a given mass is almost never the result of a single ionic species. Multiple isotopic species and fragments of different elemental composition produce ions of the same nominal mass which complicates the calibration of analyte concentration. If available, high resolution MS can be used in some cases to differentiate between such ions.

Data Acquisition

It is important that the analyte and labelled analogue ion abundance measurements are made as close together, in time, as possible in order to minimise instrumental instability effects and variation of ionic concentration such as occurs in GC–MS. When acquiring data in the SIM mode via capillary GC–MS it has been found that the standard deviation improves by a factor of 30 by reducing the sampling time per ion (two ions monitored) from 100 ms (scan time 210 ms) to 35 ms (scan time 80 ms).

6.6.5 Hints and Tips

Instrument Set-up

Care must be taken in setting up the mass spectrometer prior to analysis. A significant source of imprecision is the instability of the instrument. A long warm-up period (1–2 h) is usually necessary and the electronic signal level must be correctly set and not drift. In the case of the latest computer-controlled instruments such checks can be carried out automatically.

Variation of the mass spectrometer source pressure due to temperature ramping of the GC column whilst the analyte and isotopic analogue are eluting could lead to enhancement or suppression of the signal. It is a useful precaution to ensure that the GC temperature programme has reached an isothermal stage when recording ion abundance data.

Interferences

The presence of interferences can be detected by monitoring more than one characteristic ion of both the analyte and the labelled analogue. A change in the peak ratio of the different channels will be seen, although when using this technique sensitivity and precision can be adversely affected. It may be more satisfactory to run the sample without the labelled analogue present to detect the presence of interferences at the mass value of the labelled analogue. When using magnetic sector spectrometers problems of interferences can often be overcome by operation of the mass spectrometer at high resolution, just sufficient to resolve the interference from the analyte and labelled analogue. Tandem MS can be utilised to monitor a significant daughter ion of the relevant analyte/analogue ion.

Sensitivity and Specificity

There are a number of requirements to achieve specificity and maximum sensitivity. These are listed in Table 6.6.2. For a given analysis the items listed should be taken into account when considering how the analysis should best be carried out.

Table 6.6.2. *Sensitivity and specificity requirements of IDMS*

Sensitivity	Specificity
The number of ions detected simultaneously should be minimised	The mass of the specific ions monitored should be high
Ions measured should be of high intensity, derivatisation may be necessary	Several ions should be used
	The clean-up procedure following extraction should give as pure an extract as possible
The optimum mode of ionisation should be chosen	The material should be analysed in the absence of the labelled analogue to show
Optimise mass spectral resolution	that interferences do not occur at the
Absence of interferences	mass values of the analogue
Absence of adsorption on the GC column or elsewhere	

6.6.6 Uncertainty

The uncertainty of the measured ratio is caused by errors associated with the technique. The errors considered here are systematic and random errors.

Systematic errors can be measured and corrections made. However, random errors will lead to a loss of accuracy and precision. The following instrumental sources of error have been investigated in order to assess their significance when measuring isotope ratios.

Inlet System

Normally a direct inlet is used for solid samples, and a GC inlet for mixtures of volatile compounds. Inlet systems are discussed in Section 6.6.2. It has been found in practice that provided there is no significant separation of analyte and labelled analogue when using the GC inlet then this should not make a significant contribution to the uncertainty of the measurement.

Scanning Mode

This is discussed in Section 6.6.2. Although a systematic error is introduced when using magnetic sector instruments in the SIM mode using ESA voltage switching, this technique is widely used since it is up to 1000 times more sensitive than the magnet scanning technique.

Electrical Zero Level

Care must be taken to correctly set the electrical zero level when recording ion abundances. Failure to do so will result in large errors for masses of relatively low abundance. As stated in Section 6.6.2, for optimum accuracy and precision the amount of labelled analogue added to the sample should result in equal ion abundances for the analyte and labelled analogue. This will minimise the effect of errors in setting the zero level.

Mass Spectrometer Resolution

The resolution used should be just sufficient to resolve the analyte and labelled analogue from any interferences present. A higher resolution will result in a lower sensitivity, which will have an adverse effect on detection limits and reproducibility. Experimental work has shown that at high (1 in 10 000) resolution an apparent mass discrimination effect can take place when using the SIM mode of scanning. This is attributed to a relative defocusing of the labelled analogue ion leading to an apparently lower ion abundance measurement. This effect is not apparent at lower (1 in 1000) resolution since focusing is not so critical here.

Instrument Stability

This is discussed in Section 6.6.3 and is an important factor to be taken into account.

General Points on Accuracy and Precision

Table 6.6.3 details some significant requirements for achieving maximum accuracy and high precision. With care in use of the technique accuracy should be of the order of ±1%. A useful discussion of instrumental factors affecting precision has been published.[10]

Table 6.6.3 *Requirements for high accuracy and precision when using IDMS*

Accuracy	Precision
No known source of interference	Relatively large amounts of analyte should be used
Pure reference materials and reagents should be used	A specific ion of high intensity should be used, which is not subject to interference from column bleed ions or from other sources
There must be full equilibration between the labelled analogue and analyte	
Care in preparation of labelled analogue standard; gravimetric procedures are superior to pipetting	The instrument should be as stable as possible
If analyte-free material is available then analysis should fail to detect the component of interest	The optimum mass spectral resolution must be used
A suitable standard curve or a bracketing procedure should be used	
The amount of analyte determined should be independent of the actual concentrations of the sample and analogue	

6.6.7 Applications

The applications listed in Table 6.6.4 have been arranged, in the main, according to classes of compounds. The references provide an entry point into the literature.

Table 6.6.4 *IDMS analysis references*

Compound or class of compound	Reference
Dioxins	11, 12
PCBs	12, 13
Herbicides	14
Pesticides	15
Plasticisers	16–18
Semi-volatile toxic organic pollutants	19
Steroid hormones	20–22
Aniline	23
Benzene	24
Cholesterol	25–27

(continued overleaf)

PCBs = polychlorinated biphenyls.

6.6.8 References

1. K. G. Heumann, 'Isotope-Dilution Mass Spectrometry of Inorganic and Organic Substances', *Fresenius' Z. Anal. Chem.*, 1986, **325**, 661.
2. J. F. Pickup and K. McPherson, 'Theoretical Considerations in Stable Isotope Dilution Mass Spectrometry for Organic Analysis', *Anal. Chem.*, 1976, **48**, 1885.
3. B. N. Colby and M. W. McCaman, 'Comparison of Calculation Procedures for Isotope-Dilution Determinations Using Gas Chromatography–Mass Spectrometry', *Biomed. Mass Spectrom.*, 1979, **6**, 225.
4. E. D. Bush and W. F. Trager, 'Analysis of Linear Approaches to Quantitative Stable-Isotope Methodology in Mass Spectrometry', *Biomed. Mass Spectrom.*, 1981, **8**, 211.
5. J. A. Jonckheere, A. P. De Leenheer, and H. L. Steyaert, 'Statistical Evaluation of Calibration Curve Non-Linearity in Isotope-Dilution Gas Chromatography–Mass Spectrometry', *Anal. Chem.*, 1983, **55**, 153.
6. J. F. Sabot, B. Ribon, L. P. Kouadio-Kouakou, H. Pinatel, and R. Mallein, 'Comparison of Two Calculation Procedures for Gas Chromatography–Mass Spectrometry Associated with Stable Isotope Dilution', *Analyst*, 1988, **113**, 1843.
7. W. T. Yap, R. Schaffer, H. S. Hertz, E. V. White, and M. J. Welch, 'On the Difference Between Using Linear and Non-Linear Models in Bracketing Procedures in Isotope-Dilution Mass Spectrometry', *Biomed. Mass Spectrom.* 1983, **10**, 262.
8. D. A. Schoeller, 'Mass Spectrometry: Calculations' *J. Clin. Pharmacol.*, 1986, **26**, 396.
9. L. Dehennin, A. Reiffsteck, and R. Scholler, in 'Stable Isotopes', ed. H. L. Schmidt, H. Forestel, and K. Heinzinger, Elsevier, Amsterdam, 1982, pp. 617–622.
10. D. A. Schoeller, 'Model for Determining the Influence of Instrumental Variations on Long-Term Precision of Isotope-Dilution Analyses', *Biomed. Mass Spectrom.*, 1980, **7**, 457.
11. Y. Tondeur, W. F. Beckert, S. Billets, and R. K. Mitchum, 'Method 8290: An Analytical Protocol for the Multi-Media Characterization of Polychlorinated Dibenzodioxins and

Dibenzofurans by High-Resolution Gas Chromatography–High Resolution Mass Spectrometry', *Chemosphere*, 1989, **18**, 119.

12. J. A. Van Rhijn, W. A. Traag, P. F. Van de Spreng, and L. G. M. T. Tuinstra, 'Simultaneous Determination of Planar Chlorobiphenyls and Polychlorinated Dibenzo-*p*-dioxins and -furans in Dutch Milk Using Isotope Dilution and Gas Chromatography–High-Resolution Mass Spectrometry', *J. Chromatogr.*, 1993, **630**, 297.

13. D. W. Kuehl, B. C. Butterworth, J. Libal, and P. Marquis, 'Isotope-Dilution High-Resolution Gas-Chromatographic–High-Resolution Mass-Spectrometric Method for the Determination of Coplanar Polychlorinated Biphenyls: Application to Fish and Marine Mammals', *Chemosphere*, 1991, **22**, 849.

14. L. Q. Huang, 'Simultaneous Determination of Alachlor, Metolachlor, Atrazine and Simazine in Water and Soil by Isotope-Dilution Gas Chromatography–Mass Spectrometry', *J. Assoc. Off. Anal. Chem.*, 1989, **72**, 349.

15. H. Behzadi and R. A. Lalancette, 'Extending the Use of US EPA Method 1625 for the Analysis of 4,4'-DDT, 4,4'-DDD (TDE), and 4,4'-DDE', *Microchem. J.*, 1991, **44**, 122.

16. J. Gilbert, L. Castle, S. M. Jickells, A. J. Mercer, and M. Sharman, 'Migration From Plastics into Foodstuffs Under Realistic Conditions of Use', *Food Addit. Contam.*, 1988, **5**, 513.

17. J. R. Startin, I. Parker, M. Sharman and J. Gilbert, 'Analysis of bis(2-ethylhexyl) adipate Plasticizer in Foods by Stable Isotope-Dilution Gas Chromatography–Mass Spectrometry', *J. Chromatogr.*, 1987, **387**, 509.

18. L. Castle, J. Gilbert, S. M Jickells and J. W. Gramshaw, 'Analysis of the Plasticizer tributyl acetylcitrate in Foods by Stable-Isotope-Dilution Gas Chromatography–Mass Spectrometry', *J. Chromatogr.*, 1988, **437**, 281.

19. US EPA, 'Semi-volatile Organic Compounds by Isotope Dilution GC/MS'. Method 1625, Revision C, Office of Water Regulations and Standards, Industrial Technology Division, Woburn, MA, June 1989.

20. A. Reiffsteck, L. Dehennin, and R. Scholler, 'Oestrogens in Seminal Plasma of Human and Animal Species: Identification and Quantitative Estimation by Gas Chromatography–Mass Spectrometry Associated with Stable-Isotope Dilution', *J. Steroid Biochem.*, 1982, **17**, 567.

21. L. Dehennin, 'Oestrogens, Androgens and Progestins [progestogens] in Follicular Fluid from Pre-Ovulatory Follicles: Identification and Quantification by Gas Chromatography–Mass Spectrometry Associated with Stable-Isotope Dilution', *Steroids*, 1990, **55**, 181.

22. T. Furuta, K. Kusano, and Y. Kasuya, 'Simultaneous Measurements of Endogenous and Deuterium-Labelled Trace Variants of Androst-4-ene-3,17-dione and Testosterone by Capillary Gas Chromatography–Mass Spectrometry', *J. Chromatogr. Biomed. Appl.*, 1990, **525**, 15.

23. K. Jacob, W. Voty, C. Krauss, G. Schnabl, and M. Knedel, 'Selected-ion Monitoring Determination of Mono- and Bi-Functional Amines by Using Phosphorous-Containing Derivatives', *Biomed. Mass Spectrom.*, 1983, **10**, 175.

24. G. D. Byrd, K. W. Fowler, R. D. Hicks, M. E. Lovette, and M. F. Borgerding, 'Isotope-Dilution Gas Chromatography–Mass Spectrometry in the Determination of Benzene, Toluene, Styrene and Acrylonitrile in Mainstream Cigarette Smoke', *J. Chromatogr.*, 1990, **503**, 359.

25. A. Cohen, H. S. Hertz, J. Mandel, R. C. Paule, R. Schaffer, L. T. Sniegoski, T. Sun, M. J. Welch, and E. V. White, 'Total Serum Cholesterol by Isotope-Dilution/Mass Spectrometry: A Candidate Definitive Method', *Clin. Chem.*, 1980, **26**, 854.

26. P. Ellerbe, S. Meiselman, L. T. Sniegoski, M. J. Welch, and E. V. White, 'Determination of Serum Cholesterol by a Modification of the Isotope-Dilution Mass-Spectrometric Definitive Method', *Anal. Chem.*, 1989, **61**, 1718.

27. A. Takatsu and S. Nishi, 'Determination of Serum Cholesterol by Stable Isotope Dilution Method Using Discharge-Assisted Thermospray Liquid Chromatography–Mass Spectrometry', *Biol. Mass Spectrom.*, 1993, **22**, 247.

28. M. J. Welch, A. Cohen, H. S. Hertz, K. J. Ng, R. Schaffer, P. Van Der Lijn, and E. V. White, 'Determination of Serum Creatinine by Isotope Dilution Mass Spectrometry as a Candidate Definitive Method', *Anal. Chem.*, 1986, **58**, 1681.

29. M. J. Welch, A. Cohen, H. S. Hertz, F. C. Ruegg, R. Schaffer, L. T. Sniegoski, and E. V. White, 'Determination of Serum Urea by Isotope-Dilution Mass Spectrometry as a Candidate Definitive Method', *Anal. Chem.*, 1984, **56**, 713.

30. P. Ellerbe, A. Cohen, M. J. Welch, and E. V. White, 'Determination of Serum Uric Acid by Isotope-Dilution Mass Spectrometry as a New Candidate Definitive Method', *Anal. Chem.*, 1990, **62**, 2173.

31. M. Marlier, G. Lognay, J. Wagstaffe, P. Dreze, and M. Severin, 'Application of Radiometric and Isotope-Dilution Mass-Spectrometric Techniques to Certification of Edible Oil and Fat Reference Materials', *Fresenius' J. Anal. Chem.*, 1990, **338**, 419.

32. H. E. Hurst, R. A. Kemper, and N. Kurata, 'Measurement of Ethyl Carbamate (Urethan) in Blood by Capillary Gas Chromatography–Mass Spectrometry Using Selected-ion Monitoring', *Biomed. Environ. Mass Spectrom.*, 1990, **19**, 27.

33. J. Girault, P. Gobin, and J. Fourtillan, 'Quantitative Measurement of Clenbuterol at the Femtomole Level in Plasma and Urine by Combined Gas Chromatography–Negative-ion Chemical-ionization Mass Spectrometry', *Biomed. Environ. Mass Spectrom.*, 1990, **19**, 80.

34. W. Howald, E. D. Bush, W. F. Trager, R. A. O'Reilly, and C. H. Motley, 'Stable-Isotope Assay for Pseudoracemic Warfarin from Human-Plasma Samples', *Biomed. Mass Spectrom.*, 1980, **7**, 35.

35. J. Rosenfeld, A. Phatak, T. L. Ting and W. Lawrence, 'Mass-Spectrometric Determination of Aspirin via Extractive Alkylation', *Anal. Lett.*, 1980, **13**, 1373.

36. S. Murray and A. R. Boobis, 'Assay for Paracetamol, Produced by the *O*-de-Ethylation of Phenacetin in Vitro, Using Gas Chromatography–Electron-Capture Negative-Ion Chemical-Ionization Mass Spectrometry', *Biomed. Environ. Mass Spectrom.*, 1986, **13**, 91.

37. G. E. Von Unruh, B. C. Jancik, and F. Hoffmann, 'Determination of Valproic Acid Kinetics in Patients During Maintenance Therapy Using a Tetra-Deuterated Form of the Drug', *Biomed. Mass Spectrom.*, 1980, **7**, 164.

38. A. Van Langenhove, J. E. Biller, K. Biemann and T. R. Browne, 'Simultaneous Determination of Phenobarbital (Phenobarbitone) and *p*-Hydroxyphenobarbital and their Stable-Isotope-Labelled Analogues by Gas Chromatography–Mass Spectrometry', *Biomed. Mass Spectrom.*, 1982, **9**, 201.

39. K. K. Midha, R. M. H. Roscoe, K. Hall, E. M. Hawes, J. K. Cooper, G. McKay, and H. U. Shetty, 'Gas-Chromatographic/Mass Spectrometric Assay for Plasma Trifluoperazine Concentrations Following Single Doses', *Biomed. Mass Spectrom.*, 1982, **9**, 186.

40. B. J. Milwa, W. A. Garland, and P. Blumenthal, 'Determination of Flurazepam in Human Plasma by Gas Chromatography Electron-Capture Negative Chemical-Ionization Mass Spectrometry', *Anal. Chem.*, 1981, **53**, 793.

41. B. A. Goldberger, W. D. Darwin, T. M. Grant, A. C. Allen, Y. H. Caplan, and E. J. Cone, 'Measurement of Heroin (Diamorphine) and its Metabolites by Isotope-Dilution Electron-Impact Mass Spectrometry', *Clin. Chem.*, 1993, **39**, 670.

42. A. J. Clifford, A. D. Jones, and H. C. Furr, 'Stable Isotope Dilution Mass Spectrometry to Assess Vitamin A Status', *Methods Enzymol.*, 1990, **189**, 94.

43. M. J. Welch, A. Cohen, P. Ellerbe, R. Schaffer, L. T. Sniegoski, and E. V. White, 'Development of Definitive Methods for Organic Serum Constituents', *J. Res. Natl. Bur. Stand.*, 1988, **93**, 341.

44. D. W. Thomas, R. M. Parkhurst, D. S. Negi, K. D. Lunan, A. C. Wen, A. E. Brandt and R. J. Stephens, 'Improved Assay for α-tocopherol in the Picogram Range Using Gas Chromatography–Mass Spectrometry', *J. Chromatogr. Biomed. Appl.*, 1981, **225**, 433.

45. D. L. Hachey, S. P. Coburn, L. T. Brown, W. F. Erbelding, B. DeMark, and P. D. Klein, 'Quantitation of Vitamin B6 in Biological Samples by Isotope Dilution Mass Spectrometry', *Anal. Biochem.*, 1985, **151**, 159.

CHAPTER 7

The Analysis of Speciated Elements

7.1 Introduction

In the past most assays determined the total concentration of inorganic analytes in a sample, irrespective of their different chemical forms. Once the awareness began to spread that some forms may be harmless whereas others may be extremely toxic, analysing samples for the total concentration of some analytes became inadequate. Total concentration of such analytes yields no information on the overall toxicity. Analytes such as arsenic that are a notorious poison in some forms, are regarded as being non-toxic in others. However, mercury is regarded as being toxic in virtually all forms, but the toxicity varies according to which species is present.

In general, the organometallic forms of analytes tend to be more toxic than the inorganic forms. This is mainly due to the organic ligand providing greater permeability through biomembranes. It can therefore build up in the food chain and concentrate in the blood and reach organs such as the brain causing permanent damage. Examples include mercury, where alkylated species are substantially more toxic than the inorganic forms. In the case of tin the alkylated species are toxic to lower forms of life but are non-toxic to mammals; the inorganic form is an essential trace element. An exception to the general rule is arsenic, where inorganic As(III), *i.e.* arsenite and As(V), *i.e.* arsenate compounds, are toxic whereas most organometallic forms, *e.g.* arsenobetaine, arsenocholine, and some arsenosugars, are extremely stable and are regarded as being non-toxic. Other organic forms, such as monomethyl arsonic acid and dimethyl arsinic acid are toxic, but substantially less so than the inorganic forms.

There is also often a substantial variation in the toxicity of different oxidation states of the same species. One example is chromium. Although the mode of action of chromium poisoning is not well understood, it is known that chromium(VI) is substantially more toxic than chromium(III).

As well as supplying toxicity data, speciation can also assist in the elucidation of mechanisms of breakdown of drugs, *i.e.* which metabolites are formed *in vivo* from the parent molecule, how long the metabolites last, which organ they accumulate in (if any), *etc.* Similarly, biodegradation pathways may be studied using analytes such as tin. Butyltin derivatives are used in some antifouling paints on ships. They may then leach into the water where they enter the sediments. The organic derivative breaks down in a stepwise manner until only non-toxic inorganic tin remains.

Speciation therefore has a very large role to play in analytical chemistry. The importance of this role has only recently been realised. It is envisaged that it is a subject area that will grow considerably in the future.

Section 7.2 deals with the extraction of the speciated analyte from the sample so that it is in a suitable form for determination. Section 7.3 lists the various coupled combinations of instrumental techniques that are used for the determination of speciated elements, and Section 7.4 provides some general precautions to observe when using some of the more common combinations.

7.2 Extraction of the Analyte and Preparation of a Test Solution

In general, the same techniques are used to extract organometallic compounds as those described in Chapter 5 for the extraction of organic analytes. However, care must be taken that the speciated state of the analyte is not affected by the extraction procedure. It is obvious that acid digestion is impossible. For this reason techniques that use mild physical conditions (*e.g.* the extraction of arsenic from animal products using the enzyme trypsin and a Potter homogeniser) are preferred to more fierce methods (*e.g.* Soxhlet extraction). For biological samples, an extraction using an organic solvent (*e.g.* methanol) is fairly common. The efficiency and speed of extraction can be improved by using an ultrasonic bath. Supercritical fluid extraction is gaining in popularity as a means of extraction, because it yields high extraction efficiencies under mild conditions.

Certified matrix reference materials would provide the ideal method of validating extractions. Unfortunately, at the present time, the availability for speciated elements is extremely limited. For arsenic speciation, most of the species of interest are available commercially, but a traceable standard of roxarsone is unavailable.

The simplest method of validating the yield of an extraction technique is to compare the sum of the extracted speciated compounds with the total concentration of the element as measured by a traditional technique such as acid digestion followed by atomic absorption spectroscopy. The sum of the speciated compounds will rarely total 100% of the total concentration of the element, as extractions of organic compounds are seldom 100% efficient. However, as long as the extraction efficiency is reproducible, a correction factor can be incorporated to allow for incomplete extraction. It is advisable to always extract a sample at least three times and then combine the extracts. This will increase the extraction efficiency from, for example, 60% to 94%. If a correction factor is used this must be reported.

An alternative way of validating the extraction method is to use spiking/recovery experiments. This may be more problematic than for the extraction of conventional organic analytes, as different species may have different extraction efficiencies. It is also more difficult than for conventional organic analyses to incorporate the spike into the sample matrix in a manner that mimics the binding and solubility of the analyte. Species contained within fat may be extracted less efficiently than species bound to more lyophilic materials. If such a sample was 'spiked' the extraction yield of the spike would be higher than that of the sample since it may not enter the lipid layer.

It is important to know whether the standards used for spiking and calibration describe the concentration of the element (*e.g.* arsenic) or the species (*e.g.* arseno-betaine). In speciation, it is common for the standards to describe the concentration of the species, but this is not always so. Confusion will inevitably lead to errors.

The analysis of liquid samples presents fewer problems. If no preconcentration is required, the sample can often be centrifuged or filtered and analysed directly.

7.3 Comparison of Determination Techniques

Most analyses of speciated elements are performed using a chromatographic method to separate the species of interest, interfaced with an atomic spectroscopic detector to measure the elemental content of each species as it elutes. This coupling varies in complexity depending on the chromatographic and spectroscopic methods used.

For ease of discussion, the separation and determination components of the available coupled techniques are listed separately. However, it is imperative to consider the system as a whole. The chromatographic eluent, especially, will have a profound effect on both the method of transfer into the detector and the type of detector that can be used.

7.3.1 Separation Methods

The most common separation methods for speciated compounds are listed in Table 7.3.1. With the exception of hydride generation, these are exclusively chromatographic. High Performance Liquid Chromatography (HPLC) and the associated

Table 7.3.1 *Separation methods for the analysis of speciated compounds*

Technique	Comment
HPLC	Relatively inexpensive. Wide range of stationary phases for many applications. Easy coupling to detector. Mobile phases may be incompatible with detector
Ion-exchange chromatography	Relatively inexpensive. Applicable to many speciated compounds. Easy coupling to detector. Dissolved salts may block detector parts
GC	Slightly more expensive than HPLC. Good resolution for volatile analytes. Difficult coupling to detector
Capillary electrophoresis	Of research interest. Coupling to detector presently difficult
Hydride generation	Limited applications. Easy and inexpensive

technique of ion exchange chromatography are by far the most frequently used, but other methods do have some applications.

High Performance Liquid Chromatography

High performance liquid chromatography is one of the most common methods used for the separation of non-volatile analytes. There are published applications for a wide range of stationary phases and mobile phases for speciated compounds (see Section 7.6). Chelating phases do not tend to be used widely. However, there are many more applications for using reversed phase (octadecylsilane) in conjunction with ion-pair reagents, *e.g.* tetrabutyl ammonium phosphate, diethyl-dithiocarbamate, or 8-hydroxyquinoline. In this way, analytes with different oxidation states, such as Cr(III) and Cr(VI), can be separated.

Using an anion-exchange stationary phase with ammoniacal potassium sulfate mobile phase, arsenic can be speciated into As(III), As(V), monomethylarsonic acid, and dimethylarsinic acid. This can also be achieved using a reversed phase column with a mobile phase of sodium dodecyl sulfate in 5% methanol and 2.5% ethanoic acid.

The variety of mobile phases used to obtain good chromatographic resolution may cause problems with the method of detection. This will be discussed in detail in Section 7.4.6. As the majority of analyses using HPLC are undertaken at room temperature there are no real problems associated with the transfer line coupling the chromatograph to the detector. This is advantageous as the coupling can be made cheaply and quickly.

Ion-exchange Chromatography

Much of the ion-exchange technology is identical with that of HPLC. However, there is a growing trend to use low pressure systems incorporating flow injection micro-columns and peristaltic pumps. The cost of such systems are often very low and the ease of coupling surpasses the other techniques. Sample flow rate is usually 1–2 ml min^{-1}, which is ideal for detection by plasma instruments, but may be increased if necessary to become compatible with flame spectrometers.

Gas Chromatography

Although GC is an extremely common analytical technique for volatile analytes, its use for speciation of inorganic analytes has been relatively limited. This is because of the problems associated with transferring the analyte from the end of the column to the atom/ion source of the detector. The GC eluent is in the gas phase and is usually at an elevated temperature. The elevated temperature must be maintained all the way along the transfer line to the detector. Failure to achieve this leads to cool spots and condensation of the analyte. For analytes that are extremely volatile, *e.g.* organolead species, the transfer line may be kept relatively simple as the temperature required need not exceed 200°C. For species that have been separated using high temperature GC, *e.g.* porphyrins, the transfer line may need to be heated to temperatures of approximately 400°C. Ensuring that no cool spots exist in such a transfer line is very difficult.

The vast majority of work performed coupling GC with atomic spectrometry has been using flame atomic absorption spectroscopy (FAAS) as a detector. It has most commonly been applied for analytes present in relatively high concentrations, *e.g.* organolead species in fuel or roadside dust. The sensitivity of the analysis can be increased by placing a quartz or ceramic tube in the light beam. In this way, detection limits at the ng ml^{-1} level may be obtained.

The problem with coupling GC to plasma instruments is that the metallic transfer line may pick up radiofrequencies (it acts like an aerial). This is obviously undesirable as it will cause instability of the plasma, may cause instrument failure, and is potentially dangerous.

Derivatisation of a sample to increase volatility is not recommended for speciation studies.

Although a coupling between GC and an atomic spectrometric detector is possible, it is far more difficult, time consuming, and expensive than for LC.

Hydride Generation

Although this is not a chromatographic technique, it can be used in conjunction with a cryogenic trap to effect speciation. Arsenic(III), As(V), monomethyl arsonic acid and dimethyl arsinic acid all form hydrides. If the hydrides formed are flushed from the gas–liquid separator into a liquid nitrogen trap, the hydrides freeze. If the transfer line is then removed from the trap and allowed to warm slowly to room temperature, the hydrides will boil off at their respective boiling points, and be swept individually to the atomic spectrometer for detection. This technique is not used very frequently, but it does demonstrate that chromatographic techniques are not always necessary to obtain separation.

Capillary Electrophoresis

Capillary electrophoresis is a technique that is growing in popularity. At present the flow rate used (10 μl min^{-1}) is not readily compatible with atomic spectrometric detection, as the nebulisers of these instruments would be starved of liquid. It has therefore not really been used for speciation studies. It is probable though, that a satisfactory coupling between the two techniques will eventually be developed.

7.3.2 Methods of Detection

The most commonly used detection methods for the examination of speciated compounds are listed in Table 7.3.2. Although those using inductively coupled plasma for excitation (either atomic emission spectroscopy or mass spectrometry) give the greatest sensitivity, and so would be used in an ideal situation, FAAS is far cheaper. It also has the advantages that there are fewer problems introducing HPLC eluents than for some other techniques, and it does not require highly skilled operators.

Table 7.3.2 *Methods of detection for the analysis of speciated elements*

Detector	Comment
FAAS	Relatively inexpensive. Lack of sensitivity. Compatible with organic solvents. Does not require highly skilled operator
Electrothermal atomic absorption spectrometry	More expensive than FAAS. Higher sensitivity than FAAS. Requires specialized devices to couple on-line. Not a routine technique
Inductively coupled plasma atomic emission spectrometry	More expensive than AAS. Higher sensitivity than AAS. Longer dynamic ranges than AAS. Requires highly skilled operator
Inductively coupled plasma MS	Extremely expensive. Higher sensitivity than ICP-AES.[a] Gives higher certainty of identification. Easily fouled by LC eluents. Requires skilled operator

[a] ICP-OES = inductively coupled plasma optical emission spectroscopy.

FAAS

Although this is one of the most common techniques employed for trace inorganic determinations, its use as a detector for speciation is limited because of its lack of sensitivity. This arises from the poor transport efficiency of the sample from the nebuliser to the flame (10–15%) and the short residence time the analyte has in the light beam. These instruments are relatively cheap, simple to use, and highly skilled operators are not always necessary. They are extremely tolerant of organic solvents, except for the possible build-up of carbon, indeed nebuliser efficiency may be increased because of the decreased surface tension, viscosity, and density of such solvents when compared with water. This may lead to increased sensitivity and better detection limits.

As well as LC, GC has been coupled with FAAS. Coupling GC with atomic spectrometric detectors is not easy. The interface between the end of the chromatographic column and the atom/ion source can be especially problematic. It is true to say though, that for the large majority of speciation analyses using GC as the method of separation, flame spectroscopy is used as the detector. The sensitivity of the analysis is increased if the interface leads directly to the atom source rather than going through the conventional nebuliser/spray chamber arrangement. The use of quartz or ceramic slotted atom tubes placed in the flame are a convenient way of accomplishing this.

Electrothermal Atomic Absorption Spectroscopy

Electrothermal AAS (ETAAS) has detection limits typically 10–100 times better than the related flame technique. The instrumentation is substantially more expensive than the simpler flame spectrometers, is more difficult to operate, and may need operators with some skill and experience.

Unfortunately, ETAAS also has the drawback of being a batch technique, *i.e.* during normal operation a sample (10–100 µl) is injected, dried, ashed to remove interfering matrix ions, and then atomised into the light beam. This means that it is difficult to introduce a chromatographic eluent. Often it is possible to collect the species as a series of chromatographic fractions and then to analyse these fractions individually. This is very time consuming and needs substantial experimental work to ensure that the species of interest are in the fractions being analysed. An alternative approach is to use a thermospray or electrospray device to introduce the sample whilst the tube is maintained at a constant atomisation temperature. This may prove to be both expensive and to yield inaccurate results: for a 15 min chromatographic run, if the graphite tube is held at temperatures of 2000°C, it would wear rapidly.

The use of organic solvents is quite simple with ETAAS, provided that they can be dried evenly without spitting or frothing. The use of a mobile phase that has a high dissolved solids content (*i.e.* solvents for ion-exchange chromatography) may be problematic if a sufficiently high ash temperature, to remove the solids, cannot be used. These solids may then volatilise simultaneously with the analyte species and cause large background signals.

Inductively Coupled Plasma Atomic Emission Spectroscopy (ICP-AES)

This technique is gaining in popularity as the detection method for speciation analysis. This is because although the instruments are expensive, they have higher sensitivity and longer dynamic ranges than flame spectrometers. In addition, they also allow alternative spectroscopic lines to be monitored either simultaneously or at least far more rapidly than absorption instrumentation. The technique requires considerable operator skill to obtain optimal results. Parameters such as the gas flow rates, power, and viewing height have a substantial effect on the signal intensity and should therefore be optimised rigorously using a multivariate technique.

The effects of organic solvents on the plasma varies depending on the quality of the instrumentation. Some instruments operate readily with, for example, close to 100% methanol, without having an excessively high reflected power. Other instruments have reflected powers in excess of 40 W under the same conditions, and this is high enough to risk damage to the radiofrequency (RF) generator. In extreme cases, the plasma may be extinguished by even relatively small proportions of organic solvent (*e.g.* 30% acetonitrile). If a mobile phase containing an organic solvent is used it should be introduced in stages to give time for compensation.

The flow rates of HPLC pumps and the typical uptake rate of an ICP nebuliser are compatible (1–2 ml min^{-1}), so the interface coupling the two together can theoretically be very simple.

Gas chromatography is less simply interfaced with ICP-AES instruments, because torches tend to be enclosed so as to protect users from eye damage and damage caused by RF radiation. This often means that holes must be machined in the instruments to enable the transfer line to reach the torch. Most analytical laboratories are unwilling to do this.

ICP–MS

These instruments are often extremely expensive, although cheaper, small, benchtop models are available. The mass spectrometers are nearly all quadrupole based, although extremely expensive magnetic sector instruments are available if high resolution is required. Their use is gradually becoming more widespread in spite of their cost. The instruments are complex and require considerable operator skill to obtain reliable results. They are extremely sensitive, achieving detection limits as low as 1 ng l^{-1}, and have a linear range of 5–6 orders of magnitude. Therefore, these instruments may be used for speciation studies in most samples.

Generally these instruments share the same pros and cons as ICP-AES instruments, but they do have additional drawbacks. The interface between the plasma and the mass spectrometer can be seriously affected by organic solvents.

Inductively coupled plasma mass spectrometers are very sophisticated tools and as such should only really be used for speciation studies if the concentration of the analyte of interest is too low for any other technique to work satisfactorily.

Atomic Fluorescence Spectroscopy

This detection system has limited application in speciation studies. Atomic fluorescence has a number of drawbacks, *e.g.* scatter, but commercial instrumentation dedicated to determining specific analytes has been produced. Mercury has been speciated using GC coupled with a fluorescence detector producing limits of detection substantially below the ng ml^{-1} level. These instruments are usually cheap, but are element-specific. They require little operator skill to use, but often require an additional vapour generation stage coupled to them. It is this step that often requires a good knowledge of chemistry to ensure interferences are minimised.

Direct Current Plasma (DCP) and Microwave Induced Plasma (MIP)

These instruments are not often used as detectors for speciation studies. The DCP has a linear range of several orders of magnitude and is slightly more sensitive than FAAS. It is extremely tolerant of both organic solvents and solutions with a high dissolved solids content. They are cheaper than ICP-AES instruments, whilst still retaining the capacity of multi-line analysis. Unfortunately the DCP is regarded as being prone to matrix interferences caused by easily ionised elements. To counteract this, it is normal to add a large excess of lithium nitrate to all samples and standards. This is usually incompatible with chromatographic techniques and so the DCP has largely been ignored as a method of detection for speciation.

Microwave induced plasmas have been linked with GC for several years, but commercial instrumentation has only become available relatively recently. The

MIP has been limited to analysing gaseous samples only, as even a few microlitres of solvent have been known to extinguish the plasma. Research continues into the introduction of liquid samples, but as yet no routine methodology exists.

Other Methods of Analysing Speciated Elements

There are several techniques other than those linking chromatography with atomic spectrometric detection that may be used for speciation. These techniques are not used very frequently but mention of them should be made. Of course, HPLC may be linked with a number of detectors, including UV (although this is rarely used because of the low molar absorptivity constants of the compounds) and electro-chemical detection. Gas chromatography has been used with electron capture detectors to determine organometallic compounds of tin, arsenic, lead, and mercury. This detector is not very specific, so peaks from interfering matrix constituents may cause confusion. Electrochemical methods such as anodic stripping voltammetry, ion-selective electrodes, and polarography have also found limited use for speciation. Ion-selective electrodes do not have the sensitivity to determine analytes in anything except very polluted waters, whilst the other techniques are virtually limited to differentiating between different oxidation states of analytes. A large number of more exotic techniques have also occasionally been applied to obtain speciation data. These include X-ray fluorescence, X-ray diffraction, Auger electron spectroscopy and electron paramagnetic resonance.

7.4 Critical Factors when using Coupled Techniques

There are many general precautions and critical factors for method validation that apply to techniques such as HPLC, AAS, and ICP-AES/MS when used as standalone techniques. These are covered in the relevant sections of this book. This section deals only with the extra considerations when the techniques are coupled together. Factors that greatly influence noise, overall sensitivity, signal-to-noise ratio, and the resolution of different analytes are described.

7.4.1 Length of Transfer Lines when Coupling to LC Systems

The length of the transfer line between an LC column or flow injection valve and the sample introduction system of the detector has a very profound effect on the shape of the resulting peaks. As with any chromatographic system, the larger the dead-volume the more band broadening effects are observed. The results of an experiment comparing two lengths of transfer line between LC and ICP-MS are shown in Table 7.4.1. It can be seen that the longer transfer line yielded short, broad peaks, whilst the shorter line gave tall, narrow peaks.

The broadening caused by longer transfer lines can have adverse effects on accuracy and precision when determinations close to the detection limit are performed. Tall, thin peaks are far more simple to integrate than the short, broader ones that may easily become lost in the background noise. It is therefore desirable

Table 7.4.1 *The effect of length of the transfer line to the ICP-MS instrument*

Length of transfer line (cm)	Peak width (s)	Peak area (counts)	CV(%)
20	4.02 ± 0.16	2 736 827 ± 55 388	± 2.0
50	5.05 ± 0.13	2 695 288 ± 145 388	± 5.4

Spectrometric detector = ICP-MS, VG PlasmaQuad II, Winsford, Cheshire, SIM mode measuring m/z=208.
Sample introduction = 250 μl of 100 ng ml^{-1} Pb solution using a peristaltic pump and flow injection valve.
Transfer line = PTFE tubing, internal diameter 0.8 mm. $n = 3$; uncertainty expressed as σ.

to keep the transfer line between the chromatograph and the nebuliser as short as possible.

The distance between the nebuliser and the ion/atom source should also be considered. The results of varying this distance are shown in Table 7.4.2. It can be seen that the length of the line had minimal effect on the broadening of the peaks, but the overall signal decreased as the length of the line increased. This is probably due to the analyte aerosol condensing on the tube walls and so not being transported to the plasma. It can also be seen that the precision of the injections increases as the length of the line increases. This is probably because some of the noise from the peristaltic pump is being damped out. It can therefore be seen that even though the overall signal using a long transfer line between the nebuliser and the torch is smaller than that obtained using a shorter line, the improved precision leads to a similar limit of detection.

The length of the transfer line between the nebuliser and the torch has very little effect on peak broadening and limits of detection, whereas the transfer line between the chromatograph and the nebuliser contributes very heavily to band

Table 7.4.2 *The effect of altering the length of the transfer line between the nebuliser and the torch*

Length of transfer line (cm)	Peak width (s)	Peak area (counts)	CV (%)
10	38.6 ± 1.1	2 521 468 ± 149 548	5.9
20	38.9 ± 0.9	2 300 520 ± 82 689	3.6
40	38.4 ± 1.3	1 922 887 ± 47 667	2.5

n=6; uncertainty expressed as ± σ.
Spectrometric detector = ICP-MS, VG PlasmaQuad II, Winsford, Cheshire, SIM mode measuring m/z=208.
Sample introduction = 250 μl of 100 ng ml^{-1} Pb solution using a peristaltic pump and flow injection valve.
Meinhard nebuliser inserted into a tube (i.d. 10 mm) leading directly to the torch used in place of a conventional spray chamber.

broadening, which in turn degrades detection limits and precision. The length of this transfer line should therefore be kept to a minimum.

7.4.2 Use of the Correct Nebuliser

There are several different types of nebuliser commercially available for AAS, ICP-AES, and ICP-MS instruments. The relative advantages and disadvantages of the two major types are listed in Table 7.4.3. These are discussed more fully in Section 4.2 (Atomic Spectroscopy). When used for speciation analysis, an additional factor to consider is the fact that Meinhard nebulisers can be blocked if mobile phases with a high content of dissolved solids are used (*e.g.* ion-exchange chromatography). This gradual blocking of the nebuliser leads to excessive signal drift until, when blocking is complete, no signal is observed. An example is the speciation of arsenic using an anion-exchange column where a mobile phase of 0.1 mol l^{-1} potassium sulfate is used. This has a dissolved solids content of approximately 1.75% m/v and its use rapidly leads to nebuliser blockage. In cases such as this, the use of a nebuliser that is far more tolerant of high solids, *i.e.* a V groove nebuliser is to be recommended.

Table 7.4.3 *The two most common types of nebuliser*

Nebuliser	*Comment*
V groove nebuliser (*e.g.* Babington, Ebdon, and De Galan)	Have to be pumped, therefore increased noise Do not block easily. Robust
Concentric nebulisers (*e.g.* Meinhard)	Venturi action, therefore decreased noise. Fragile and easily blocked

The V groove nebulisers have to have the sample pumped to them and so they pick up a lot of noise from the peristaltic pump. Other than that, they do not suffer from any of the drawbacks that the concentric nebulisers do. They tend to be far less fragile, do not block so easily, and are inert.

There are other types of nebuliser available. The use of ultrasonic nebulisers is becoming more common because of their improved noise properties. These nebulisers also have the advantage of more sample reaching the ion source. This means that detection limits are often improved in comparison with conventional nebulisers. Thermospray and electrospray devices for sample introduction to atomic spectrometric detectors have also been used, but their use has not yet become widespread for speciation studies. It is anticipated that research into this area will, in the future, yield an efficient, commercial coupling between HPLC and ICP instrumentation.

A small air bleed into the nebuliser is often used to compensate for the different flow rates of the HPLC (1–2 ml min^{-1}) and the uptake rate of the flame spectrometer (5–8 ml min^{-1}). This air bleed often improves peak shape and decreases noise.

7.4.3 The Spray Chamber

There is a wide variety of spray chambers available commercially for ICP instrumentation. Of these, the two most popular are the Scott type double pass and the single pass spray chambers. The volume of such spray chambers is typically 110 ml for the double pass and 70 ml for the single pass. The large surface area within the spray chamber and the large volumes may lead to peak broadening and memory effects. Serious memory effects (appearing as tailing on peaks) have been observed when using a double pass spray chamber for lead analysis when an ammonium acetate mobile phase of pH 5.5 was used. This may be due to the increased volume, but is more likely to be due to the very large internal surface area.

Many research laboratories have constructed in-house spray chambers that have a reduced internal volume. Unfortunately, these too have some limitations. The results for different spray chambers used for the FIA-ICP-MS analysis of lead are shown in Table 7.4.4. Several conclusions can be drawn from these data. It is evident that although the cyclone spray chamber yielded by far the highest overall signal, the precision was not improved (*i.e.* the noise characteristics were worse). This produced a limit of detection (LOD) comparable to that of the Scott double pass spray chamber. It can be seen that in general, the larger volume spray chambers produced more peak broadening. It can also be seen that although the reduced volume double pass chamber produced narrow peaks, the sensitivity and precision were considerably poorer than the other spray chambers and, consequently, a much poorer limit of detection was obtained.

It should be noted that although the reduced volume spray chambers give rise to narrow peaks, they have a much higher noise level, which is undesirable for ultratrace work. For the results tabulated the single pass spray chamber gave optimum performance, producing narrow peaks with excellent precision and sufficient sensitivity.

The results also demonstrate that the largest overall peak magnitude does not necessarily yield the best detection limits. It must be stressed that although virtually any spray chamber will suffice for the determination of quite concentrated analytes, the choice of spray chamber is vital if determinations close to the detection limit are to be performed.

Most spray chambers have a water-filled cooling jacket surrounding them. This is especially desirable when mobile phases containing organic solvents are being used. It assists in the general stability of the system by decreasing the solvent loading of the plasma. If even less solvent is required, then the jacket can be filled with isopropanol, which can be taken down to approximately −15°C. At these temperatures the solvent within the spray chamber becomes much more viscous and so less of it reaches the plasma.

7.4.4 Coupling Lines for GC

As described earlier, the coupling between GC and plasma instruments is not easy and relatively few workers have attempted it. Far more work has been performed using FAAS as the method of detection because it is much simpler. Specially

Table 7.4.4 *A comparison of spray chambers*

Type of spray chamber	Peak width (s)	Peak area (counts)	CV (%)	LOD($3\sigma_{n-1}$) ($\mu g\, l^{-1}$)
Scott double pass	50.33 ± 2.34	854 921 ± 22 463	2.6	7.5
Single pass	43.50 ± 0.91	426 276 ± 2 792	0.7	2.0
Mini double pass (10 m)	43.50 ± 2.99	108 134 ± 11 361	10.5	20.5
Cyclone	47.62 ± 0.75	2 121 380 ± 56 002	2.6	7.3

Spectrometric detector = ICP-MS, VG PlasmaQuad II, Winsford, Cheshire, SIM mode measuring m/z = 208. Sample introduction = 250 μl of 100 ng ml^{-1} Pb solution using a peristaltic pump and flow injection valve. n = 6; uncertainty expressed as ±σ.

designed transfer lines have to be constructed that have uniform heating along their length. These transfer lines usually have to be made in-house and ensuring that they have no cool spots can be very problematic. It should be held at the same temperature as the column, which for high temperature GC, may be in excess of 400°C. Obtaining temperatures such as this requires equipment which supplies a large amount of power. These have special safety requirements such as sufficient lagging around all hot parts, *etc.* If the GC is coupled to an ICP, positioning the transfer line in the torch (nebulisers and spray chambers are not required) can also be difficult as it has to be close enough to the plasma so that no dead-volume or cool spots occur, but not too close so that the end of the column melts or picks up RF radiation. If the latter does occur, the transfer line can act as an aerial and can direct the RF to electrical circuits within the GC. This could obviously cause serious damage.

To prevent RF leakage, most commercial instruments have the torch enclosed in a box. For coupling with GC, holes usually have to be cut in this and the surrounding casing. After this has been done, checks should be made to ensure that RF radiation is not escaping.

7.4.5 Mass Spectrometer Cones and Lens Stacks

The sample and skimmer cones as well as the lens stack that focuses the ions into the mass spectrometer, can become clogged with carbon. This leads to excessive signal drift until the cones become completely clogged, and no signal is obtained. The problem can be partially overcome by the addition of oxygen (approx. 3% v/v) into the nebuliser gas flow. However, the lens stack still requires cleaning after virtually every day's use. The use of the oxygen bleed potentially has one other drawback. If too much oxygen is admitted, it will vastly accelerate the wear on the sample cone and will rapidly render it unusable.

The cones are usually made of nickel, although platinum ones are available. The platinum cones are obviously extremely expensive, but unlike those made of nickel are not attacked by mobile phases containing sulfate or phosphate. If the MS is coupled to an ion chromatograph, a nickel sample cone will become worn very rapidly if large concentrations of sulfate/phosphate have been used in the chromatography.

7.4.6 The Use of Organic Solvents

One minor problem that may be encountered with LC–AAS is that the organic solvents used produce a build-up of carbon in the burner slot. This can be removed by gently rubbing a spatula or other non-flammable item over the burner, although care must be taken not to damage the burner head. Failure to clean the head will cause gaps to appear in the flame, and in turn, to decreased sensitivity. Alternatively, the flame may be extinguished, so that the burner may be cleaned properly using an ultrasonic bath.

For ion chromatography, in which mobile phases with large amounts of dissolved solids (high salt content) may be employed, the nebuliser and burner system of these instruments may occasionally block, especially after extended use. It may therefore be necessary to have a washout period at the end of each chromatographic run. This should delay or perhaps even prevent the salting up of the burner. The washout period is no real hardship as it is not time consuming and it prevents excessive instrumental drift.

The plasmas of some ICP spectrometers are also extremely prone to instability or possibly even extinction if organic solvents are introduced. The choice of either the mobile phase of the HPLC method, or the solvent used for extraction of the samples must therefore be considered carefully before attempting any analysis. Many of the more modern spectrometers have free-running RF generators, which can cope more readily with organic solvents. Unfortunately, many of the older instruments use crystal-tuned RF generators, which take a considerable amount of time to retune from aqueous to organic introduction. In cases such as this, the organic solvent should be introduced gradually, *i.e.* if a methanol/water (3:1) mobile phase is to be used, then initially only 10% methanol should be introduced, and then the proportion should be increased gradually until the final proportion is obtained. In this way, the generator can retune before the plasma becomes unstable or even extinguished.

The presence of organic solvents in the plasma often increases the level of reflected power. There are several ways to decrease the amount of reflected power produced by the organic solvents. These include altering the impedance by changing the load, changing the position of the load coil, manually retuning the generator, or using a method to decrease the amount of solvent reaching the plasma. These methods include using a chilled spray chamber, using Peltier coolers between the spray chamber and the torch, and using tubular membrane driers, which remove the organic solvent through porous membranes via a counter flow of argon. These methods have all been used successfully. By using these methods, solvents such as acetonitrile may be introduced at levels which otherwise would lead instantly to plasma extinction.

Table 7.4.5 shows the maximum percentage of some common solvents that can be introduced into a plasma without causing excessively high reflected powers (>45 W, an amount that may cause damage to the generator after extended use). The amount of solvent admissible depends on its volatility and, to a lesser extent, the number of carbon atoms per molecule it contains. For example, the boiling point of propan-2-ol is relatively high (82.4°C), so more may be used than for the more volatile methanol (65°C) or propanone (57°C). Propanone also contains three carbon atoms per molecule, so this limits its use to only very low proportions (<20%). Acetonitrile has two carbon atoms per molecule and a relatively high boiling point (81.6°C), but the amount that can be introduced is less than expected. Therefore, factors such as surface tension, viscosity, and density (the parameters that control nebulisation efficiency) must also play a role.

Addition of Peltier coolers or drier tubes, using higher forward powers, or other modifications increases the amount of organic solvent admissible. It must also be mentioned that high levels of organic solvents make the plasma less energetic. This is why many workers use higher incident powers to run their analyses.

Table 7.4.5 *The maximum amount of solvent admissible before reflected power exceeded* 45 W

Solvent	Amount of solvent (% υ/υ in water)
Propan-2-ol	100
Methanol	75
Propanone	20
Acetonitrile	50

Spectrometric detector = ICP-MS, VG PlasmaQuad II, Winsford, Cheshire, SIM mode measuring $m/z=208$.
Spray chamber = double pass. Forward power = 1500 W.

7.5 General Precautions and Tips

It is imperative that the chromatography and the detection system should be considered as a whole and not as separate entities. If an existing chromatographic method is transferred onto ICP technology in an attempt to obtain increased sensitivity, the mobile phase may prove to be incompatible. In this case, a different chromatographic system will have to be developed that is compatible with detection by plasma systems.

It is also important to know whether the standards describe the concentration of the analyte (*e.g.* arsenic) or the species (*e.g.* arsenobetaine). In speciation, it is common for the standards to describe the concentration of the species, but this is not always so. Confusion will inevitably lead to errors.

The importance of matrix matching cannot be overemphasised. If a sediment is to be speciated for tin, the Certified Reference Material PACS-1 (supplier LGC) is available. Unfortunately, not all sediments have the same chemical composition. A procedure using glacial ethanoic acid is very efficient at extracting the tin species from PACS-1, which has a very high organic composition, *e.g.* humic material, but it is considerably less efficient at extracting them from sediments that contain significant proportions of silica or alumina. In cases such as this, it is impossible to assume that simply because a procedure works for one specific sample, it will work for all samples. If assumptions like this are made, serious underestimates of the true content can be made.

Although spiking experiments can be performed, these are often less than informative because different analyte species have different extraction efficiencies. Losses can occur in virtually all steps of a workup procedure. Some animal-based samples may be defatted. This may lead to losses of some lipid-based species. Even a simple filtering step may lead to some losses. If losses are occurring somewhere in the sample preparation procedure, and there is doubt as to exactly where, spiking of sub-samples before each step of the procedure will isolate the sources of error. This is a time-consuming exercise, where even a simple three-step procedure will lead to several samples being analysed.

For optimal performance, the transfer line between the liquid chromatograph and the spectroscopic detector system should be as short as possible to prevent band broadening. Conversely, the transfer line to the torch has very little effect on the overall signal. If this transfer line has a wide enough diameter, signal intensity is not affected and there is no band broadening.

The nebuliser/spray chamber assembly should be chosen so that blockage by mobile phases with high salt contents does not occur and so that the dead-volume/ surface area within the spray chamber does not cause broadening or tailing of peaks. If broadening does occur, it results in a loss of precision and a degradation of detection limits.

Cooling of the spray chamber, tubular membrane driers, or Peltier coolers are almost a prerequisite if organic solvents are being used, to prevent excessive solvent loading of the plasma. Even then, the plasma may have to be modified to allow successful introduction, *e.g.* by using higher incident powers, an oxygen bleed, retuning the generator, or changing the impedance.

7.6 References

Books

1. 'Lead, Mercury, Cadmium and Arsenic in the Environment', ed. T. C. Hutchinson and K. M. Meema, John Wiley, Chichester, UK, 1987.
2. P. J. Craig and F. Glockling, 'The Biological Alkylation of Heavy Elements,' Royal Society of Chemistry, London, 1988.
3. R. M. Harrison and S. Rapsomanikis, 'Environmental Analysis using Chromatography Interfaced with Atomic Spectroscopy', Ellis Horwood, Chichester, UK, 1989.
4. 'Metal Speciation in the Environment', ed. J. A. C. Broekaert, S. Gücer, and F. Adams, Springer Verlag, Berlin, 1990.
5. M. Bernhard, F. Brickman, and P. Sadler, 'The Importance of Chemical Speciation in Environmental Processes', Springer Verlag, Berlin, 1987.
6. 'Trace Element Speciation: Analytical Methods and Problems', ed. G. E. Batley, CRC Press, Boca Raton, Fl, 1989.

Reviews

7. L. Ebdon, S. Hill, A. P. Walton, and R. W. Ward, 'Coupled Chromatography–Atomic Spectrometry for Arsenic Speciation – Comparative Study'. *Analyst*, 1988, **113**, 1159.
8. K. Robards, P. Starr, and E. Patsalides, 'Metal Determination and Metal Speciation by Liquid Chromatography', *Analyst*, 1991, **116**, 1247.
9. L. Ebdon, S. Hill, and R. W. Ward, 'Directly Coupled Chromatography–Atomic Spectroscopy, Part 1 Directly Coupled Gas Chromatography–Atomic Spectroscopy, A Review' *Analyst*, 1986, **111**, 1113.
10. L. Ebdon, S. Hill, and R. W. Ward, 'Directly Coupled Chromatography–Atomic Spectroscopy, Part 2 Directly Coupled Liquid Chromatography–Atomic Spectroscopy, A Review.' *Analyst*, 1987, **112**, 1.
11. R. F. Zainullin and V. G. Berezkin, 'Flame Photometric Detectors in Chromatography: Review'. *Crit. Rev. Anal. Chem.*, 1991, **22**, 183.

12. S. J. Hill, M. J. Bloxham, and P. J. Worsfold, 'Chromatography Coupled with Inductively Coupled Plasma Atomic Emission Spectrometry and Inductively Coupled Plasma Mass Spectrometry, A Review'. *J. Anal. At. Spectrom.*, 1993, **8**, 499.
13. N. P. Vela and J. A. Caruso, 'Potential of Liquid Chromatography–Inductively Coupled Plasma Mass Spectrometry for Trace Metal Speciation'. *J. Anal. At. Spectrom.*, 1993, **8**, 787.
14. Y. K. Chau and P. T. S. Wong, 'Recent Developments in Speciation and Determination of Organometallic Compounds in Environmental Samples'. *Fresenius' J. Anal. Chem.*, 1991, **339**, 640.
15. W. Lund, 'Speciation Analysis – Why and How?', *Fresenius' J. Anal. Chem.*, 1990, **337**, 557.
16. R. J. A. Van Cleuvenbergen, W. D. Marshall, and F. C. Adams, 'Speciation of organolead compounds by GC–AAS', in 'Metal Speciation in the Environment' (NATO ASI Ser., *G 23*) ed. J. A. C. Broekaert, S. Gücer, and F. Adams, Springer Verlag, Berlin, 1990, pp. 307–337.
17. B. L. Sharp, 'Pneumatic Nebulisers and Spray Chambers for Inductively Coupled Plasma Spectrometry. A Review. Part 1 Nebulisers', *J. Anal. At. Spectrom.*, 1988, **3**, 613.
18. B. L. Sharp, 'Pneumatic Nebulisers and Spray Chambers for Inductively Coupled Plasma Spectrometry. A Review. Part 2 Spray Chambers', *J. Anal. At. Spectrom.*, 1988, **3**, 939.

CHAPTER 8

Techniques Suitable for Both Inorganic and Organic Analytes

8.1 Absorption and Emission of Ultraviolet and Visible Radiation

8.1.1 Overview

Ultraviolet and visible absorption spectrophotometry, fluorescence, and chemiluminescence are used for trace analysis. Although the discussion in this section is concerned with solution measurements, which account for the great majority of determinations, the techniques may also be applied to solids and gases.

8.1.2 Scope of UV/Visible Absorption Spectrophotometry

Spectrophotometry is widely employed in automated methods of analysis in, for example, clinical and water analysis. Fixed wavelength, variable wavelength, and diode-array UV detectors are widely used in HPLC. To some extent, visible spectrophotometry is also used in HPLC, when metal ions are separated and undergo post-column reaction with a chromogenic reagent.

In general, the absorption of radiation in the UV/visible (UV/VIS) region results in the excitation of electrons, and the absorbing species may be categorised according to the types of electronic transitions that occur.

Absorptions Involving π, σ, and n Electrons

Species in this category include organic molecules and ions containing groups of atoms that contain unsaturated bonds and/or lone-pair electrons, *chromophores*. These species may also contain suitably positioned polar groups, *auxochromes*, which, although they do not themselves absorb, enhance the absorption of the chromophores.

Some inorganic oxy-anions containing unsaturated bonds and/or lone-pair electrons also absorb in the UV. Table 8.1.1 lists some chromophoric groups.

Table 8.1.1 *Examples of common chromophoric groups*

Group		Approximate λ_{max} (nm)[a]
Alkene	>C=C<	177
Conjugated diene	>C=C–C=C<	220
Conjugated polyenes	$H(HC=CH)_6H$	360
Azo	–N=N–	340
Carbonyl	>C=O	279
Azomethine	>C=N–	190
Nitroso	$-CH_2-N=O$	300
Thioketone	>C=S	205
Sulfone	$>SO_2$	180
Sulfoxide	>S=O	210
Benzene	C_6H_6	205, 255
Carboxyl	RCOOH	204
Benzoic acid	C_6H_5COOH	230, 270
Nitro	NO_2	270
Nitrobenzene	$C_6H_5NO_2$	250, 280, 330
Nitrite	–ONO	219, 350
Nitrate, organic	$-ONO_2$	270
inorganic		194, 303

[a] The exact position depends on the solvent and the presence of other groups.

Absorptions Involving d and f Electrons

The majority of transition metal ions absorb UV or visible radiation. For ions of the first and second transition-metal series, 3d and 4d electrons respectively are responsible for the absorption, and for lanthanide elements, the absorptions arise from transitions of the 4f and 5f electrons.

Elements of the first and second transition series have more than one oxidation state and these are usually coloured. However, their visible absorption bands are weak, and very broad, often 100 nm or more, and significantly influenced by the chemical environment of the ion. The low intensity of the absorption bands makes them unsuitable for trace analysis, unless charge transfer is involved. Charge-transfer spectra are discussed later.

Most lanthanide and actinide elements absorb radiation in the visible or, more often, UV region. Unlike the transition metals of the first and second series, the corresponding electronic transitions involve f electrons, which are largely unaffected by ligand fields. The absorption bands are narrow but the absorption intensities are too low to make them useful in trace analysis.

Absorptions due to Charge Transfer

Charge-transfer absorptions are often very intense, making them particularly suitable and important for trace inorganic analysis. Examples of inorganic complexes include the thiocyanate complex of iron(III), the 1,10-phenanthroline complex of iron(II), and the tri-iodide complex I_{3-}. The charge-transfer absorption bands arise when the photon absorbed causes an electron to be transferred from one ion to the other and occurs because of the ease of oxidation and reduction of the species comprising the complex. With metal ions in high oxidation states and with ligands that are oxidisable, as in the iron(III) thiocyanate complex, the charge-transfer bands correspond to transitions involving the reduction of the metal ion from Fe(III) to Fe(II). The converse is true of a lower oxidation-state metal ion and a reducible ligand, as exemplified by the red complex iron(II) tris(1,10-phenanthroline).

Most of the colorimetric reagents used for the detection and determination of trace metal ions are ligands that form complexes having intense charge-transfer bands.

8.1.3 Quantitative Applications of UV/Visible Spectrophotometry

The quantitative applications of absorption spectrophotometry depend on the combined Beer and Lambert–Bouger relationships known colloquially as Beer's Law. This law concerns the relationship between the intensity of radiation falling on an absorbing substance and the intensity of the radiation transmitted, expressed as in Equation (1):

$$\log_{10}(I_0/I) = A = \varepsilon cl \tag{1}$$

where
I_0	=	intensity of incident radiation
I	=	intensity of transmitted radiation
A	=	absorbance (formerly called optical density)
ε	=	molar absorptivity (l mol^{-1}cm^{-1}) formerly called molar extinction coefficient)
c	=	concentration (mol l^{-1})
l	=	path length of substance (cm)

The molar absorptivity is the absorbance that a 1 mol l^{-1} solution of the absorbing species would have if it were possible to measure it using a 1 cm pathlength.

An indication of the limiting concentration for quantitation may be obtained as follows. If we substitute the appropriate values into Equation (1), assuming a cell of pathlength 1 cm used to measure the absorbance of a solution of an absorbing substance, and that an absorbance value of 0.01 represents the lower limit of reliable measurement, then the lowest limit of quantitative determination becomes:

(i)   with a molar absorptivity of 10^4 l mol^{-1}cm^{-1}, equivalent to a concentration of 2×10^{-6} mol l^{-1};

(ii) and for a molar absorptivity of 10^5 l mol^{-1}cm^{-1}, equivalent to a concentration of 2×10^{-7} mol l^{-1}.

These values are of practical interest because the first is a fairly typical value for a colorimetric reagent, and so gives an indication of the concentration limit to be expected in a normal colorimetric method; the second figure is the theoretical limit for molar absorptivity and so indicates the approximate limiting lower concentration capability of absorption spectrophotometry.

Absorbances are additive: if a solution contains two or more non-interacting absorbing species, then the sum of the absorbances is given by Equation (2):

$$A = A_1 + A_2 + \ldots + A_n$$
$$= l(\varepsilon_1 c_1 + \varepsilon_2 c_2 + \ldots + \varepsilon_n c_n) \tag{2}$$

It should be noted that Beer's Law is a limiting law and only applies to dilute solutions and for monochromatic radiation.

Deviation from Beer's Law

When the absorbance of the components of a set of standard solutions of different concentrations does not vary linearly with concentration, the set is said to deviate from Beer's law. What usually occurs is that the graph of absorbance against concentration curves towards the concentration axis increasingly with increasing concentration (*i.e.* negative deviation). There are several possible causes for this behaviour and only those of chemical importance or that are under the control of the analyst are considered here. It must also be pointed out that small deviations are tolerable for analytical purposes, provided that they are reproducible.

Deviations may occur for one or more of the following reasons.

(i)   The radiation is not sufficiently monochromatic, and ε varies over the spectral range of the radiation used. The broader the spectral bandwidth, the more likely this effect is to occur. The effective bandwidth of a monochromator depends upon the dispersion of the prism/grating as well as the width of the entrance and exit slits.

(ii) Radiation reaching the detector contains a significant fraction of stray light. Stray light is radiation of a different wavelength from the instrument setting and which may not have passed through the sample. This is most often a problem at wavelengths below about 230–240 nm, and is the reason that tungsten lamps must not be used for wavelengths below 350 nm. The deviation caused by stray light is similar to that caused by a large bandpass. Instrument design including the use of non-reflecting inner surfaces, light baffles, stray radiation filters, and double monochromators is important in reducing stray light.

(iii) Deviations may be due to concentration effects. Species in solution may be interacting, and the interaction is dependent on concentration. This is a common cause of deviation and can arise in several ways. One of these is by the dissociation or association of the species being measured. Obviously, the product of dissociation or association will have a different ε value from the analyte and so cause deviation. Such a deviation can be noticed with, for example, copper(II) in ammonia solution when a number of different ammino-aquo copper(II) species may co-exist.

(iv) Dust or turbidity in a solution will cause light-scattering, which is recorded as an absorbance, and causes similar deviations.

(v) There may be problems establishing a calibration relationship because the analytical system is sensitive to temperature or the signal has a limited lifetime. The former factor is not usually very significant unless the absorption is due to an equilibrium between two species since, in this case, the position of the equilibrium will be temperature-dependent.

The time factor is often important when the method depends on the formation of a charge-transfer complex. Charge-transfer spectra are related to the redox reaction within the complex whose absorbance is being measured. It is not uncommon for the reaction to proceed in a time period ranging from minutes to hours and so accurate calibration and reliable analytical data can be obtained only by establishing time conditions for the period between the colour reaction and the measurement of the absorbance. The decomposition of the coloured complex in this way will, like all reactions, also be temperature-dependent.

8.1.4 Photoluminescence and Chemiluminescence

Radiation emitted in the UV/VIS region may also be used for qualitative and quantitative analysis. Photoluminescence is the term used for the emitted radiation from samples that are in the excited state as a result of absorbing photons; it includes both fluorescence and phosphorescence. Fluorescence is the only one of these that has much application in trace analysis. Chemiluminescence arises when an excited state species, produced during a chemical reaction, loses its excess energy by emission of radiation. This technique is limited in its scope but does have the advantage of simple instrumentation.

Types of Fluorophore

There are far more molecules that absorb UV/VIS radiation than the number that exhibit fluorescence. Molecules that have fused aromatic rings and have a rigid structure normally fluoresce. The fact that all absorbing molecules do not fluoresce may be an advantage when dealing with mixtures of chemicals. Figure 8.1.1 shows the types of structures that fluoresce.

Choice of Excitation and Emission Wavelengths

A simple means of determining the conditions for fluorescence measurements is to record the absorbance spectrum of the analyte to determine λ_{max}. The λ_{max} value that provides greatest selectivity for the analyte can then be selected; the emission wavelength (fluorescence wavelength) is generally about 30 nm higher than the selected λ_{max} value. Table 8.1.2 lists some typical values.

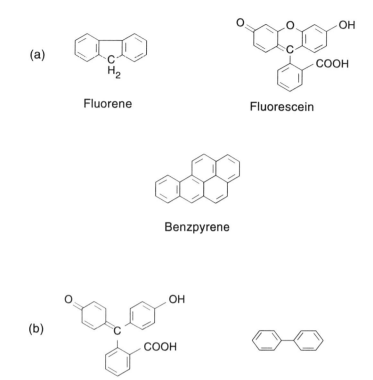

Figure 8.1.1 *Structures of molecules that (a) fluoresce and (b) do not fluoresce*

Table 8.1.2 *Excitation and emission wavelengths for selected fluorophores*

	Excitation λ_{max} (nm)	Emission λ_{max} (nm)
Anthracene	360	402
Naphthalene	290	330
p-Terphenyl	295	338
Benzpyrene	520	545
Rhodamine B	562	573

Quantitative Aspects of Fluorescence

The advantage of fluorescence techniques as compared with absorbance is their inherent sensitivity, with detection limits being up to three orders of magnitude smaller than those for absorption spectroscopy. Selectivity and linearity are two other features that may be better than in absorption. Precision and accuracy may be a factor of 2 to 5 times lower than absorption.

$$I_f = I_0 \, \Phi_f \, (1 - \exp -\varepsilon cl) \qquad (3)$$

where I_f =   intensity of the emitted light
        I_0 =   intensity of the exciting radiation
        Φ_f =   quantum efficiency
        εcl =   absorbance of the solution

At low concentration (absorbance less than 0.02) the fluorescent intensity is proportional to concentration. This is because of the approximation made in solving Equation (3) to give:

$$I_f = I_0 \, k \, c \qquad (4)$$

where k = a constant for the system. Increasing the incident intensity (I_0) increases the fluorescent intensity. Increasing the incident intensity for absorbance measurements does not have the same effect since it is the ratio I/I_0 that is proportional to absorbance. In luminescence the background signal is lower and so sensitivity is increased.

Fluorescence may not vary linearly with concentration for reasons other than the approximate nature of Equation (4). At high concentration self-quenching may occur; this is not usually a problem in trace analysis. This occurs as a result of collisions between excited state molecules leading to a radiationless transfer of energy. Self-absorption is another reason: this is when the emission band overlaps with the absorption band. Changing the excitation wavelength may avoid this condition. See Section 8.1.7.

8.1.5 UV/Visible Absorption Spectrophotometry in Practice

This section is concerned mainly with those features of practical analytical methodology that must be addressed in order to obtain reliable analytical data in spectrophotometry. Features of analytical instruments are mentioned only when they have a particular bearing on the topic under consideration because, except on the occasion of selecting a new instrument, they are outside the control of the analyst.

Spectrophotometers may be set to read in absorbance or transmission mode. Since the quantitative applications of analytical spectrophotometry make use of Beer's Law, normal practice is to derive analytical data from absorbance. This section assumes absorbance is being measured.

Analytical Determination

When working close to the lowest concentration that can be measured with a given method, it is important to know the lowest concentration that can be recorded, known as the detection limit, and the lowest concentration that can be reliably measured, the determination limit. Both these limits are multiples of the standard deviation calculated from measurements made on a large number of reagent blanks. Three times the standard deviation of the blank or three times the signal-to-noise ratio may be used as the detection limit. The determination limit is several times the detection limit, normally ten times the standard deviation of the blank. It should be remembered that it is wrong to assume that using a reagent blank in the reference cell and measuring the analyte absorbances against it will cancel out the effect of even a high blank.

As in all trace analysis one of the most important considerations is cleanliness and several precautions need to be taken to prevent inadvertent contamination of the sample or the analytical system. This is particularly important in trace metal analysis when very common metals such as aluminium or iron, or widely distributed metals such as lead or zinc, are being determined.

Reagents of suitably pure grade giving low blank values must be employed, and blanks should be run to check reagent suitability every time a new bottle is opened. Many chromogenic reagents are commercial dyestuffs and show considerable variations from batch to batch, so that full calibration functions must be prepared each time a new bottle of colorimetric reagent is opened. It is advisable, in fact, to extend this practice to all chromogenic reagents unless experience has proved this to be unnecessary. Trace impurities may cause high blanks, and organic reagents other than dyestuffs are not always well characterised.

Organic solvents used in UV spectrophotometry, including HPLC, must have low absorbances in the regions they are to be used, and the purchase of those grades, specially purified, is strongly recommended. The extra expense over normal analytical reagent grades is more than compensated for by the savings made in spectral quality.

Calibration. When establishing a calibration relationship, it is usual to measure the absorbance of each of the standards used and then to draw or calculate the 'line of best fit'. Since each measurement, including those made in the preparation of the standards, has a variance it is advisable to prepare, at least in duplicate, a set of solutions of a chemical calibrant of known concentration. For each of the solutions three or four absorbance measurements should be made. The mean values for each point should be determined and these used to establish the exact relationship between absorbance and concentration. The same practice should then be applied to the samples.

Absorption spectrophotometry is capable of quite high precision (coefficient of variation <0.1%) though the techniques of high-precision spectrophotometry are not practical for routine work. However, one lesson from high-precision operations can often be applied in routine work. There is a tendency for calibrations to be prepared over too wide a range of concentrations, perhaps to save making two or more calibration curves or because a series of samples might contain a few with particularly high or low absorbances. Wherever possible, the working range of a method should be confined. The absorbances of the chemical calibrants should 'bracket' those of the samples; the closer the values the more reliable will be the measurement.

Temperature. The temperature dependence of the solution being measured should be examined during the method validation. It is good practice to thermostat the cell holder to control the temperature to within ±0.5°C. In the range 20–30°C, the temperature coefficient of the molar absorptivity of the 350 nm peak in potassium dichromate in acid solution is $-0.05\%°C^{-1}$. If the cell compartment is not thermostatted there may be significant changes in the temperature within the sample cell during a series of measurements.

Sample Cells. Cells come in a variety of shapes and sizes but the path length most often used is 10 mm. For the UV region high quality silica should be used whereas optical glass can be used in the visible region. Disposable plastic cells may be used for some applications. On its passage through the sample cell (or cuvette), radiation will be lost by reflections from the surfaces, and may also be lost through absorption by the cell material. Pathlength variability and non-parallelism of end windows can introduce errors. The cells must be matched over the wavelength range for which they are intended. They must be matched for wall thickness and for optical surface and so require a high level of quality control. Using the same cell orientation for all measurements is one way of minimising errors.

Before they are put into use, a new set of cells should be checked to ensure that they do match both in absorbance and in pathlength. The latter is checked by ensuring that a solution of moderately high absorbance, say about 0.5, gives identical absorbance readings in the series of cells.

Cells must be kept clean using methods that are no harsher than absolutely necessary. Plastic cells can be cleaned only with well-diluted detergents, glass cells may be cleaned with detergents and acids but not alkalis, and fused silica cells are treatable with detergents, strong oxidants, acids, or alkalis. New silica cells, described as 'matched', may need immersion in dilute alcoholic potassium hydroxide before they give negligible absorbance and become matched, possibly due to a very thin coating of a silicone grease. Finally cells should be rinsed with high purity water and left to drain. Rinsing with isopropyl alcohol (propan-2-ol), followed by distilled water is another method of cleaning.

Always handle cells by the opaque sides. Touching the optical faces of the cells with fingers must be avoided and it is recommended that any contamination to the faces is removed as soon and as thoroughly as possible, finally polishing them with a clean, dry optical cloth. At low wavelengths, below about 250 nm, cells that appear to be clean may cause false absorptions due to fingerprints or traces of grease *etc.* on the optical faces. Clean, dry cells should be stored in small, well-sealed clean containers, and if they are unlikely to be used for a prolonged period, they should be wrapped in lens tissue.

For some measurements, cell position can introduce an unacceptable uncertainty. To reduce the uncertainty the same cell is used for calibration and measurement. The cell should be placed in the spectrometer and all the filling, emptying, and cleaning operations carried out without removing the cell from the spectrometer. Great care is required if these procedures are necessary for the assay.

In automatic equipment, the methods must often be adapted for use with shorter pathlength cells. It can be shown that the 'theoretical' minimum error for a measurement occurs at an absorbance of either 0.43 or 0.8 depending on the type of detector used. It is good practice to select, if possible, the cuvette size and associated analytical conditions so that the expected sample absorbances fall within the range 0.2–0.8.

8.1.6 Colorimetric Procedures

A typical colorimetric procedure for a trace metal may contain all the following steps after the sample solution has been prepared, although not necessarily in this sequence:

oxidation or reduction;
addition of masking agents;
pH adjustment;
addition of buffer;
addition of colorimetric reagent;
dilution to known volume.

The solution will then be set aside for a specified time interval to allow the colour to develop and then the absorbance will be measured, again usually within a stated time. When the coloured complex is extracted into an organic solvent, the instructions for the extraction process will also be stated.

It is important to follow the instructions exactly, and to ensure that the various reagents are added in the specified order and that the solution is properly mixed between each addition. Oxidation or reduction required in the method, for example when forming a complex of iron(II) or copper(I), may not proceed to completion except under the conditions specified. Likewise, masking reactions and the colour reaction itself may not proceed to completion unless the exact instructions are followed.

It must be remembered that the analytical system is complex and that many competing reactions are possible and will occur. Hydrogen ions will compete for the chromogenic reagent with the metal ions being determined. Hydroxyl ions, or ammonia, or amines present, may compete with the colour reagent by acting as ligands for the metal ion. For this reason, there is always an optimum pH range for the desired reaction. Masking agents are clearly sources of complexing ligands and may, in sufficient concentration, begin to combine with the determinand. Finally, by their nature, buffers may contain weak bases or the anions of weak acids, both of which are electron donors and hence ligands, and which, once again could, in sufficiently high concentration, combine to some extent with the metal sought and reduce the quantity that is available to react with the colour reagent.

It is important therefore never to use larger amounts of any of the reagent solutions than is necessary, and to measure them reasonably accurately to ensure that reaction conditions are both reproducible and will, in a series of tests and standards, be the same from one solution to another. Another reason for using any reagent at concentrations no higher than necessary is that the ionic strength of the system should be kept as low as possible, since high ionic strengths have adverse effects on the stabilities of metal complexes and hence on the analytical results. Some electronic transitions are sensitive to the ionic strength of the medium and hence this will be reflected in the observed absorbance.

Metal complexes showing charge-transfer absorption bands require protecting from light, which will promote the charge-transfer redox reactions. It is always good practice to protect solutions of coloured metal complexes from direct light and some methods clearly state conditions for colour development in reduced light or complete darkness.

Metal solutions containing less than about 1 µg ml^{-1} are often unstable, the instability increasing with decreasing concentration. Such solutions need to be used as soon as possible after preparation. Dilute standards need frequent checking or replacement and sample solutions should be measured soon after preparation. Losses occur by adsorption or ion exchange at glass surfaces and are much less likely to take place in silica or plastic apparatus. Losses are also less common from moderately acidic solutions. Some metals are more prone to be lost than others. These include silver, arsenic, antimony, and bismuth and in some cases lead and mercury. Surface losses of metal are not predictable and this may be because they depend on many factors including the history of the apparatus. One possible solution to the problem is to work with plastic apparatus whenever analytical conditions permit, but care must still be exercised. Some solutes may be absorbed by low density polyalkenes making it difficult to remove the solute by washing. However, they may be released later into other solutions that contain the same solute. It is good practice to reserve apparatus fabricated from these materials for particular solutions.

8.1.7 Photoluminescence and Chemiluminescence Analyses

The number of applications of photoluminescence to organic and biochemical species increases annually. Some of the most important areas of analyses are in food products, pharmaceuticals, clinical samples, and natural products. The applications of chemiluminescence are fewer because of the small number of reactions that produce luminescence.

The precautions necessary for absorption spectrophotometric analysis also apply to luminescence spectroscopy but there are some additional points. Raman emission may occur as an unwanted contribution to fluorescent emission. This problem can be overcome either by changing the experimental conditions or by mathematical treatment of the data. The pH dependence of the emission may be different from that in absorption because of the different electron configuration of the excited state species. Luminescence is also quenched by oxygen and by some metal ions.

For complex mixtures it may be necessary to measure the excitation–emission matrix, which allows the plotting of the emission intensity at all combinations of excitation and emission wavelengths in a single three-dimensional graph. Synchronous spectroscopy is a simplification of this technique.[1,2] In this procedure both the excitation and emission wavelengths are scanned with a constant interval $\Delta\lambda$ between them; typically $\Delta\lambda$ values of 50–100 nm are used. This procedure generates a single two-dimensional slice through an excitation–emission matrix surface. In some instances setting $\Delta\lambda$ to zero sharpens the emission lines.

Mathematical treatment of the data can also be used when two species have overlapping spectra; a technique known as H-point standard addition method (HPSAM) has been used in absorption spectrophotometry and spectro-fluorimetry.[3–6] This method may also be used if the interference is from Raman scattering.

Fluorescence Cells

Fluorescence cells are similar to cells used for absorption spectrophotometry except that at least three of the faces are polished. Care of the cells is as for those used for absorption.

Inorganic Species

There are two ways fluorescence techniques are used to detect inorganic analytes:

(i)   formation of fluorescent chelates;
(ii) quenching action of an inorganic species on a fluorophore

A limiting factor is that many transition-metal chelates are paramagnetic and thus are unlikely to fluoresce. Non-transition metal chelates are generally colourless but are fluorescent, thus making fluorimetry complementary to absorption spectro-photometry.

Organic Species

Formation of fluorescent derivatives is possible if the analyte is not a natural fluorophore. One of the problems is that fluorescent quantum yields vary within a set of derivatives, and may be pH-dependent.

Chemiluminescence Analyses

Instrumentation is simple but the analysis has to be well controlled because the emission varies with time. The emission signal tends to rise fairly rapidly to a maximum and then decay exponentially. The time over which the signal is integrated has to be controlled as well as the temperature and pH of the reaction.

Several metal ions affect the chemiluminescence of reactions, *e.g.* the reaction of luminol with hydrogen peroxide or oxygen in alkaline solution. The increase or decrease in chemilumiscence permits the determination of such ions at the trace level. Similarly organic species can be determined in this way.

8.1.8 Checking and Maintaining Instrument Performance

Checks on wavelength and photometric accuracy should be made at regular intervals, the time between checks depending on the frequency of use of the instrument.

Calibration Standards

Calibration standards are more likely to be used if they are easy to use and require minimum preparation. The standards must also simulate the normal use of the instrument. Solid reference materials are the most convenient to use since they are usually stable and insensitive to temperature change. The disadvantage is that they have fixed shape and are difficult to keep clean. Alternatives are solid reference materials which are used in solution. These are available either as a solid which the analyst dissolves in an appropriate solvent or already in solution in a sealed cuvette.

Wavelength Standards

When determining some trace organic substances, or rare earth ions, a shift in the wavelength scale can have a significant effect on the validity of the measurement. However, if the analysis involves measuring the absorption of a broad absorption band at its peak maximum this is usually relatively insensitive to a small change in the wavelength scale, see Figure 8.1.2. In addition, the wavelength setting should not be changed during the analysis of the samples and standards.

The filters most often used to check the wavelength scale consist of soda-glass doped with holmium oxide or with a mixture of rare earth oxides called, didymium oxide. These have several disadvantages since they are difficult to produce reproducibly. The bands are broad and subject to variation, depending on the method of manufacture. A great improvement has been achieved by using a single rare earth oxide in a fixed concentration in a monocrystalline host material. These 'McCrone filters' cover the region 253–795 nm and have sharp bands. These filters need recalibration at intervals of 2–3 years.

Solutions of holmium oxide in perchloric acid or of samarium perchlorate are available. Samarium perchlorate-filled cuvettes cover the region 235–559 nm. Holmium oxide in perchloric acid covers the region 240–640 nm. One disadvantage is that the band position depends on bandwidth. However, there is no temperature variation in the 20–30°C range for bandwidths of 0.1–1.0 nm.

It should be noted that the bandwidth used for the calibration should be the same as that used for the samples. Where an instrument is fitted with a constant energy mode, which causes the slit width to vary to compensate for loss of energy, the bandwidth should also be checked.

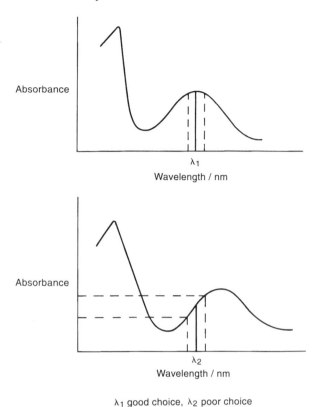

λ_1 good choice, λ_2 poor choice

Figure 8.1.2 *Selection of* λ_{max} *value*

Absorbance Standards

In addition to the general requirements for standards, absorbance standards ideally should have a fixed absorbance, which is independent of wavelength, be unaffected by stray light, be non-fluorescent, and be easy to calibrate.

Neutral glass filters are convenient to use but have several disadvantages. They can only be used in the visible region and may suffer from reflectance errors, which depend on the geometry of the incident radiation. They require regular recalibration. The temperature dependence is wavelength and absorbance dependent.

Metal on quartz filters suffer some of the problems of glass filters but are less dependent on temperature variation. Nickel chromium alloy on silica can be used from 200 to 800 nm; chromium on fused silica is also available.

A number of liquid standards are available from various sources. Liquid standards need also to be insensitive to pH and temperature changes and obey Beer's Law. A set of standards prepared from pure metallic Cu, Ni, and Co dissolved using a mixture of nitric and perchloric acids covers the range 298–794 nm. This suffers from having sharp peaks, so wavelength settings need to be very precise. The solutions also have a significant temperature dependence.

Ampoules of solutions of Co and Ni in a mixture of perchloric and nitric acids are also available as certified standards. There are four bands in the range 302–678 nm as shown in Figure 8.1.3. The reference values of absorbance are quoted for a particular bandpass and the wavelength settings need to be correct to ±0.5 nm. Temperature coefficients are lower than for potassium dichromate solutions.

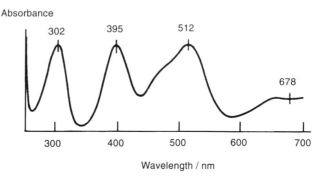

Figure 8.1.3 *Spectrum of Co/Ni in a perchloric acid/nitric acid mixture*

Crystalline potassium dichromate is also available and the absorbance of 10 concentrations of the salt in 0.001 mol l^{-1} perchloric acid are recorded. Solutions of potassium dichromate in sealed cuvettes are also available. Perchloric acid is preferred to sulfuric acid because it has no tendency to form mixed complexes with Cr(VI) species. Higher concentrations of perchloric acid also produce interfering species. The distilled water must be free of reducing impurities. Distilling the water from alkaline KMnO$_4$ may be necessary. Solutions of potassium dichromate have a large temperature dependence.

Bandwidth

Changes in spectral bandwidth can affect both the band position and the absorbance values. Peak heights are generally constant if the bandwidth of the instrument is no more than one-tenth of the effective bandwidth of the absorption peak. The difference between the observed and the true absorbance is proportionally greater at high absorbance values. Samarium perchlorate solutions can be used to check effective bandwidth.

Stray Light

Stray light is any radiation which reaches the detector but has not passed through the sample. This is usually only a problem at the extremes of the wavelength scale of the instrument. Low absorbance values or false peaks are a result of stray radiation (see Figure 8.1.4). If the stray light is slit-width-dependent this suggests the sample compartment is not light-tight. Solutions of potassium iodide in distilled water can be used for measuring stray light below 260 nm. Saturated lithium carbonate solutions are available, which may be used to check for stray light below 225 nm. Vycor filters are useful for a quick check but suffer from a slow wavelength cut-off.

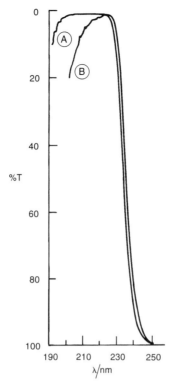

Spectrometer A can be used 195–200 nm
Spectrometer B can be used below 220 nm

Figure 8.1.4 *Effect of stray light on absorption bands*

Fluorescent Standards

Because of variation in source intensity, photomultiplier sensitivity, and other instrument variables, it is necessary to adjust fluorimeters from day to day. Solid standards are available which can be routinely used to calibrate and monitor the performance of fluorimeters. Standards used are anthracene, naphthalene, ovalene, *p*-terphenylbutadiene and rhodamine B. The intensity standards are embedded in a poly(methyl methacrylate) matrix. Another common standard is a solution of quinine sulfate in perchloric acid.

8.1.9 References

1. J. B. F. Lloyd, 'Synchronized Excitation of Fluorescence Emission Spectra', *Nature Phys. Sci.*, (*London*), 1971, **231**, 64.
2. M. Salgado, C. Bosch, F. Sanchez-Rojas, and J. M. Cano-Pavon, 'Derivative and Synchronous Scanning Spectrofluorimetry. Basic Principles and Applications', *Quim. Anal.*, 1986, **5**, 374.
3. F. Bosch-Reig and P. Campins-Falco, 'H-Point Standard-additions Method. Fundamentals and Application to Analytical Spectroscopy', *Analyst*, 1988, **113**, 1011.

4. M. J. Cardone, 'H-Point Standard-additions Method. Fundamentals and Application to Analytical Spectroscopy', *Analyst*, 1990, **115**, 111.
5. F. Bosch-Reig and P. Campins-Falco, 'H-Point Standard-additions Method. Fundamentals and Application to Analytical Spectroscopy. Reply', *Analyst*, 1990, **115**, 112.
6. P. Campins-Falco, J. Verdu-Andres, and F. Bosch-Reig, 'Development of the H-Point Standard Addition Method for the Use of Spectrofluorimetry and Synchronous Spectrofluorimetry', *Analyst*, 1994, **119**, 2123.

Selected Bibliography

7. 'Trace Analysis – Physical Methods', ed. G. H. Morrison, Interscience Publishers, New York, 1965.
8. J. R. Edisbury, 'Practical Hints on Absorption Spectrometry, Ultra-violet and Visible'. Hilger & Watts, London, 1966.
9. D. A. Skoog and J. L. Leary, 'Principles of Instrumental Analysis', Harcourt Brace Jovanovich College Publishing, Philadelphia, PA, 4th edn, 1992.
10. UV Spectrometry Group, ed. C. Burgess and A. Knowles, 'Techniques in Visible and Ultra-violet Spectrometry, Vol. 1, Standards in Absorption Spectrometry', Chapman and Hall, London, 1981.
11. UV Spectrometry Group, ed. A. Knowles and C. Burgess, 'Techniques in Visible and Ultra-violet Spectrometry, Vol. 3, Practical Absorption Spectrometry', Chapman and Hall, London, 1984.
12. H. H. Willard, L. L. Merritt, Jr., J. A. Dean, and F. A. Settle, Jr., 'Instrumental Methods of Analysis', Wadsworth Publishing, Belmont, CA, 7th edn, 1988.
13. 'Luminescence Applications in Biological, Chemical, Environmental, and Hydrological Sciences', ed. M.C. Goldberg, *ACS Symp. Ser.* 383, 1989.
14. 'Molecular Luminescence Spectroscopy, Methods and Applications', Part 1, ed. S. G. Schulman, John Wiley, Chichester, UK, 1985.
15. 'Molecular Luminescence Spectroscopy, Methods and Applications', Part 2, ed. S. G. Schulman, John Wiley, Chichester, UK, 1988.
16. 'Molecular Luminescence Spectroscopy, Methods and Applications', Part 3, ed. S. G. Schulman, John Wiley, Chichester, UK, 1993.

8.2: Electrochemical Techniques

8.2.1 Overview

Electrochemical techniques, sensors, and electroanalytical techniques can be used for the determination of trace amounts of a wide range of analytes (usually metals and organometallics). All electrochemical techniques use at least two electrodes and a meter with which to measure a potential or a current and in some cases an applied voltage. The analytes will most often be in solution (usually aqueous). There are an enormous variety of electrochemical techniques but they can be subdivided into two groups.

(i) *Potentiometry*: the measurement of emf at zero current. Measurements using ion-selective electrodes, which includes pH measurements, are potentiometric techniques.

(ii) *Voltammetry*: the application of a voltage function and measurement of the associated current, which includes polarography, amperometry, and coulometry. The sub-group of voltammetry known as polarography is any voltammetry performed with a mercury electrode. Anodic stripping voltammetry for the analysis of trace cations is an example of this group.

Other electrochemical techniques which fall outside the definitions given here but are useful in trace analysis are biosensors and amperometry. These are discussed briefly at the end of this section. There are other electrochemical techniques that are so rarely used (or not at all), that it is not within the scope of this book to discuss them. These include some of the sub-groups of voltammetry, coulometry, and conductimetry.

For metal ion analysis, electrochemical techniques are potentially the cheapest and quickest method of carrying out a determination; this is when they are compared with instrument techniques such as atomic absorption spectroscopy (AAS) and inductively coupled plasma (ICP)-MS. Depending upon the type of technique, they can be very specific and extremely sensitive. The portability and non-destructive nature of some of the electrodes is also an advantage, suiting them to on-site environmental analysis and the analysis of flowing industrial processes. However, electrodes have a reputation for poor stability which is often deserved. As the device has to be in physical contact with the sample, they are prone to fouling and interference. Accordingly, few laboratories make use of the full range of electrochemical techniques available. However, ion-selective electrodes (potentiometry) and anodic stripping voltammetry are popular for the analysis of metals in water. Historically, many analysts had an aversion to using electro-chemistry, but nowadays the commercial instruments in use tend to be simple to use and do not require a full understanding of the theory involved.

The advantages and disadvantages of electrochemical techniques are summarised in Table 8.2.1. It should be noted that these are generalisations and may not apply to every electrochemical technique since the number of options is so vast.

Unfortunately, despite the large number of obvious advantages of using electrochemical techniques, the suspicion of unreliability has resulted in their slow uptake.

8.2.2 Potentiometric Methods

Introduction to Potentiometric Techniques and Ion-selective Electrodes (ISE)

To carry out a potentiometric determination, there needs to be a potential difference between two half cells as in a galvanic cell. Any cell arrangement of an indicator electrode and reference electrode can be used. These are dipped into a solution and connected to a (high impedance) voltmeter. No current is applied to the system, *i.e.* it is a completely passive technique. Additionally, virtually no current is drawn from the system.

The *voltmeter* measures the potential between the two electrodes which, after a suitable calibration, will equate to the concentration of the selected analyte. The

Table 8.2.1 *Advantages and disadvantages of electrochemical techniques*

Advantages	Disadvantages
Cheap instruments	Doubts about reliability
Portable	In some case requires highly skilled analyst
No moving parts, suitable for on-site analysis	Lack of specificity (analytes of similar charge and size may interfere)
Reduces need for transporting samples because of on-site possibilities	
Quick analysis time, usually a few minutes or less. This does not include the necessary calibrations, controls, and blanks	
Possibility of high sensitivity, can be comparable to that of existing elemental techniques such as AAS and ICP-MS	
Can be made very small	

indicator electrode is an electrode specific to the analyte being measured. The most important aspect of this electrode is the membrane which is in contact with the analyte solution and separates the analyte from an internal reference solution. The *reference electrode* is usually a calomel electrode (Hg/Hg_2Cl_2) or a silver/silver chloride electrode. The function of the reference electrode is to provide a constant potential with the minimum of drift against which the potential (which will be affected by the concentration of the analyte in the solution) of the indicator electrode is compared. Thus the concentration of the analyte can be established after a calibration has been performed.

Ion-selective electrodes is the generic term for all commercially available electrodes used as part of a potentiometric technique. The functioning of an ISE is based on the selective passage of charged species from one phase to another leading to the creation of a potential difference. There are several sub-groups that come under this heading:

glass electrodes (including pH electrodes);
solid state electrodes (including single crystal electrodes);
neutral carrier membrane electrodes;
ion-exchange electrodes;
gas sensing probes;
enzyme electrodes.

The simplest system, with an electrode costing about £300, costs about £1000. Alternatively, dedicated systems can be purchased, with autosamplers and data processing (excluding electrodes), for approximately £10 000. The electrodes would be purchased according to the requirements of the analyst.

The use of an ISE system may simply involve dipping the electrodes in the test solution and reading a signal from the voltmeter (prior calibration with test solutions having been made). Some commercially available devices have some form of internal calibration procedure. However, the only way to ensure adequate calibration is to use a range of chemical standards. Ion-selective electrodes are reasonably sensitive. They can be calibrated to about 10^{-5} mol dm^{-3} (equivalent to about 0.19 ppm of fluoride ions). Calibration ranges can be extended down to the region of 10^{-8} mol dm^{-3} by the use of ionic buffers. All the analytes to which ISE are applicable must be either ions or molecules that ionise.

Common applications in the use of ISE are the determination of halides, cyanide, sulfide, heavy metal ions, nitrate, borate, carbonate, lithium, sodium, potassium, ammonium, and calcium ions. They can be made specific to pharmaceuticals and food additives such as sulfonamides.

Like most analytical techniques, ISE work better in 'clean' solutions without interferences. They are ideally suited to analyses in flowing streams such as rivers or industrial processes, as they can be adapted to almost any size, shape, or material in order to fit in with their location. Microelectrodes with built-in reference electrodes have been used for measuring K^+ and other cations in cells, in the spinal cord, and in the brain.

Comparison of Techniques

In the following paragraphs the different ISE are discussed and the most important ones are also summarised in Table 8.2.2. This Table shows the pertinent aspects of the membrane and to which analytes it is applicable. However, it is quite likely that, when purchasing electrodes, they will be classified in terms of their respective analytes and not their membranes.

Glass Electrode

This type of electrode has a glass membrane permeable to H^+, Na^+, K^+, Li^+, and NH_{4+}. To alter the membrane selectivity elements and compounds are added to the glass. These adaptations can take the form of adding Li_2O and Al_2O_3 to the glass membrane for the analysis of sodium. Modified glass electrodes have been used, for example, for the analysis of sodium and potassium in wines. By far the most popular use of a glass electrode is for the measurement of pH which is probably the most frequently performed electrochemical analysis. Although not usually considered as trace analysis, this can measure hydrogen ion concentration at low levels. It is the most robust and reliable of the ISEs and is fairly easy to use. It is accurate over a range of about pH 0.5–9, because it is the activity of the proton that is measured. The linearity of the response will vary at very low or very high pH. Below 0.5 the protons are hydrated (acid error) whilst at high pH ions such as sodium, potassium, and ammonium interfere (alkaline error).

Solid State Electrodes

Solid state electrodes are so called because they do not have an electrolyte in the indicator electrode. They are ionic solids that have a low solubility product to

Table 8.2.2 Manufacturers' detection limits for different analytes

Analyte	Membrane	Detection limit ($\mu g\ kg^{-1}$)
NH_3	Gas permeable	0.01
NH_4^+	PVC	0.01
Ba^{2+}	PVC	10
Br^-	Solid state	0.4
Cd^+	Solid state	0.01
Ca^+	PVC	0.02
CO_2	Gas sensing	4.0
Cl^-	Solid state	1.8
Cl^-	Liquid membrane	0.18
Cl_2	Solid state	0.01
Cu^{2+}	Solid state	0.3
CN^-	Solid state	0.03
F^-	Solid state	0.01
BF^-	Liquid membrane	0.6
I^-	Solid state	0.02
Pb^{2+}	Solid state	0.2
NO_3^-	PVC	0.08
NO_x	Liquid membrane	0.08
O_2	Gas permeable	0.2
ClO_4^-	Liquid membrane	0.7
K^+	PVC	0.04
Ag^+	Solid state	0.01
Na^+	Glass	0.02
S^{2-}	Solid state	0.003
SCN^-	Solid state	0.3

avoid dissolution of the membrane and to ensure a response that is stable with time. The membrane is either a single or a mixed crystal. It is this that gives them their specificity. The lanthanum fluoride crystal membrane is an example of a solid state membrane that is used for F^- determination. Other solid state electrodes are available (for example silver halide crystal for the analysis of a particular halide), but these are rarely employed. The sulfide electrode has been used for the detection of sulfate-reducing bacteria, and for the determination of acetylenes, amine, and alkaloids following their reaction with carbon disulfide. It unfortunately suffers interference from dissolved oxygen. The nitrite electrode is occasionally used for environmental analysis of ground waters, but is subject to interference from perchlorate and chloride ions (these can be removed by precipitating with silver sulfide).

Single Crystal

Single crystal electrodes have a single crystal as the membrane. The most common of these electrodes is the lanthanum fluoride electrode which is used for the (highly selective) analysis of fluoride ions as the F^- ion can migrate through the solid crystal. Although generally more selective, single crystal membrane electrodes are more susceptible than glass membranes to interferences from oxidising agents.

The fluoride electrode is not as widely used as the glass electrode, but does find some applications. The fluoride electrode can be used as a sensor in high temperature and pressure environments. It has also been used for routine screening for fluoride in the urine of industrial workers. Fluoride in tap water can be measured at levels down to about 10^{-7} mol dm^{-3} and fluoride inhibitors (such as aluminium ions) can also be determined at similar concentrations. It should be remembered that interfering aluminium will mask fluoride ions when carrying out standard fluoride analyses.

Neutral Carrier Membrane Electrodes

These electrodes use designer synthetic polymers for the membrane in which an active ingredient is impregnated. Neutral liquid membranes do not contain the ion to be detected but a polymeric species whose geometry offers selectivity to a particular ionic species, due to the size and shape of the cavities in the added species. An example is the potassium electrode, using valinomycin to transport the ion across a hydrophobic membrane. This electrode is frequently used to determine blood potassium levels. Similarly, the nonactin ammonium electrode is used for the analysis of ammonium. Solid membrane electrodes use a polymer such as PVC as the membrane into which the neutral carrier is introduced in a suitable solvent. These electrodes do perform functions that would be difficult to do with either glass electrodes or solid state electrodes. The neutral carrier membrane electrodes are used for mainly clinical applications.

Electrodes Based on Ion Exchange

The operation of ion-exchange membranes is virtually identical to that of the neutral carrier membranes, although useful applications are rare. Both solid and liquid membranes can be used.

Gas Sensors

These are not technically electrodes but are actually electrochemical cells containing the reference electrodes. For this reason they are usually termed sensors or probes. Consequently they do not need an additional reference electrode. They are usually used for the analysis of dissolved gases. They are conventional sensors whose response is modified by quantitative interaction between the dissolved gas and the analyte specific to the electrode. There are two types of permeable membranes that can be used for such sensors.

(i)   *Microporous membranes* are made from hydrophobic polymers such as polypropylene and PTFE. The gas molecules move through the pores.

(ii) *Homogeneous membranes* are made from relatively dense polymeric substances such as silicone rubber. The gas molecules diffuse through the membrane.

Typical gases that can be detected by these techniques, with their detection limits, are shown in Table 8.2.2. The ammonia probe is probably the most widely used gas-sensing membrane electrode, used to analyse fresh water, sewage, and effluent.

Enzyme Electrodes

There is only really one commercially available example of the use of enzyme electrodes in potentiometry, the urea enzyme electrode, although others appear in the literature. There are many such electrodes in the field of amperometry.

Urea is detected following its breakdown to ammonium by an enzyme incorporated into the membrane. However, there are a number of problems with interferences from cations in the media in which urea might be measured (such as blood). This has resulted in the limited uptake of this technique for routine analysis.

8.2.3 Potentiometric Techniques: Hints and Tips

Quality of Electrodes

The largest source of error when using ISE originates from the reference electrode, and the difficulty of obtaining a suitable calibration standard identical to the sample.

As crystal membranes age, the contact surface will become pitted and cracked. This results in carry-over unless care is taken to clean the surface thoroughly. The polymer-based electrodes have a limited shelf-life; the manufacturer should provide details. If the electrode is not to be used for some time then drying and storing in cold conditions should be considered. Otherwise the electrode could be left clamped vertically to avoid damage.

It is possible to repair some of the electrodes (the pH electrode is difficult to fix). It will depend, for the most part, on the level of damage to the membrane. Both the membrane and the electrolyte can be replaced or topped up. Indeed, it is usually recommended that this be done at regular intervals to ensure the continued performance of the electrode.

Measurement Conditions

The most important aspect in the use of ISE is to ensure reproducibility. The reproducibility of solid state membrane electrodes depends on the pre-conditioning. The temperature and stirring rates need to be carefully controlled. In addition, most manufacturers recommend the addition of an 'ionic strength adjustment buffer' (ISAB). This is added to the analyte solution and ensures that the electrode is working under the most favourable conditions possible with respect to both pH and ionic strength. Along with each electrode, the manufacturer normally supplies information that enables the analyst to make the best use of the electrodes. This includes:

(i)   temperature range;
(ii)   pH range;
(iii) recommended ISAB;
(iv) concentration range (including detection limit);
(v)   recommended reference electrode;
(vi) interferences.

The most critical is probably (vi). 'Ion-selective electrodes' is, in most cases, a misnomer. An electrode will merely have a preference for one particular ion. Other ions in the solution will (at particular concentrations) compete and interfere. The extent of the interference can be calculated if the selectivity coefficient of the ion is known. The selectivity coefficient can be determined experimentally and is specific to each electrode and analyte solution. Examples of interferents include halides with cyanide, silver, copper, and mercury with other metals of similar charge and size. Cyanide can also leach chloride from silver chloride electrodes.

The response time (the time taken for a system to equilibrate after the start of the analysis) will vary between electrodes. In addition, the presence of interfering ions and the concentration of the analyte will have an effect. Generally gas sensors take the longest to equilibrate (up to 1 min is quoted by one manufacturer).

If measurements are made in a flowing system then the distance between the electrodes should be minimised.

Miniaturisation not only provides a size advantage but also improves the signal-to-noise ratio.

Frequent calibration is necessary, at least at the beginning of each measurement session. The calibration medium should be as similar as possible to the medium in which the sensor will be used.

Analytes

The range of ISE that are commercially available is summarised in Table 8.2.2, along with the membrane type and detection limit. This is compiled from the literature supplied by different manufacturers of ISE. Other information on the use of the electrode should be supplied by the manufacturer.

8.2.4 Voltammetric Methods

Introduction

Voltammetry encompasses a wide range of different techniques. Whereas potentiometry involves setting up a galvanic cell with no voltage applied, voltammetry relies on the application (and variation) of a voltage to a system and the measurement of the resulting current. A voltammogram is established, which relates the voltage applied to the current (or a function of the current). By calibrating with solutions of known concentration, the current–voltage profile can be used for the quantitative analysis of analytes in a solution. The different voltammetric techniques depend on the way in which the voltage is applied. In addition, the techniques will vary in such aspects as the type of electrode used and the diffusion and deposition of analytes.

The two most important techniques for trace analysis are pulsed voltammetry and stripping voltammetry. As was stated in the overview, polarography is simply voltammetry performed with a mercury electrode.

The reason most voltammetric techniques are not applicable to trace analysis is because the current due the analyte (the faradaic component) gets dwarfed by the current required to charge the double layer at the electrode. This is what gives rise to a sloping baseline in polarography. Many of the techniques that are used are designed to get around this problem of double layer charging. Some of the techniques not covered are rejected because they cannot achieve the detection limits required in trace analysis.

Commonly, when purchasing voltammetry equipment, one complete instrument will be capable of performing all the tasks outlined in this section, and others besides. Indeed, manufacturers of voltammetry equipment are likely to be the best source of methods for voltammetric techniques. However, like any other method derived from the literature, these should be checked and thoroughly validated in the analyst's own laboratory under the appropriate conditions of use.

The most relevant trace analytical voltammetric techniques are, *stripping voltammetry* and *pulse voltammetry*. These are discussed in Sections 8.2.5 and 8.2.6.

Analytes

There is considerable scope for voltammetric techniques in the analysis of anions, cations, and organics. However, in practice, it is used mainly for the analysis of trace metals in waters. The limits of determination as found by one manufacturer are shown below in Table 8.2.3. Obviously these figures have to be taken with some skepticism having been probably attained under the most favourable conditions. However, they do illustrate the scope of voltammetric techniques.

Substrate/Matrix

To obtain reliable trace analysis results, it is preferable if the metal cations of interest are in aqueous solvents. However, metals have been analysed in foods and beverages, electroplating baths, pharmaceuticals, and biological fluids.

Instrumentation

The execution of voltammetry usually involves the assembly of three electrodes. These will be the working electrode (*e.g.* dropping mercury electrode), the

Table 8.2.3 *Applications and detection limits using voltammetric techniques*

Element	Limit of determination ($\mu g\ kg^{-1}$)
Al	5
As	1
Ag	5
Bi	0.05
Cd	0.05
Co	0.05
Cr	0.02
Cu	0.05
Fe	0.5
Hg	1
Mo	0.05
Mn	2
Ni	0.05
Pb	0.05
Pt	0.0001
Se	1
Sn	0.5
Tl	1
V	1
Zn	0.05

reference electrode (*e.g.* the saturated calomel electrode or silver/silver chloride electrode), and the secondary or counter electrode (*e.g.* a gold electrode). The reason for the need for three electrodes is that the calomel and other reference electrodes will only sustain a stable potential (*i.e.* work properly) under low polarisation conditions. This is established by putting the reference electrode in a high impedance circuit and allowing a current to flow through the secondary electrode. An exception can be when microelectrodes are used, where low currents mean that sometimes there is no need for a secondary electrode.

Materials

Mercury was the first standard material for voltammetric electrodes and still the most commonly used material for metal ions. This electrode has good reduction kinetics and forms stable conducting amalgams. However, mercury is toxic, messy, and requires skilled handling to avoid contamination. Mercury drop electrodes are also unsuitable for measuring organic analytes, as organic solvents are electrically resistant (the electrode is also obviously unsuitable for measuring mercury). Materials that are now used for voltammetric electrodes are carbon and the noble metals, such as platinum or gold. Other materials are sometimes used for specialist applications *e.g.* copper is used for the analysis of nitrates. Carbon paste and platinum replacements have unfortunately yet to achieve the precision of the hanging mercury drop, which is why the older technique still finds some use.

Microelectrodes

A whole field has grown up in voltammetry which uses 'microelectrodes'. Essentially these are very small electrodes that substitute for the classical mercury and other electrodes. Microelectrodes will mainly substitute for the working electrode. When working in a restricted space, or for other reasons when miniaturisation is required then the reference and counter electrodes might also be microelectrodes. If this is the case then a three-electrode system is required. The theory and processes of voltammetry remain the same but, because of the low volumes and currents involved, microelectrodes possess special properties.

There are advantages in miniaturising most electrochemical sensors. With a small, spherical electrode the migration of ions around it is uniform rather than the diffusion gradient produced when ions move towards a planar surface (as seen with larger electrodes). This gives a steady state current, and therefore a more confident measurement. As was mentioned earlier, a lower current will also negate the need for the secondary electrode unless the reference electrode is also miniaturised.

8.2.5 Stripping Voltammetry

This is a technique that combines the preconcentration stage with the end-determination. The good detection limits seen in stripping voltammetry techniques is due to the accumulation of the analyte on one of the electrodes. Until the last few years, this technique was restricted to Anodic Stripping Voltammetry, used mainly for the analysis of metal ions in water. Now, however, Cathodic and Adsorptive Stripping Voltammetry have expanded the applications to include many more ions and organic species, and creating great interest in biological and pharmaceutical analysis.

The technique is attractive as it is accurate, precise, easily automated, and has detection limits comparable to ICP-MS at a tenth of the cost. It is not, however, a universal technique, and many analyses suffer from interferents. In particular, dissolved oxygen and chlorine affect the stripping stage, and it is very difficult to produce a reference solution with exactly the same dissolved oxygen as the sample solution; this makes it difficult to provide an accurate background correction. In addition, other aspects such as stirring rates need to be reproducible.

Anodic Stripping Voltammetry

Anodic stripping voltammetry remains the electrochemical technique most widely applied to trace analysis. As the analyte is preconcentrated in the technique, limits of determination can be in the region of parts per trillion. Saline water makes an ideal electrolyte, and so it is the technique of choice for the analysis of sea-water. When using a mercury electrode it is extremely good for metals that form amalgams with mercury, *e.g.* cadmium, lead, and zinc.

Anodic stripping voltammetry works by the reduction of the metal ion at a fixed potential (related to the ion of interest) at the (usually mercury) working electrode (which will, in effect, be the cathode). During this time the analyte will accumulate on the working electrode. After a fixed period, the voltammetric part of the analysis takes place. The voltage is made more positive, the metal is oxidised back to the ion, and it is stripped off the working electrode. The integral of the current–time curve or voltammogram peak height (the current) is then measured and compared with the calibration data.

Cathodic Stripping Voltammetry

Anodic stripping voltammetry at a mercury electrode depends on the ability of a metal ion to be reduced to the metal and form an amalgam with the mercury. The technique is therefore limited to elements such as bismuth, cadmium, copper, indium, lead, and zinc. Cathodic stripping voltammetry involves accumulation on the electrode of anions that are subsequently oxidised. Consequently, analytes such as halides, sulfide, phosphate, arsenate, and arsenite can be analysed by cathodic stripping voltammetry. All the potentials for anodic stripping voltammetry are reversed. In addition, because the working electrode needs to be positive with respect to the analyte, this will often preclude the use of mercury because of its low oxidation potential (i.e. it will be oxidised too easily). Possible electrodes are platinum, gold, and carbon.

Cathodic stripping voltammetry can suffer more from interferents, but these are generally easily removed: adsorbent organics and potential metal ligands can be removed by UV photolysis, and competing metal ions by optimising the operating parameters. It is limited to operating at a pH of less than 9; in highly alkaline conditions analyte ions can co-precipitate with calcium and magnesium hydroxides which may be present.

Adsorptive Stripping Voltammetry

Adsorptive stripping voltammetry widens the technique to include organic analytes. It has been applied to surface-active organics, *e.g.* detergents and some oil components, as well as many pesticides, pharmaceuticals, and growth stimulants. It is only useful for compounds of low polarity. By using complexing agents metal ions have been analysed down to levels of 0.1 ppb. The main drawback is the need to add an electrolytic reagent, which increases the risk of contamination and can alter the speciation equilibrium of the analyte.

8.2.6 Pulsed Voltammetry

There are a few techniques which fall into this category that are applicable to trace analysis: these are *normal pulse voltammetry* and *differential pulse voltammetry*. Each of these techniques relies on the same principles, i.e. that the voltage is applied and varied in a series of pulses which may or may not be superimposed upon a ramp function. The resulting current or function of the current is sampled. As has been mentioned earlier, the major problem of voltammetry that prevents its use for trace analysis is the relative contribution to the current signal from the double layer charging. Double layer charging occurs every time the potential on the electrode is changed (which happens in all forms of voltammetry). When a voltage ceases to be applied, the double layer charging current will decrease exponentially, whereas the signal current decreases at the slower rate of $t^{1/2}$. Pulsed voltammetry reduces the relative impact of the double layer charging by sampling the current a discreet time after the voltage has been applied when the signal/charging ratio has swung in favour of the signal current. The pulse can be applied in several ways. Modern instruments apply the voltage as a pulsed ramp (the voltage will be increased over an individual pulse), and the differential of the current/voltage curve is plotted to give a sharp, easily measurable peak, *i.e.* 'differential-pulse voltammetry'. This improves selectivity, and gives sensitivity down to a recorded 10^{-11} mol dm^{-3} for the analysis of riboflavin (more typically around 10^{-8} mol dm^{-3}).

The voltage application and current sampling lend themselves to computer control, and chemometric techniques can be applied to the data from a complex sequence of ramps and pulses.

8.2.7 Voltammetry: Hints and Tips

For all stripping techniques each stage of the process must be precisely timed. In practice this means the use of reliable automated equipment.

It is not necessary to run the deposition step in stripping voltammetry until all the metals have been deposited on the electrode. However, the calibration and analysis have to be done under identical conditions of time, stirring, temperature, and voltage.

Precise thermostatting is necessary, as small changes in solution temperature can cause large variations in the reading. A 1°C increase has been found to give a 7% higher determination.

8.2.8 Bioelectroanalysis

Biosensors take many forms including many methods of detection [optic, mass (piezoelectric), fluorescence/luminescence]. However, by far the most popular form of biosensor is the electrochemical type. This technique incorporates an enzyme to aid in the detection of an analyte. The enzyme converts the analyte into a species that can be analysed electrochemically. An example already discussed is the urea molecular-selective electrode. The urea is converted to ammonium and detected with either an ammonia electrode or a pH electrode.

In principle, enzymes can be attached to any kind of potentiometric or amperometric electrode. However, enzyme activity tends to be reduced by such a process. Most amperometric sensors are based on the monitoring of oxygen (which may be produced or consumed), hydrogen peroxide, or other reduced or oxidised species. Commercially available potentiometric applications are limited to the urea electrode. Microelectrodes for *in vivo* use have been produced for the analysis of substances such as glucose and catecholamines.

The major problem with biosensors is their lack of robustness and instability. Immobilized enzymes have a working life of only 1–8 months, and are easily disabled by fouling agents in the sample solution and mild acids, alkalis, or corrosive agents. There is much research interest in methods of immobilising enzymes on electrodes.

Although biosensors are theoretically highly selective, there are still many potential interferents. Chief amongst these are enzyme inhibitors or activators, such as heavy metals, present in trace amounts. In addition, there is the familiar problem of competing analytes. These can be removed by the addition of complexing agents such as EDTA (ethylenediaminetetraacetic acid). Impurities in the enzyme preparation can also lead to false high readings. All enzyme reactions are pH and temperature dependent, and so the pH must be precisely controlled. This is particularly difficult in the case of sensors based on the pH electrode, since the signal depends on fluctuations in pH.

The glucose electrode is the most common enzyme-based voltammetric biosensor. Glucose is broken down in the presence of the enzyme glucose oxidase to give hydrogen peroxide which gives rise to the current measured. The current is directly proportional to the glucose concentration.

Biosensors based on gas-permeable membranes, *e.g.* the urea electrode, show much greater selectivity. However, they suffer from slow dynamic response times: it can take up to 45 min for an ammonia sensor to reach equilibrium.

Being selective, rapid, cheap, and reversible, biosensors are an attractive technique. They find use in industrial monitoring, the food industry, the determination of a variety of environmental pollutants, and, most importantly, clinical applications. Enzyme inhibitors, such as heavy metals, drugs, and pesticides, have been determined at trace levels (0.5 μg kg^{-1} for organophosphates). Bacteria can also be assayed by the depletion of oxygen. As the technology of their construction improves, biosensors are likely to become a valuable weapon in the analyst's arsenal.

8.2.9 References

General Electrochemical Techniques

1. T. Riley and C. Tomlinson, 'Principles of Electroanalytical Methods', John Wiley, Chichester, UK, 1987.
2. D. B. Hibbert, 'Introduction to Electrochemistry', Macmillan Press, Basingstoke, UK, 1993.
3. C. M. A. Brett and A. M. O. B. Brett, 'Electrochemistry: Principles, Methods and Applications', Oxford University Press, Oxford, 1993.
4. D. R. Crow, 'Principles and Applications of Electrochemistry', Chapman and Hall, London, 4th edn, 1994.
5. W. F. Smyth and M. R. Smyth, 'Electrochemical Analysis of Organic Pollutants', *Pure Appl. Chem.*, 1987, **59**, 245.
6. M. D. Ryan and J. Q. Chambers, 'Dynamic Electrochemistry: Methodology and Application', *Anal. Chem.*, 1992, **64**, 79R.
7. P. M. Bersier, J. Howell, and C. Bruntlett, 'Advanced Electroanalytical Techniques versus Atomic Absorption Spectrometry, Inductively Coupled Plasma Atomic Emission Spectrometry and Inductively Coupled Plasma Mass Spectrometry in Environmental Analysis', *Analyst*, 1994, **119**, 219.

Potentiometry and Ion-selective Electrodes

8. A. Evans, 'Potentiometry and Ion Selective Electrodes', John Wiley, Chichester, UK, 1987.
9. J. Koryta, 'Theory and Applications of Ion Selective Electrodes', *Anal.Chim.Acta*,1988, **206**, 1.
10. J. D. R. Thomas, 'Potentiometric Anion Sensing: Prospects and Limitations', *Anal. Proc.*, 1990, **27**, 117.

Biosensors

11. M. A. Arnold and M. E. Meyerhoff, 'A Review of Biosensors', *CRC Crit. Rev. Anal. Chem.*, 1988, **20**, 148.

Voltammetry and Polarography

12. T. Riley and A. Watson, 'Polarography and other Voltammetric Methods', John Wiley, Chichester, UK, 1987.
13. F. W. Smyth, 'Voltammetric Determination of Molecules of Biological Significance', John Wiley, Chichester, UK, 1992.
14. 'Analytical Voltammetry', (Vol. XXVII of Wilson and Wilson's Comprehensive Analytical Chemistry), ed. M. R. Smyth and J. G. Vos, Elsevier, Amsterdam, 1992.
15. J. Wang, D.-B. Luo, P. A. M. Farias, and J. W. Mahmoud, 'Adsorptive Stripping Voltammetry of Riboflavin and other Flavin Analogues at the Static Mercury Electrode', *Anal. Chem.*, 1985, **57**, 158.
16. R. Kaldova and M. Kopanica, 'Adsorptive Stripping Voltammetry in Trace Analysis', *Pure Appl. Chem.*,1989, **61**, 97.
17. S. Daniele, M.-A. Baldo, P. Ugo, and G.-A. Mazzocchin, 'Determination of Heavy Metals in Real Samples by Anodic Stripping Voltammetry with Mercury Micro-electrodes', *Anal. Chim. Acta*, 1989, **219**, 19.
18. A. R. Fernando and J. A. Plambeck, 'Effects of Background Electrolyte and Oxygen on Trace Analysis for Lead and Cadmium by Anodic Stripping Voltammetry', *Anal.Chem.*, 1989, **61**, 2609.
19. C. M. G. van den Berg, 'Adsorptive Cathodic Stripping Voltammetry of Trace Elements in Sea Water', *Analyst*, 1989, **114**, 1527.
20. D. W. M. Arrigan, 'Voltammetric Determination of Trace Metals and Organics after Accumulation at Modified Electrodes', *Analyst*, 1994, **119**, 1953.
21. L. Mart, 'Minimisation of Accuracy Risks in Voltammetric Ultratrace Determination of Heavy Metals in Natural Waters', *Talanta*, 1982, **29**, 1035.

8.3 Bioanalytical Techniques

8.3.1 Overview

There are a wide variety of methods and techniques for trace analysis that can broadly be described as 'bioanalytical'. These include immunoassays, the use of enzymes as analytical reagents, analyses of enzymes, and assays for DNA. This vast range of techniques available may at first sight seem bewildering to the analyst contemplating venturing into this area. The available literature tends to be specialised and sometimes assumes a great deal of prior knowledge, and it is easy to become lost in a maze of biochemical terms and acronyms. This section is aimed at those analytical chemists who are considering using a bioanalytical alternative to current methods.

Advantages of Bioanalytical Techniques

The potential benefits of using biological techniques are considerable.

Economy. Bioanalytical methods are generally extremely rapid, cheap, and easy to use. A great many samples can be tested in one batch.

Sensitivity and Specificity. In many cases the sensitivity and selectivity of such techniques are very good.

Suitability for Screening Analyses. Although they cannot rival classical techniques in terms of quantitation, they are ideally suited to screening large numbers of routine samples. Positive results can then be confirmed using one of the more precise, but generally more laborious, techniques described elsewhere in this book. Such screens are ideal for analyses such as the determination of environmental pollutants in ground water samples.

Disadvantages of Bioanalytical Techniques

Bioanalytical techniques do have some inherent limitations which may preclude their use for certain applications. These are not always apparent from the literature, as they apply more to commercial kits and spot tests than to the types of optimised and dedicated systems used in specialist biochemical laboratories.

Detection Limits. Limits of detection for test kits, particularly those relying upon a colour change, are often not as good as expected. This may be due to a variety of reasons: the test extract may already be coloured or contain particulates, giving increased background noise; or the spectrophotometer or other measurement device supplied with the kit may be of a lower specification than those used in research laboratories.

Robustness. Enzymes are not robust reagents. Some kits have very short shelf-lives, and are extremely susceptible to variations in temperature and/or pH. Enzymes will not remain active in organic solvents, and so their use is excluded for sample extracts in organic solvents. A few percent of methanol in water is about the limit.[1,2] There has been much research interest in encapsulating enzymes within micelles in a two-phase system with water-immiscible solvents.

Availability. Some of the reagents used are difficult to obtain. For example, the antibodies needed for immunoassays must be produced. If not available commercially, this means the laboratory must have access to an animal house.

Classification of Techniques

The techniques covered within this section are split into three categories.

(i) *Enzyme/substrate reactions*. These include the analysis of substrates using enzymes as reagents, the analysis of enzymes using substrates as reagents, and the analysis of substances that inhibit or activate such reactions.

(ii) *Immunological reactions*. The analysis of compounds by monitoring their reaction with a specific antibody, or *vice versa*.

(iii) *DNA reactions*. There are three types, the replication of DNA strands to produce measurable quantities, DNA 'fingerprinting', and the use of DNA probes.

It must be remembered, however, that classifications of bioanalytical techniques tend to be both arbitrary and occasionally misleading. Frequently, techniques will be developed for particular applications which do not conform to a convenient type and cannot be described by any of the conventional labels.

8.3.2 Enzyme/Substrate Reactions

Comparison of Techniques

Enzymes are proteins with the ability to catalyse specific biological reactions. They display a specificity seldom found in non-biological systems. Thus, any compound that produces a distinctive reaction in the presence of an enzyme can be assayed, and its concentration calculated. Alternatively the concentration of enzymes can be determined.

The range of analytes (substrates) that undergo enzyme catalysed reactions is vast. Table 8.3.1 lists six types of enzyme, with examples of their use in analysis. As indicated enzymes cover a range of possible chemical reactions, *e.g.* oxidations, dimerisations, *etc.* Any compound that is subject to any of the types of reactions indicated can be assayed using enzymes, provided that a suitable enzyme can be found for the catalysis and that a method can be devised for monitoring the progress of the reaction.

Enzymes are described in terms of their activity, quantified by the International Unit (U). The definition of 1 U is the amount of enzyme capable of converting 1 micromole of substrate per minute under specified conditions. The specific activity is then defined as the activity (in U) per milligram of enzyme. It is fortunate that the specific activity of most enzymes is high and very little is needed to catalyse each reaction; because enzymes are difficult to extract, purify, and store they tend to be expensive.

The kinetics of enzyme catalysed reactions are described by the Michaelis–Menten (Briggs–Haldane) equation.[3] This describes the rate of the reaction in terms of the concentration of the substrate. As can be seen from Figure 8.3.1, there comes a point where increasing the substrate concentration gives no further increase in the reaction rate; this is because the enzyme present is working at its

Table 8.3.1 Classification of enzymes

Type of enzyme	Reactions catalysed	Example of analytical use
Oxidoreductases	Oxidations and reductions	Alcohol dehydrogenase for the analysis of ethanol $$C_2H_5OH + NAD^+ \xrightarrow{\text{Enzyme}} CH_3CHO + NADH + H^+$$
Transferases	Transfer of groups from one molecule to another	Hexokinase for the analysis of glucose $$\text{Glucose} + \text{adenosine -5'- triphosphate (ATP)} \xrightarrow{\text{Enzyme}} \text{Glucose-6-phosphate} + \text{adenosine - 5'- diphosphate (ADP)}$$
Hydrolases	Hydrolysis	Urease for the analysis of urea $$\text{Urea} \xrightarrow{\text{Enzyme}} \text{Ammonia} + \text{carbon dioxide}$$
Lyases	Remove groups from molecules	Difficult to monitor reaction; not often used for analysis
Isomerases	Intramolecular rearrangement	Phosphoglucose isomerase for the determination of fructose $$\text{Fructose} \xrightarrow{\text{Enzyme}} \text{Glucose}$$
Ligases	Condense two molecules together	Difficult to monitor reaction; not often used for analysis

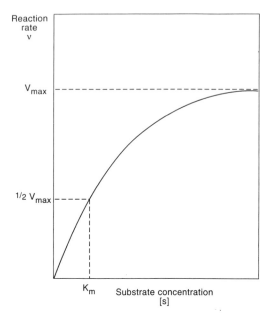

Figure 8.3.1 *A Michaelis–Menten plot*

maximum activity. The substrate concentration which gives a reaction rate of half of the maximum is referred to as the Michaelis constant, K_m.

K_m is related to the reaction rate (v) and substrate concentration [S] by the Michaelis–Menten equation (Equation 1).

$$v = \frac{V_{max}\ [S]}{K_m + [S]} \tag{1}$$

There are a number of simplifying assumptions that allow Equation (1) to be used for the quantitation of either enzyme or substrate concentration from the measured reaction rate.

To Quantify the Substrate Concentration. If the substrate concentration is negligible compared with K_m *i.e.* when the analyte is at trace levels, then the reaction rate is directly proportional to the substrate concentration. The measurement of the initial reaction rate (and thus the initial substrate concentration) can be calculated by measuring the difference in rate between two different times and extrapolating back. This process is very easily automated. As the assumption holds best for trace levels of substrate, the method is ideal for trace analysis, and is the most widely used application of enzyme/substrate reactions for trace level work. At higher substrate concentrations the simplifying assumption does not hold. Some analyses merely involve measuring the amount of product a fixed period of time after adding the enzyme. This is compared with the amount produced by a standard substance. Other analyses require, for example, measuring the absorbance after a fixed period of time after adding the enzyme and until the rate of change with time is constant. The linear graph of absorbance against time is extrapolated back to the time of adding the enzyme.

To Quantify the Enzyme Concentration. In some cases the enzyme is the analyte. Examples are the monitoring of body functions by tracking enzyme levels, and the monitoring of food treatment processes such as pasteurisation or freezing and thawing.[4] In these cases, if K_m is negligible compared with the substrate concentration, *i.e.* if the system is deliberately overloaded with substrate, then the initial reaction rate is directly proportional to the initial enzyme concentration.

To Quantify Enzyme Inhibitors or Activators. Apart from measuring substrates or enzymes, enzyme/substrate reactions can be used to assay a variety of substances that inhibit or activate enzymes. This extends the scope of the method to other common analytes such as metal ions and insecticide residues.[5-7] For example, fluoride in teeth can be measured down to ng ml^{-1} levels by inhibition of esterases. To quantify such substances the reaction rate is measured as under the heading 'To Quantify the Enzyme Concentration', and then the reaction repeated in the presence of an extract containing the inhibiting/activating analyte. The assumptions about reaction kinetics must be modified depending upon the nature of the activation/inhibition.[3]

There are three factors that are critical to ensuring the reproducibility of enzyme catalysed reactions: temperature, pH, and timing. The activity of an enzyme is highly pH and temperature dependent; both of these must be controlled very precisely. The time at which measurements are taken must also be precisely controlled. The availability of well-characterised material for use as analytical standards is also necessary.

If temperature, pH, and timing are sufficiently well controlled, the reliability of quantitative enzyme/substrate reactions then depends largely upon the analyst's knowledge of the mechanism and kinetics of the reactions, and particularly of conditions under which any simplifying assumptions will no longer hold. However, a great many enzyme/substrate reactions are intended to be either qualitative or semi-quantitative. In these cases it is not necessary to measure the reaction rate; the analyst merely needs an indication of whether or not the reaction has occurred (usually a colour change). In these cases a knowledge of reaction kinetics is rarely required.

Methods of Monitoring Enzyme/Substrate Reactions

In order for enzyme/substrate reactions to be of analytical use, it is necessary to devise methods of monitoring them. For many spot tests and screening kits it is only necessary to determine whether a reaction has occurred, but for quantitation a property that changes in proportion to the amount of analyte present has to be monitored.

The ideal method of monitoring a reaction would be extremely selective, sensitive, easy to automate, and non-intrusive (*i.e.* will not have an effect upon the reaction rate). Some of the possible options are listed in Table 8.3.2.

Table 8.3.2 *Methods of monitoring enzyme/substrate reactions*

Monitoring method	Comments	Examples of applications
Spectrophotometry	Usually does not affect reaction. Cheap, easy, and a variety of applications	Monitoring coloured product
Spectrofluorimetry	Usually does not affect reaction. More selective and sensitive than spectrophotometry, but fewer applications	Reactions involving nicotinamide adenine dinucleotide
Chemiluminescence	Totally external. Does not affect reaction. Very selective and sensitive but applications rare	Use of luminol to follow reactions involving hydrogen peroxide
Polarimetry	Does not usually affect reaction. Has been slow and tedious in past, but instrumentation now improved	Only useful for optically active compounds
Volume of gas evolution	Totally external. Does not affect reaction. Very tedious. Has only recently developed sensitivity to match other techniques	Reduction in pressure on oxidation of glucose with O_2
Polarography	Need to introduce electrode; may affect the reaction. Slow, difficult, and not easily automated	Using the oxygen electrode to follow the oxidation of glucose with O_2
Ion-selective electrodes	Need to introduce electrode; may affect reaction	Glass electrode to follow reactions involving pH changes. Ammonia electrode to follow urea hydrolysis
Thermistor	Need to introduce probe; may affect reaction	Most reactions produce temperature changes
Radiolabelling	Difficult to apply to rapid, routine reactions	Labelled isotopes of any compound can be produced

If none of the reactants or products are easily measurable, it is always possible to design quite complicated schemes and secondary reactions in order to obtain a product that can be measured. For example, for the enzyme catalysed reaction of glucose and oxygen as shown in Equation (2),

$$\text{Glucose} + \text{Oxygen} \xrightarrow{\;\beta\text{ - D - glucose oxidase}\;} \text{Gluconic acid} + \text{Hydrogen peroxide} \quad (2)$$

it would be possible to measure, amongst other things:

  (i)   the decrease in oxygen with an oxygen electrode;

(ii)   the increase in hydrogen peroxide with a platinum electrode;
(iii) the increase in hydrogen peroxide by reaction with *o*-dianiside to produce a coloured product, with homovanillic acid to form a fluorescent product, or with luminol to produce chemiluminescence.

It is vital, if using a secondary indicator reaction, that the overall kinetics are unaffected, *i.e.* the indicator reaction must *not* be the rate determining step.

Immobilisation of Enzymes: Biosensors

In addition to test kits where an enzymatic catalyst is added to the sample solution, the enzyme can also be bonded or immobilised onto a solid support, *e.g.* the well in which the reaction takes place, or a plate that is dipped into solution. In principle, any enzyme can be bound to a carrier by chemical bonding, adsorption, or polymerisation into plastic. They can also be incorporated into membranes. There are some technical difficulties in immobilising enzymes, but if these can be solved then the test procedure is straightforward.

Enzymes immobilised on insoluble supports continue to act as catalysts until they are destroyed mechanically or chemically; in fact, their stability is often increased by immobilisation. An additional advantage of immobilisation is that it opens the way to continuous processing, whereby test solutions are passed through a reactor containing immobilised enzyme. The extent of the enzymatic reaction is controlled by the rate of flow through the reactor since this determines the time the enzyme is in contact with the substrate.

In some cases there is only a slight change in the catalytic activity following immobilisation, while other bound enzymes exhibit much lower activities, sometimes as low as 1% of their optimum. The particular method employed to immobilise the enzyme affects the catalytic efficiency of the bound form. One requirement is that it leaves the catalytic centre of the enzyme accessible to the substrate.

Developments in the field of membrane technology have resulted in a different system of performing continuous enzyme reactions. This system, known as the membrane reactor, combines the fast reaction rate of the free enzyme with its continuous use. The enzymatic reaction is performed on one side of a porous membrane, the permeability of which allows the reaction products to pass through but prevents passage of the enzyme. Thus the products are continuously removed, while the enzyme is retained to act on fresh substrate flowing into the reactor.

The immobilised enzyme method can be combined with several measurement techniques such as potentiometric, polarographic, and microcalorimetric sensors. The ideal design is an integrated probe or sensor that can be dipped straight into a sample solution and provide an immediate result. Devices known as biosensors include examples using immobilised enzymes. They are very promising, but as yet have not come into use other than in research laboratories. Technical difficulties such as gumming up of the enzyme surface by sample constituents, long reaction times due to bonding of the enzyme to the carrier, *etc.*, have so far only been partially overcome.

Critical Aspects and Practical Advice

Enzymes are not robust reagents. Even when purchased in the form of kits, care must be taken to obtain valid results. Some of the key factors to consider are listed below.

Storage of Enzymes. Enzymes only have a limited shelf-life; 2 months is typical. They should be stored in a refrigerator. Long-term storage of some enzymes in the freezer is preferable, but repeated freezing and thawing is inadvisable as it can denature some enzymes. Manufacturers' guidelines should therefore be carefully followed for each individual enzyme.

Dilute solutions of enzymes and substrate should also be stored in a refrigerator because of chemical instability and the possible growth of micro-organisms. The enzyme activity should be checked regularly (every 10 days). In general, pure enzymes are more stable than enzymes in tissue extract, which often contain components (*e.g.* proteins) that decrease the stability of the enzymes. Buffer and substrate solutions should be stored in clean, sterilised, dark bottles, especially phosphate acetate buffer, amino acid, and sugar solutions.

All reagents required for enzymatic assays, including stock buffer solutions and redistilled water, should be kept covered, otherwise contamination may occur by airborne bacteria or chemical compounds. Deionised water is not a suitable reagent due to the presence of organic impurities such as softening agents from the exchange resin.

If an analytical procedure is used frequently, it is advisable to premix reagent solutions in order to cut down on the use of the pipette in the determination. This considerably increases the precision of the procedure. Furthermore, it enables even unskilled personnel to perform enzymatic analyses. Various premixed commercial reagents of satisfactory quality are also available, particularly for the determination of the catalytic activity of enzymes. In general the reagents and procedures are adapted to a specified sample matrix. If other sample types are to be analysed, the procedure using premixed reagents should be checked for reproducibility.

Temperature of the Reaction. The importance of precise temperature control has already been stressed. Enzymes are heat-sensitive proteins and lose their catalytic capabilities when the three-dimensional structure is disrupted by heat. As any home-brewer will know, denaturing occurs rapidly at temperatures over about 37°C. If it is necessary to quench a reaction at a given time this can be done by heating to ~35°C. Other materials present need to be stable at this temperature.

Like all chemical reactions, enzyme catalysed reactions are sensitive to temperature changes. The temperature coefficient varies from one enzyme to another. This temperature dependence is particularly critical when checking the activity of enzymes.

pH of the Reaction. Enzyme reactions are also influenced by the pH of the medium. An enzyme may exist in several ionic forms, and the ionic form that predominates is a function of pH. Only one of these forms may be catalytically active, or one form may be more active than the others. The pH optimum for different enzymes varies over a broad range, from pH 2 for pepsin to pH 10 for arginase. Some enzymes exhibit a broad optimal range extending over several pH units, others have a very sharp pH optimum. The pH optimum of an enzyme, like the temperature optimum, may vary depending on the substrate being used and on other experimental conditions (temperature, ionic strength of the buffer, *etc.*).

Timing. The other major factor, which is critical to obtaining reproducible results, is precise timing. In order to quantify reactions with any degree of certainty rate measurements must be taken at exactly the same times.

Potential Interferents. Although enzyme catalysed reactions are usually described as completely selective, this is not always the case. Commercial test kits, in particular, are often designed to respond to a whole class of compounds rather than just one individual compound. In these cases, any compound belonging to the targeted class (and some that are merely vaguely similar) might interfere.

Other potential interferents are the very activators and inhibitors that some tests are designed to measure. Metal ions *etc.* are common in the laboratory environment, and can have a marked effect upon reaction rates.

With bioanalytical screening kits the question has to be asked 'Is an interference critical?' The answer in most cases is that interferents producing false positive identifications are acceptable, as the result will then be confirmed by an independent method. However, false negatives are unacceptable, as the sample will be reported as 'none detected'. Kits should therefore be designed so that if anything does go wrong the default result is 'positive'. The analyst should carefully consider possible inhibitors, conditions that may denature the enzyme, and other potential interferents to determine whether they would give a 'false positive' or 'false negative' result.

8.3.3 Immunoassays

Immunoassay techniques are based on the specific reaction between an antibody and a large protein molecule (its *antigen*). An antibody is a compound that is naturally produced (*raised*) within an animal to react with a foreign substance. Each antibody is designed to target one particular antigen.

In principle large protein molecules having a relative molecular mass exceeding 1000 can be used to raise antibodies. A molecule used to raise antibodies is termed an *immunogen*. An immunogen can be the antigen itself or, as is the case in most instances, a molecule that has the same functional group as the antigen, but is larger. Antibodies raised to such an immunogen can therefore also target the antigen. Antibodies can then be extracted for use as analytical reagents. Antibodies may be raised as *monoclonal* (specific to one antigen) or *polyclonal* (specific to a class of antigens), depending upon whether they are taken directly from an animal's serum (polyclonal) or produced by cells in culture (monoclonal). Polyclonal antibodies are the more sensitive, but are less specific and are limited to the supply from one animal. When the animal dies, so does the source.

Monoclonal antibodies are more specific and have an unlimited source, but are less sensitive, and also more expensive than polyclonal antibodies.

This production of antibodies obviously requires specialist facilities, and one of the main restrictions upon the use of immunoassays is currently that there are few suppliers of antibodies, and that the properties of those raised tend to vary from batch to batch.

Antibodies are therefore highly specific reagents for detecting antigens. If the reaction between the two can be monitored, then antibody–antigen reactions can be extremely selective and sensitive analytical tools.

Applications requiring the analysis of large proteins are obviously rare. However, analytes that are too small to act as antigens (called *haptens*) can be derivatised by binding them to a large protein (*e.g.* bovine serum albumin). Antibodies can then be raised to this bound derivative. This extends the scope of immunoassays to analytes such as pharmaceuticals and pesticides.

Labels for Immunoassays

As with enzyme/substrate reactions, it is necessary to devise some method of monitoring the reaction. In immunoassays, this consists of derivatising (tagging) either the antibody or the antigen with a label that can be easily detected.

The success of immunoassays depends upon the use of methods of labelling, which allow the detection of extremely small numbers of labelled molecules. The primary attribute of a label must be that it is detectable at very low concentrations. Four of the most popular types of label are compared in Table 8.3.3.

RadioImmunoAssay (RIA). Producing radioactive isomers of antibodies or antigens was one of the first methods used to monitor immunoassays. Radioactivity can be detected using simple methods. It has great sensitivity (concentrations down to 10^{-12} mol dm^{-3}) and is independent of environmental factors. Radiolabelling procedures are relatively simple as many labelled compounds are commercially available. Radiolabelling does not normally affect reaction kinetics.

However, radioactive labelling has drawbacks. Labelled antigens show batch-to-batch variation and have shelf-lives of 2 months or shorter (depending upon the half-life of the radioisomer). Preparation of the labelled antigen involves risks, it should be remembered that these are cumulative.

Once the antibody–antigen complex is formed, separation of reacted from unreacted labelled compounds is essential, and so a heterogeneous system is required. Such a separation process is difficult to automate and has limited the development of RIA. In recent years several 'non-centrifugation' separation methods have been developed, but none of these systems have been widely applied.

Using Enzymes as Labels. As already described, enzymes are catalysts and so can be detected at very low levels by any of the techniques listed in Section 8.3.2. There are some difficulties in binding an enzyme to another protein, *e.g.* the

Table 8.3.3 *A comparison of immunoassay labels*

	Enzymes	Radiolabels	Chemiluminescence	Fluorescence
Shelf-life of kits	>6 months	<2 months	>6 months	>6 months
Cost of reagents	inexpensive	expensive	inexpensive	inexpensive
Cost of detector	inexpensive	expensive	expensive	expensive
pH of detection step	critical	not critical	critical	critical
Detection limit	10^{-10} to 10^{-11} mol dm^{-3}	10^{-9} to 10^{-12} mol dm^{-3}	10^{-9} to 10^{-12} mol dm^{-3}	10^{-9} to 10^{-14} mol dm^{-3}
Type of analysis	(semi)-quantitative	quantitative	quantitative	(semi)-quantitative
Hazards	some carcinogenic reagents	radiation	none	none
Automation	easy	difficult	difficult	difficult
Matrix effects at detection step	yes	no	yes	yes

enzyme must not become deactivated in the process, but these difficulties have been overcome for the 20 or so enzymes that are used as labels. Enzymatic labelling is now the most frequently used method of monitoring immunoassays. However, it must be remembered that enzymes are not robust, and there can be any number of substances in the sample solution that may adversely affect the enzyme activity.

The most popular labels are phosphatases, peroxidases, and galactosidases, as these are conveniently assayed by a final incubation with chromogenic substrates (Table 8.3.4). Peroxidase is the cheapest of these enzymes and several chromogens are available which produce colours that are suitable for visual determinations. However, most of these chromogens have been shown to be carcinogenic or mutagenic. A relatively safe alternative is tetramethyl benzidine.[8] Peroxidase is inactivated by polystyrene, and some tissue cells which may be present in the sample have an interfering, inherent peroxidase activity.

Table 8.3.4 *Enzyme labels for immunoassays which give spectrophotometric substrate reactions*

Enzyme label	Substrate for detection reaction	Optimum detection wavelength (nm) (pH-dependent)
Peroxidase and glucose oxidase	2′,2-Azinodi-3-ethybenzothiazoline sulfonate (ABTS)	410
	o-Phenylenediamine (OPD)	492
	5-Aminosalicylic acid	500 (pH 1)
		450 (pH 5)
	3,3′,5,5′-Tetramethylbenzidine	450
Phosphatase	4-Nitrophenyl phosphate	405
Galactosidases	Fluorescent adducts	Substrate-dependent

Glucose oxidase can be measured with the same chromogens as peroxidase but as it has a lower specific activity the assays tend to be less sensitive. Alkaline phosphatase and its conjugates are very stable and the colorimetric and fluorimetric substrates that produce sensitive assays are safe chemicals. The main disadvantage is that the purified calf intestine enzyme, preferred because of its high specific activity, is expensive.

Luminescent Labels. Fluorescent labels can prove extremely sensitive, with some techniques having the potential to detect a single molecule. Compared with isotopic labelling fluorescent labels are relatively inexpensive, the lifetime of the kit is considerably longer, and the fluorescence can be measured on relatively simple fluorimeters. Fluorescence immunoassay is, however, more limited in scope than radioactivity due to a number of factors, such as pH dependence and the presence of quenchers of fluorescence in the biological material.

To obtain the maximum sensitivity, molecules that are used as fluorescence labels should emit radiation that is distinguishable from the background fluorescence. The detection limit using conventional fluorescence is 10^{-9} to 10^{-10} mol dm^{-3}. Use of europium complexes as labels can take limits down to 10^{-12} to 10^{-13} mol dm^{-3}, as their high wavelength shift and long relaxation times allow any fluorescent emission to be easily distinguished from interferents.[12] Another technique to improve sensitivity is the use of secondary fluorescence. For example, if the antigen is labelled with fluorescein *and* the antibody is labelled with rhodamine B, the resulting antibody–antigen complex undergoes energy transfer. The fluorescein is excited and the fluorescent emission from the fluorescein label then excites the rhodamine B label. If the rhodamine B emission is measured, there is no interference from unreacted labelled antigen or antibody, *i.e.* rhodamine B only emits radiation when it is complexed with a fluorescein-labelled antigen. This allows detection limits to be as low as 10^{-14} mol dm^{-3}.

Because background and interference problems can be decreased using separation assays (especially if a washing procedure is incorporated) solid phase techniques have been widely applied to fluorescent immunoassays, although some plastics can contribute to the background fluorescence and scattering effects.

Chemiluminescence (the use of a chemical reaction that produces luminescence) offers lower detection limits still as there is a theoretical background of zero, although errors can occur due to both physical and chemical effects. Chemiluminescence measurements are insensitive to turbidity, but they can be affected by light-absorbing or fluorescent molecules present in the sample. Temperature, pH, and time are also factors that should be standardised.

Other Labels. Potentially, anything that is easily detected at low concentration can be used as an immunoassay label. Amongst others, metal ions and free radicals have been tried.

Immunoassay Techniques

There are a great number of variants of immunoassay procedures. Analysts tend to develop their own protocols for their particular problem by applying general principles in a flexible manner. They may well find that kits developed for specific applications bear only passing resemblance to the convenient classifications found in the literature.

Immunoassays usually require much simpler sample preparation than those required for chromatographic techniques. These requirements must be verified rather than assumed. Control experiments are necessary to quantify both matrix effects and antibody cross-reactivity.

In all of the examples to be given the following notation is used.

Ab   Antibody
Ab*   Labelled antibody
Ag   Antigen (usually the analyte)
Ag*   Labelled antigen

Non Competitive Immunoassay in Solution. In concept this is the simplest of immunoassay techniques. The reaction proceeds as shown in Equation (3).

$$Ab^* \text{ (in excess)} + Ag \xrightarrow{\hspace{2cm}} [Ab^*\text{–}Ag] + \text{excess } Ab^* \qquad (3)$$
$$\text{(analyte)}$$

The reaction occurs in solution. The large $[Ab^*\text{–}Ag]$ product precipitates out of solution. The reaction can therefore be monitored by following the reduction of Ab^* concentration in solution.

In order to measure the labelled antibody in solution, it is necessary to remove any labelled precipitated product that would interfere. This is usually achieved by the use of a centrifuge. Such techniques, which require the separation of the product from the reactants, are referred to as 'heterogeneous' immunoassays. These suffer from being difficult to automate.

Competitive Immunoassay in Solution. In this case an additional quantity of labelled antigen is premixed with the antibody. The scheme proceeds as shown in Equation (4).

$$\text{(a) } Ag^* + Ab \xrightarrow{\hspace{2cm}} [Ab\text{–}Ag^*]$$
$$\text{(b) } Ag + [Ab\text{–}Ag^*] \xrightarrow{\hspace{1.5cm}} [Ab\text{–}Ag^*] + Ag^* + [Ab\text{–}Ag] + Ag \quad (4)$$
$$\text{(analyte)} \hspace{4cm} \text{(measured)}$$

If there is insufficient antibody in solution to complete the reaction (*i.e.* the total antigen is in excess) then there is competition between Ag and Ag^* for binding to the antibody. The more Ag in the sample, the less Ag^* ends up in the precipitate. In order to quantify the analyte, the measured, labelled complex is compared with that formed from a blank assay, *i.e.* without the addition of the analyte.

The amount of Ag^* in the precipitate can therefore be measured, and related to the concentration of analyte Ag in the sample extract (Figure 8.3.2). In order to ensure that quantitation is based upon the linear portion of the curve, it is necessary to run at least five or six calibrants along with each batch of samples analysed.

In order to measure the Ag^* in the precipitate, it is again necessary to completely separate the precipitate from any Ag^* remaining in solution (heterogeneous immunoassay). Labelled antigens are also more expensive than labelled antibodies because of the difficulty in labelling antigens.

An exception to the rule that the precipitate needs to be separated from the solution is in the case where the antigen is labelled with an enzyme. The enzyme label can be designed such that it is deactivated when $[Ab\text{–}Ag^*]$ is formed. Any label detected is therefore only due to the free Ag^* in solution, which can again be related to the analyte Ag concentration. This is one example of the enzyme labelled immunoassays known as Enzyme Mediated Immunoassay Techniques (EMIT)[‡]. Such methods are referred to as 'homogeneous' immunoassays.

[‡] EMIT is also an acronym for Enzyme Multiplied Immunoassay Technique. This is one example of an area where bioanalytical nomenclature can be confusing, and highlights the importance of fully explaining the mechanism of techniques.

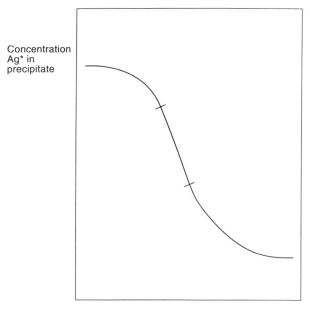

Concentration
Ag* in
precipitate

Analyte (Ag) concentration
in solution

Figure 8.3.2 *Relationship between labelled antigen and analyte antigen concentrations in competitive binding*

Solid Phase Immunoassays. There are considerable advantages in immobilising either the antibody or the antigen on a solid support. It can facilitate the automation of an immunoassay, or enable it to be performed in a fast, routine manner. The reagent is usually immobilised either on the inside of the reaction well or on a plate that is dipped in and out of solution.

One example of a competitive binding scheme is shown in Figure 8.3.3. A well is coated with the antibody. Analyte Ag and an excess of the labelled form Ag* are then added to the well; they compete for the Ab binding sites. The unbound antigen is then washed off, leaving Ag and Ag* bound to the plate.

If the label is an enzyme then the well can be incubated with a substrate and/or chromogen solution to quantify the bound Ag* [*e.g.* glucose 6-phosphate dehydrogenase gives a colour change when nicotinamide–adenine dinucleotide phosphate (NADP) solution is added]. This type of technique is known as Enzyme Linked ImmunoSorbent Assay (ELISA). The term 'ELISA' is one which is used freely (and sometimes inaccurately), but is usually taken as referring to any competitive solid phase assay that uses an enzyme label. It is one of the most popular immunological techniques in use due to its simplicity and rapidity.

Derivatisation with Magnetic Species. If antibodies or antigens are derivatised with magnetic species it provides an easy method of separating them from

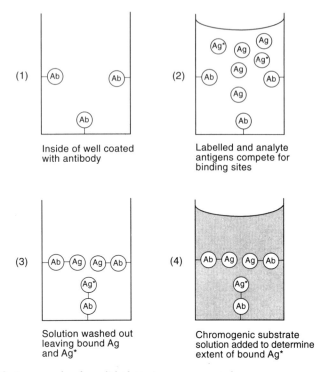

(1) Inside of well coated with antibody

(2) Labelled and analyte antigens compete for binding sites

(3) Solution washed out leaving bound Ag and Ag*

(4) Chromogenic substrate solution added to determine extent of bound Ag*

Figure 8.3.3 *An example of a solid phase immunoassay scheme*

solution; an alternative to solid phase techniques. The magnetic derivatisation is combined with another labelling method, usually enzyme labelling.

Sandwich Immunoassays. All of the techniques described so far have been for the determination of antigens. There are some cases where the antibody is the analyte. One way round this problem is to raise antibodies to the antibody.

The antibody is immobilised onto a solid support using its antigen (Figure 8.3.4). Antibodies (Ab_2*) can then be raised to the resulting large molecule (the new antigen), and can be labelled.

This type of sandwiching gives extremely strong and selective binding arrangements. It relies upon the analyte having two binding sites (one to the antigen on the solid support and one to the newly raised antibody), and therefore has the potential to discriminate between the analyte and similar compounds containing only one binding site such as precursors, metabolites, and fragments.

Critical Aspects and Practical Advice

Non-specific Binding. One problem frequently encountered when setting up an immunoassay is non-specific association. For example, in a solid phase immuno-assay the antigen may adsorb to the walls of the coated plate as well as to the bound antibody. If this is a problem it is necessary to block the plate with an

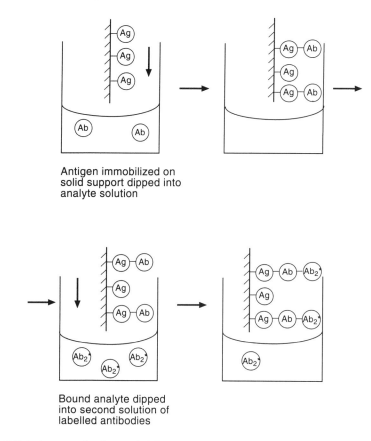

Figure 8.3.4 *An example of a sandwich immunoassay*

unreactive substance (bovine serum albumin, glycine, or milk powder) to avoid non-specific binding of subsequent reagents to the well walls.

As a general precaution against non-specific binding, wetting agents such as non-ionic detergents, Tween 20, and Triton-X are added to diluents and washing buffers. These do not interfere with the antigen–antibody reactions but prevent hydrophobic interaction between added proteins and the solid phases.

Batch-to-Batch Variations of Solid Phase Kits. Plates from various manufacturers and even from the same batch may vary in ability to bind the antibody or antigen. Certain microtitre plates and multiple cuvette systems suffer from 'edge effects' which are caused by inconsistent antigen binding characteristics. In these cases if the wells at the edges of the plate must be used this should only be as controls.

The plates must produce uniform adsorption or coating of the antigen or antibody. Before deciding on a regular supplier of microtitre plates, it is advisable to check the reproducibility of a few sample plates. This can be performed by coating all the wells with the same amount of antigen and then incubating with the enzyme-linked antibody. Following the incubations, washing, and the addition of the enzyme substrate, the colour developed in each well should not vary more than ±5% from the mean value.

Optimising the Conditions for Antigen/Antibody Binding to the Solid Phase. Adsorption of the antigen and antibody onto the solid phase is a physical process and generally does not involve any covalent interaction with the solid surface. The optimal conditions for coating the wells will depend on the concentration of the antigen or antibody, the incubation time for coating, and the pH.

Preliminary experiments are necessary to establish optimum binding conditions for a particular antigen or antibody. For many proteins a concentration of 1–10 μg ml^{-1} in carbonate/bicarbonate buffer (pH 9.8) is sufficient; some users prefer saline or phosphate buffered saline. For proteins, incubation times of 2–3 h at room temperature are adequate but, for convenience, in many instances the coating step is carried out at 4°C overnight.

Although adsorption of antibodies or antigens is simple, the adsorbed antibodies may undergo denaturation or losses may occur during the washing stages. Use of proteins covalently attached to the appropriate surface can overcome these problems.

Obtaining Reproducible Washing Steps. Some analytes (proteins or other molecules) can be lost during the washing step, so this step needs to be standardised in order to improve the reproducibility of the assays. A standard procedure for washing the microtitre wells is to flood the wells with the wash buffer, leave to soak for 1 min, then shake off the wash buffer and repeat the process three times. It is important that the wells are never allowed to dry completely and that the washing step follows the incubation step without a break. Thorough washing is necessary to ensure no carry-over of reagents occurs.

Length of Incubations. When the sensitivity is not critical, the optimum time and the temperature for ELISA incubations are often adopted on the grounds of convenience. Generally, incubations at room temperature or 37°C are used. Incubations at 37°C can reduce the length of the procedure. However, in some cases elevated temperature can drastically reduce assay sensitivity, and incubations at room temperature may be essential.

For some applications or with some antibodies, the signal strength can be increased by using longer incubation times. For rapid semi-quantitative assays the incubation times can be shortened to between 30 min and 1 h. It is less easy to reduce the length of other steps; *e.g.* typically over 2 h is required for blocking steps.

Dispensing Reagents. Immunoassay procedures involve a large number of steps, which require accurate and precise dispensing of reagents. This is normally carried out using various types of digital pipettes. Incorrectly adjusted pipettes will produce a consistent bias from the true value.

Various types of pipetting equipment are available and the choice will depend on such factors as the number of assays required, the budget, and the degree of quantitation required.

For assays requiring processing of only one or two microtitre plates, Eppendorf (air displacement) multishot pipettes can be used, which allow quick transfer of reagents. If a larger number of plates need to be processed (*e.g.* when optimising the parameters of the assays) multichannel pipettes, which consist of 8 or 12 channels, are ideal. Automated dispensing, washing, and reading equipment is also available for routine assays. The advantage of using the automated equipment is that improved repeatability and reproducibility is achieved and it forgoes the need for skilled personnel.

Before dispensing, the pipette tip must be wetted by suction and expulsion. Although the pipette tip is made of water repellent material this step is essential, in particular with viscous solution or solutions that have differing wetting properties towards the plastic.

Although pipette tips are disposable, care must be taken that there is no splashing or spilling of liquid onto the main body of the pipette; this is to avoid the risk of carry-over. This is a particular danger with air displacement pipettes: with positive displacement pipettes the plunger, which is in contact with the liquid, is also disposable. A fresh pipette tip should be used for each sample. Suction of liquid into the shaft of the pipette (*e.g.* by rapid aspiration of the solution or formation of bubbles in the pipette) or laying down the pipette with a filled tip must be avoided. Aggressive liquids (*e.g.* concentrated acids) and those with high vapour pressure should not be measured with a plunger pipette.

The accuracy of plunger pipettes should be checked gravimetrically at regular intervals.

Timing of Breaks. The timing of most stages of an immunoassay must be assumed to be critical until proven otherwise. Incubations, reaction quenching, *etc.*, must be performed in a reproducible manner. If it is not possible to perform the analysis in a single session, the analyst must plan carefully the stages at which the assay can be left.

Antigen-coated plates can be stored after the blocking step either in phosphate buffered saline containing 0.02% sodium azide at 4°C for 1 week, or indefinitely at −20°C after removing the blocking solution. Storage conditions vary from method to method, but stabilising the plate in a sugar solution is common.

Increasing the Sensitivity. If the amount of antigen/antibody bound to the plate is too low to produce a response, the use of amplification techniques or changing the solid support to nitrocellulose sheets improves the response. Nitrocellulose can bind 1000 times more protein per unit surface area and can be purchased either as sheets or as a filter sealed to the bottom of a microtitre plate.

8.3.4 DNA Analysis

Introduction: Principles of Use

DNA analysis is another bioanalytical technique which is based upon the specific reaction between two biochemical molecules: in this case, a complementary pair of DNA sequences. The following properties of DNA make it suitable for use as an analytical reagent:

(i)   DNA is a double-stranded molecule;
(ii)   each strand is complimentary to the other;
(iii)   short DNA sequences can be chemically synthesised;
(iv)   a DNA sequence can be induced to bind to its complementary partner;
(v)   DNA can be labelled and visualised by using a number of methods.

The reaction between two complementary DNA strands is therefore analogous to the antibody-antigen reaction described in Section 8.3.3.

Many methods and variations of genetic analysis are available. Most of these utilise as their basis either gene probes or one of a growing number of amplification-based methodologies. The most common amplification method is the polymerase chain reaction.

The DNA Probe. A DNA probe is a sequence of labelled DNA. The labels used are as those described for immunoassay in Section 8.3.3, with radiolabels or, increasingly, luminescent labels being the most popular. The assay then proceeds in an analogous manner to immunoassays (Figure 8.3.5). Target DNA (the analyte) is first purified from the sample in question. It is then typically immobilised in a single-stranded form on a membrane support made of either nitrocellulose or nylon. The labelled DNA probe is then allowed to bind to its complementary

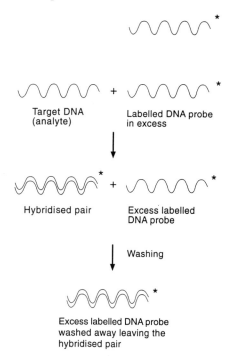

Figure 8.3.5 *DNA probe analysis*

partner in a process known as hybridisation. Unbound probe is removed by a number of washing steps and the hybridised probe remaining is visualised by a method appropriate to the label used.

The specificity of the DNA probe can be modulated at the washing stage by progressively increasing the temperature or decreasing the salt concentration of the wash. This has the effect of removing more and more closely related labelled sequences until only those with a perfect match remain. The harshness of the washing step is referred to as the stringency.

In a variation of this technique, the DNA is first cut into specific pieces using restriction endonucleases. The fragments produced are separated by size using agarose gel electrophoresis and then transferred to the membrane support in a process known as 'Southern transfer'. Detection of specific fragments using the DNA probe is then the same as above but yields additional information in terms of the size of the detected DNA sequences.

The Polymerase Chain Reaction (PCR). Polymerase chain reaction is a technique for amplifying the target molecule before analysis. The reaction occurs in distinct stages (Figure 8.3.6):

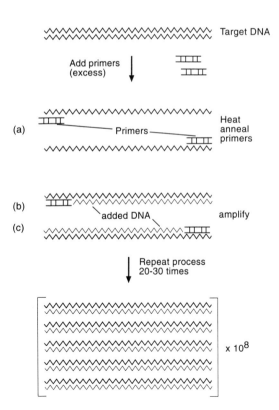

Figure 8.3.6 *Schematic representation of a polymerase chain reaction*

(a)   two oligonucleotides ('primers') are induced to specifically bond ('hybridise') at either end of a selected DNA target;

(b)   a twin of the original DNA target is synthesised between the two primers;

(c)   the twin strands of DNA are separated, *i.e.* the original DNA strand has been cloned.

This process can be repeated in the region of 20–30 times, giving a many-million-fold amplification of a specific piece of DNA if the process is 100% efficient. Because of the exponential nature of the amplification, minor differences in experimental conditions or starting materials give major differences in the amount of end product. In practice, this means the technique is generally used for qualitative analysis. The product may be identified using electrophoresis on an agarose gel, or by a DNA probe.

Each stage of the PCR is performed at a regulated temperature, so that the experiment consists of repetitive heating and cooling. This is usually achieved using an apparatus known as a thermal cycler.

Laboratory practice needs to be scrupulous to avoid contamination of samples with DNA from previous reactions or other samples and thereby the generation of false positive results. Rigorous controls need to be included. Negative controls check for spurious background bands and DNA contamination, whilst positive controls verify the efficiency and specificity of the PCR.

Reasons for Using DNA Analysis

In particular circumstances, genetic approaches may be superior in terms of speed, sensitivity, and specificity over other available techniques. They may be the only choice if other methods are not available. The results produced by such methods may be qualitative, semi-quantitative, or quantitative depending on the procedure used and the method of application.

Specificity. In principle, it is possible to select regions of an organism's genetic material for use as DNA probes or PCR targets that are as specific or as general as required. This means that, given sufficient effort, it should be possible to develop a method to distinguish DNA from any group of related species, any individual species within such a group, or any individual within a species.

Sensitivity. For trace analytical measurements a DNA probe alone is unlikely to be sufficiently sensitive. A PCR-based method may be more suitable. Such methods are, in principle, capable of detecting single cells in a vast excess of sample matrix and competing material. The limiting factor is likely to be the isolation of the analyte from the sample matrix. Each sample presents its own set of problems. The detection limits may vary by many orders of magnitude depending upon the matrix.

Speed. DNA probes can generally provide results within 1–2 days. However, this time can vary depending upon the precise nature of the probe/detection system being utilised and, perhaps more significantly, the amount of sample processing required before the analysis can be performed. The PCR protocols in general take between 3 and 4 h, although this can vary. Sample processing and the analysis of the labelled product can add significant amounts of time to the analysis. It should be possible to complete the whole analysis for the shorter protocols within 1 working day, although this may increase to several days where more processing is required.

Quantitation. It is possible to produce quantitative results from both DNA probe and PCR procedures. Such results are relatively easy with DNA probes, since the inclusion of suitable controls and dilution series allows the comparison of signals from the different probes and samples (*i.e.* the construction of a calibrated response). However, PCR is rather more complicated, since the exponential nature of the amplification allows significant differences in product yield to result from relatively small differences in amplification efficiency. Such problems may be overcome by the use of a number of types of mimic (experimental controls using DNA with minor differences from the target), which can be included in the reactions to control for variations that may occur between tubes and samples and even between different areas of the thermal cycler. Even when taking these precautions, quantitation is not simple. The inclusion of dilutions of the mimics and a knowledge of the reaction kinetics is required if reliable results are to be achieved. Procedures are continually being developed allowing other routes to quantitation, such as the on-line analysis of product either directly or indirectly by the co-generation of a coloured product.

Applications of DNA Analysis

DNA analysis is considered as an alternative to existing bioanalytical techniques rather than to classical chemical analytical methods. It is still mainly used in specialist bioanalytical research centres, but commercial kits are now being produced, which make the technology available to the practising analyst.

There are many cases where it may be necessary or advantageous to analyse for DNA. These include the detection of micro-organisms in food or water samples; the species of origin of meat products, hair, or skin samples; and the differentiation of individual people for forensic purposes.

Detection of Microbiological Contamination. The traditional method of microbial assay is by using viable cell counts. This requires incubation for a period of time, which may be overnight or up to 2 weeks. This inevitably produces a delay in providing a result. In addition all it demonstrates is the presence of viable cells (those capable of independent growth). Immunoassay techniques detect non-viable cells, but problems may be encountered with non-specific binding of antibodies to other matter within a sample. However, DNA methodology avoids both of these problems. Unless a high microbial presence is suspected, PCR is the method of choice for genetic-based detection, giving benefits in speed over viable counts as well as specificity.

Detection of Adulterated Food. The detection of an adulterating species of, *e.g.* meat in a highly processed sample, can be difficult since the structure of the constituents may be broken down. The protein fragments may not respond to the antibody making the analysis difficult using antibodies as a means of identification.

A number of specific advantages over the use of antibodies are offered by DNA probes, particularly where a quantitative result is required. The presence of DNA is uniform within most tissues (unlike proteins, the expression of which may be tissue-specific). Even if the DNA in a sample becomes degraded during processing it will still usually respond to DNA probes.

Forensic and medical applications. Additionally, DNA methods can be used for the forensic identification of an individual, *e.g.* to prove their presence at the scene of a crime, or for paternity testing. Medical applications are also found in areas such as prenatal diagnosis and monitoring of disease, particularly of viruses such as HIV. In such cases, reliability, rather than speed, is likely to be the most important consideration.

The ability of DNA-based techniques to analyse samples with such a high degree of specificity is the major reason why this form of analysis is becoming more widespread. Other techniques do not offer the same degree of discrimination and consequently require a number of independent tests to be carried out to achieve a comparable level of confidence.

Sources of DNA Probes and PCR Methods

The choice of technique for a particular application also depends upon its availability and cost. The easiest way to perform a genetic analysis is to buy a commercial kit, available from an increasing number of companies. If this is not possible, scientific publications offer a valuable source of information on DNA probe and PCR-based methods that may be utilised. Such systems may or may not have been fully validated and care must be taken to ensure that these work under the circumstances in question. Most research groups are willing to supply clones detailed in publications; oligonucleotides can be obtained from companies and research institutes for relatively little cost. However, access to a laboratory set-up to perform molecular biology is required since some form of biological containment and certain items of specialist equipment may be required.

Critical Aspects and Practical Advice

Control Experiments. As with all forms of analysis it is essential to perform the correct positive and negative controls. For species determination purposes, positive and negative controls, representative of the species between which differentiation is required, need to be included to ensure sufficient levels of discrimination are being achieved. Similar controls need to be used with PCR methods.

Contamination. The avoidance of false positive results is the greatest concern with PCR. The contamination of reactions with trace amounts of DNA and, particularly, amplified product from previous reactions must be avoided. Strict quality control procedures must be followed. Precautions taken normally include the use of dedicated positive displacement pipettes or tips containing filters to prevent cross-contamination arising from the generation of aerosols; the physical separation of areas for handling the reagents used for pre- and post-PCR samples; and the inclusion of multiple negatives to ensure the efficacy of the precautionary procedures.

Since many forensic samples are human in origin, great care needs to be taken to avoid contamination of the sample with DNA from the operator during handling and processing, particularly when the analysis involves one of the amplification-based technologies.

Optimising the Specificity of the Reaction. With DNA probes, it may be possible to increase the specificity by adjusting the temperature or pH of the washing steps. For PCR amplifications specificity can again be modulated by increasing the temperature or decreasing the time of the priming step. General reaction optimisation is an important part of any PCR. This can be affected particularly by magnesium ion concentration.

8.3.5 References

Enzyme/Substrate Reactions

1. R. Zacs, 'Enzymes in Organic Solvents – Properties and Applications', *J. Biotechnol.*, 1988, **8**, 259.
2. 'Biomolecules in Organic Solutions', ed. A. Gomez Puyou, CRC Press, London, 1992.
3. L. Stryer, 'Biochemistry', W. H. Freeman, New York, 3rd edn, 1988.
4. D. W. Moss, 'Measurement of Enzyme Activity', *Scand. J. Clin. Lab. Invest.*, 1989, **49** (Suppl. 193), 20.
5. M. E. Leon-Gonzalez and A. Townshend, 'Determination of Organophosphorus and Carbamate Pesticide Standards by Liquid Chromatography with Detection by Inhibition of Immobilized Acetylcholinesterase', *J. Chromatogr.*, 1991, **539**, 47.
6. F. Schubert, R. Renneberg, F. W. Scheller, and L. Kierstein, 'Plant Tissue Hybrid Electrode for the Determination of Phosphate and Fluoride', *Anal. Chem.*, 1984, **56**, 1677.
7. C. Tran-Minh, P. C. Pandey, and S. Kumaran, 'Studies on Acetylcholine Sensor and its Analytical Application based on the Inhibition of Cholinesterase', *Biosens. Bioelectron.*, 1990, **5**, 461.
8. E. S. Bos, A. A. van der Doelen, N. van Rooy, and A. H. W. M. Schuurs, '3,3',5,5'-Tetramethylbenzidine as an Ames Test Negative Chromogen for Horse Radish Peroxidase in Enzyme Immunoassay', *J. Immunoassay*, 1981, **2**, 187.
9. J. J. Kulys, 'Amperometric Enzyme Electrodes in Analytical Chemistry', *Fresenius' Z. Anal. Chem.*, 1989, **335**, 86.
10. M. A. Arnold and M. E. Meyerhoff, 'Recent Advances in the Development and Analytical Applications of Biosensing Probes', *CRC Crit. Rev. Anal. Chem.*, 1988, **20**, 149.

11. D. Hawcroft, 'Diagnostic Enzymology', John Wiley, Chichester, UK, 1987.

Immunoassays

12. E. P. Diamandis and T. K. Christopoulos, 'Europium Chelate Labels in Time Resolved Fluorescence Immunoassays and DNA Hybridization Assays', *Anal. Chem.*, 1990, **62**, 1149A.
13. D. S. Hage, 'Immunoassays', *Anal. Chem.*, 1993, **65**, 420R.
14. 'Principles and Practice of Immunoassay', ed. C. P. Price and D. J. Newman, Stockton Press, New York, 1991.
15. A. Truchaud, B. Capolaghi, J. P. Yvert, Y. Gourmelin, and M. Bogard, 'New Trends for Automation in Immunoassays', *Pure Appl.Chem.*, 1991, **63**, 1123.
16. L. Fukal and J. Kas, 'The Advantages of Immunoassay in Food Analysis', *Trends Anal. Chem.*, 1989, **8**, 112.
17. D. Levieux, 'Immunoassays Applied to Meat Products', *Analusis*, 1993, **21**, M36.
18. R. Niessner, 'Immunoassays in Environmental Analytical Chemistry: Some Thoughts on Trends and Status', *Anal. Methods Instrum.*, 1993, **1**, 134.
19. J. M. van Emon and V. Lopez-Avila, 'Immunochemical Methods for Environmental Analysis', *Anal. Chem.*, 1992, **64**, 78A.
20. G. Guebitz and C. Shellum, 'Flow Injected Immunoassays', *Anal. Chim. Acta*, 1993, **283**, 421.
21. D. A. Armbruster, R. H. Schwarzhoff, E. C. Hubster, and M. K. Liserio, 'Enzyme Immunoassay, Kinetic Microparticle Immunoassay, RadioImmunoassay and Fluorescence Polarization Immunoassay Compared for Drugs-of-Abuse Screening', *Clin. Chem.*, 1993, **39**, 2137.
22. P. Ekins and F. Chu, 'Multi-Analyte Micro-Spot Immunoassay', *Anal. Proc.*, 1993, **30**, 488.

DNA Analysis

23. G. H. Keller and M. M. Manak, 'DNA Probes', Macmillan, London, 1989, ISBN 0-935859-63-2.
24. 'Essential Molecular Biology: A Practical Approach', ed. T. A. Brown, 1991, vols. 1 and 2, ISBN 0-19-963111-5.
25. 'PCR Technology: Principles and Application for DNA Amplification', ed. H. A. Erlich, Stockton Press, New York, 1989, ISBN 0-333-48948-9.
26. 'PCR: A Practical Approach', ed. M. J. McPherson, P. Quirke, and G. R. Taylor, IRL Press, Oxford, 1991, ISBN 0-19-963196-4.
27. 'PCR Protocols: A Guide to Methods and Application', ed. M. A. Innis, D. H. Gelfand, J. J. Sninsky, and T. J. White, Academic Press, New York, 1990, ISBN 0-12-372181-4.
28. 'DNA Cloning', ed. D. M. Glover, IRL Press, Oxford, 1995, vols. 1 and 2, ISBN 0-199634-77-7 and 0-199634-79-3.
29. R. W. Old and S. B. Primrose, in 'Principles of Gene Manipulation. An Introduction to Genetic Engineering', ed. N. G. Carr, J. L. Ingraham, and S. C. Rittenberg, Blackwell Scientific, Oxford, 4th edn, 1990, ISBN 0-632-02608-1.
30. J. Sambrook, E. F. Fritsch, and T. Maniatis, 'Molecular Cloning: A Laboratory Manual', Cold Spring Harbor Laboratory, Cold Spring Harbor, NY, 2nd edn, 1989, ISBN 0-87969-309-6.
31. C. R. Newton and A. Graham, 'PCR', Bios Scientific, Oxford, 1994.

Applications: Environmental Analysis

32. B. M. Kaufman and M. Clower, 'Immunoassay of Pesticides', *J. Assoc. Off. Anal. Chem.*, 1991, **74**, 239.
33. L. Quillien, 'Detection of Vegetable Protein Products in Foods', *Analusis*, 1993, **21**, M39.
34. X. Drouet, 'Determination by Enzyme Immunoassays of Contaminants in Agricultural Products, Feeds and Foodstuffs', *Analusis*, 1993, **21**, M32.
35. B. Dunbar, B. Riggle, and G. Niswender, 'Development of Enzyme Immunoassay for the Detection of Triazine Herbicides', *J. Agric. Food Chem.*, 1990, **38**, 433.
36. P. C. C. Feng, S. J. Wratten, S. R. Horton, C. R. Sharp, and E. W. Logusch, 'Development of an Enzyme Linked Immunosorbent Assay for Alachlor and its Application to the Analysis of Environmental Water Samples', *J. Agric. Food Chem.*, 1990, **38**, 159.
37. J. C. Hall, R. J. A. Deschamps, and M. R. McDermot, 'Immunoassays to Detect and Quantitate Herbicides in the Environment', *Weed Technol.*, 1990, **4**, 226.
38. F. Jung, S. J. Gee, R. O. Harrison, M. H. Goodrow, A. E. Karu, A. L. Bruan, Q. X. Li, and B. D. Hammock, 'Use of Immunochemical Techniques for the Analysis of Pesticides', *Pestic. Sci.*, 1989, **26**, 303.
39. J. M. van Emon, J. K. Selber, and B. D. Hammock, 'Immunoassay Techniques for Pesticide Analysis', in 'Analytical Methods for Pesticides and Plant Growth Regulators, Vol.XVII', ed. J. Sherma, Academic Press, San Diego, CA, 1989.

Applications: Biomedical and Clinical Analysis

40. P. Mayersbach, R. Augustin, H. Schennach, D. Schoenitzer, E. R. Werner, H. Wachter, and G. Reibnegger, 'Commercial Enzyme Linked ImmunoSorbent Assay for Neopterin Detection in Blood Donations Compared with RIA and HPLC', *Clin. Chem.*, 1994, **40**, 265.
41. A. O'Rorke, M. M. Kane, J. P. Gosling, D. F. Tallon, and P. F. Fottrell, 'Development and Validation of a Monoclonal Antibody Enzyme Immunoassay for Measuring Progesterone in Saliva', *Clin. Chem.*, 1994, **40**, 454.
42. G. A. Chiabrando, F. E. Zalazar, M. A. J. Aldao, and M. A. Vides, 'Rapid and Sensitive Sandwich Immunoassay for Low Concentrations of Albumin in Human Urine', *Clin. Chim. Acta*, 1994, **225**, 155.
43. H. Brailly, F. A. Montero-Julian, C. E. Zuber, S. Flavetta, J. Grassi, F. Houssiau and J.VanSnick, 'Total Interleukin-6 in Plasma Measured by Immunoassay', *Clin. Chem.*, 1994, **40**, 116.
44. E. Ishikawa, S. Hashida, T. Kohno, K. Hirota, K. Hashinaka, and S. Ishikawa, 'Principle and Applications of Ultrasensitive Enzyme Immunoassay (Immune Complex Transfer Enzyme Immunoassay) for Antibodies in Body Fluids', *J. Clin. Lab. Anal.*, 1993, **7**, 376.
45. S. M. Yie, E. Johansson, and G. M. Brown, 'Competitive Solid Phase Enzyme Immunoassay for Melatonin in Human and Rat Serum', *Clin. Chem.*, 1993, **39**, 2322.
46. D. Collins, D. J. Wright, M. G. Rinsler, P. Thomas, S. Bhattacharya, and E. B. Raftery, 'Early Diagnosis of Acute Myocardial Infarction with use of a Rapid Immunochemical Assay of Creatine Kinase MB Isoenzyme', *Clin. Chem.*, 1993, **39**, 1725.
47. M. Pergande and K. Jung, 'Sandwich Enzyme Immunoassay of Cystatin C in Serum with Commercially Available Antibodies', *Clin. Chem.*, 1993, **39**, 1885.
48. N. Fujimoto, J. Zhang, K. Iwata, T. Shinya, Y. Okada, and T. Hawakawa, 'One-Step Sandwich Immunoassay for Tissue Inhibitor of Metalloproteinases-2 Using Monoclonal Antibodies', *Clin. Chim. Acta*, 1993, **220**, 31.

49. C. Larue, C. Calzolari, J. P. Bertenchant, F. Leclercq, R. Grolleau, and B. Pau, 'Cardiac-Specific Immunoenzymometric Assay of Troponin I in the Early Phase of Acute Myocardial Infarction', *Clin. Chem.*, 1993, **39**, 972.
50. M. Hirvonen, S. Koskinen, and H. Tolo, 'Sensitive Enzyme Immunoassay for the Measurement of Low Concentrations of IgA', *J. Immunol. Methods*, 1993, **163**, 59.

Applications: Pharmacological and Toxicological Analysis

51. J. R. Patrinely, O. A. Cruz, G. S. Reyna, and J. W. King, 'Use of Cocaine as an Anaesthetic in Lachrymal Surgery', *J. Anal. Toxicol.*, 1994, **18**, 54.
52. F. Moriya, K. M. Chan, T. T. Noguchi, and P. Y. K. Wu, 'Testing for Drugs of Abuse in Meconium of Newborn Infants', *J. Anal. Toxicol.*, 1994, **18**, 41.
53. R. C. Meatherall and R. J. Warren, 'High Urinary Cannabinoids from a Hashish Body Packer', *J. Anal. Toxicol.*, 1993, **17**, 439.
54. R. G. Morris, N. C. Saccoia, B. C. Sallustio, L. K. Fergusson, S. Mangas, and C. Kassapidis, 'Experiences with the Enzyme Multiplied Immunoassay Cyclosporine Specific Assay in a Therapeutic Drug Monitoring Laboratory', *Ther. Drug Monitor.*, 1993, **15**, 410.
55. P. T. Feldsine, M. T. Falbo-Nelson, and D. L. Hustead, 'Polyclonal Enzyme Immunoassay Method for Detection of Motile and Non-Motile Salmonella in Goods: Comparative Study', *J. Assoc. Off. Anal. Chem. Int.*, 1993, **76**, 694.
56. M. Wermeille, E. Moret, J. P. Siest, S. Ghribi, A. M. Petit, and M. Wellman, 'Determination of Dimethindene in Human Serum by Enzyme Linked ImmunoSorbent Assay,' *J. Pharmacol. Biomed. Anal.*, 1993, **11**, 619.
57. M. Hennies and W. Holtz, 'Enzyme Immunoassay for the Determination of Bovine Growth Hormone Using Avadin–Biotin–Peroxidase Complexes', *J. Immunol. Methods*, 1993, **157**, 149.

CHAPTER 9

From Signal Processing to Reporting

9.1 Signal Processing

9.1.1 Overview

Almost every piece of laboratory equipment now manufactured contains some form of microprocessor control or manipulation of the measuring process itself and the production of the results of that process. This ranges from the simple laboratory balance to laboratory robots, and from chromatographic integrators to laboratory information management systems dealing with analytical information for a major organisation or indeed an entire continent. Procedures and methods for ensuring the reliability of the data and the results produced in processes where a large degree of automatic manipulation of raw and derived data is involved are receiving increasing attention throughout the analytical community.

The measurement system of most analytical instruments involves detection of a change in a parameter, generally a physical one such as radiation intensity in spectroscopy, which is detected by some form of transducer. The signal from the transducer, normally an analogue electrical change in current or voltage, can be directly processed by electronic circuitry. There are two main goals of signal processing in analytical chemistry. Firstly to distinguish the signal resulting from the analyte from other signals produced both by the instrument itself (noise) and by the other components of the sample matrix. Secondly to quantify the signal from the analyte. The purpose of the analysis determines which of these goals is given priority although the first objective is generally a requirement for the second one to be achieved.

The manner in which the signal is processed is crucial to the reliability of the end result. Rigorous extraction and clean-up procedures may have been used in preparing a sample for chromatographic determination and considerable time and expertise may have been spent in optimising chromatographic conditions; but all this effort will be wasted if the signal processing procedure contains the wrong rules for deciding the baselines of the chromatograms. The fundamental problem with such automated systems is that the processes in which such errors can occur are not directly apparent to the analyst and therefore can go unnoticed.

Indeed the increasing complexity of data processing systems makes it less likely that they will be questioned by the analyst or even thoroughly checked, whilst at the same time this increasing complexity increases the probability of errors within the system. The complexity of the sophisticated software makes it less likely that the analyst has the knowledge to thoroughly check the system and reliance has to be placed on a computing expert who, however able, may not understand the chemistry of the analytical procedure. Use of standard data sets should be encouraged. These provide a means of checking the algorithms.

There have also been disturbing trends in recent years for an increasing number of people carrying out analytical measurements to accept with little question the data coming out of 'black boxes'. It is essential that all staff receive appropriate training before they carry out any analyses. Problems arise because pressure of work in some laboratories inhibits a questioning approach to analytical work, especially the more routine work; and the increasing complexity of equipment means that many analysts do not have the knowledge to check out the operation of their equipment in any detail. The implementation of some form of internal quality control is necessary to ensure that the data produced are fit for their intended purpose.[1] This is usually done by use of reference materials and duplicate analyses.

9.1.2 Good Automated Laboratory Practice (GALP)

In general accreditation and certification schemes concentrate on the management structure and procedures, test methods, staff training, *etc.* and pay little attention to the possibility of errors arising within the automated methodology and data handling processes themselves. This situation and its effects on the possibility of the corruption of data for Good Laboratory Practice (GLP) purposes was recognised and investigated by the US Environmental Protection Agency (EPA) who published their primary findings in 1990. They found that:

(i)   there were serious gaps in the security of the system and in data validation and documentation.
(ii)   laboratory information systems did not offer software that met all the requirements of GLP and computer hardware did not assure data integrity.
(iii)   commercial laboratory staff expressed a need for guidance in protecting the integrity of computer resident data and supported the proposal for a single source of guidance.
(iv)   the main sources of risk to data integrity that are present in automated financial systems exist also in automated laboratory systems.

The first finding can be overcome by the adoption and enforcement of correct procedures but the second identifies the weaknesses of computerised systems and overcoming them requires the adoption of specialised practices for the assessment and utilisation of hardware and software. These are set out in 'Good Automated Laboratory Practice', the EPA's recommendations for ensuring data integrity in automated laboratory operations with implementation guidance.[2]

The guidance is built on six principles:

(i)   *Data*: the system must provide a method of assuring the integrity of all entered data.

(ii)   *Formulae*: the formulae and decision algorithms employed must be accurate and appropriate.

(iii)   *Audit*: an audit trail that tracks data entry and modification to the responsible individual is a critical element in the control system.

(iv)   *Change*: a consistent and appropriate change control procedure capable of tracking the system operation and application software is a critical element in the control process (*i.e.* changes to data, software, and operation procedures should only be made by authorised staff and according to written procedures).

(v)   *Standard Operation Procedures*: control of even the most carefully designed and implemented system will be thwarted if appropriate user procedures are not followed.

(vi)   *Disaster*: consistent control of a system requires the development of alternative plans for system failure, disaster, and unauthorised access. (Disaster means unusual events and system stresses.)

The guidance discusses each of the 83 GALP recommendations.[2] An advisory note on the application of the requirements in the UK by laboratories operating to GLP requirements has been produced by the GLP Authority at the Department of Health.[3]

It is always preferable that quality is designed into the operating system of a laboratory from the start rather than 'documented in' at a later stage. A well-designed system will require the minimum of documentation necessary to verify that the system is maintained and operated correctly. Even where a quality system is added to an existing automated operation it may well be that not all of the requirements of GALP need necessarily be adopted by non-GLP laboratories. Nevertheless, the guidance set out in the sections on equipment, laboratory operations, software documentation, and operational records/logs are in the main generally applicable to any laboratory operating automated equipment or equipment with data handling capabilities. Similar guidance is produced by NAMAS,* which reflects the requirements of ISO 9001.[4,5] In addition, the NAMAS Accreditation Standard M10, requires that computers and automated test equipment should meet the requirements for all laboratory equip- ment set out in Section 6 of that document.[6] Most of the requirements relate to proper documentation of the history of the equipment which, in practice, means that when faults are discovered their source can be readily identified and rectified.

The normal quality control procedures in the laboratory, not the documentation requirements, are the main way in which faults and errors in the system will be identified, but proper records are essential if time is not to be wasted in dealing with the problem. Of particular importance, with automated systems for analysis and for data handling, are changes made to the software. If changes are made,

---

*NAMAS, National Measurement Accreditation Service, now replaced by UKAS, the United Kingdom Accreditation Service.

tests must be carried out to ensure the validity of the data produced. In addition, separate, individual changes can be made to sophisticated software without a full appreciation of their capability to interact, sometimes only in particular instances. In order to deal with such situations it is essential that changes are fully recorded in a standardised manner.

9.1.3 Signal Integrity

The first requirement in measurements that are derived from instruments is that the signal that is being measured arises from the analyte and is not corrupted by signals arising from other components present in the sample or matrix, or by signals arising from within the instrument and its associated systems. The second requirement is that the signal gives the information that is required. The signal should be sufficiently large to measure the analyte with the sensitivity and accuracy required. These requirements are not independent of each other and the main objective of signal processing is to optimise the signal to meet them.

The main types of signal processing carried out in an analytical laboratory fall into two broad categories, the resolution and integration of chromatograms and the resolution and measurement of intensity of spectrograms. In order to effect better separation of components before spectroscopic identification and quantification, chromatographic and spectroscopic techniques are often used in sequence. If this entails any transfer of data between the instrument and the data processing system this may result in the corruption of data which can be difficult to identify and rectify.

Most detectors used in laboratory instruments measure the intensity of some physical parameter. Changes in the intensity of the parameter generally occur across some cavity containing the chemical component of interest and therefore the intensity of the signal from the detector is a function of the concentration of that chemical component in the detector cavity. In both chromatography and spectroscopy, the intensity or size of the signal from the detector has to be calibrated against known concentrations of the substance in order to obtain a quantitative measure. However, in addition, in chromatography, the total integration of the signal with respect to time enables the amount of a substance passing through the column to be determined.

Many detectors, used in analytical instruments, have a linear relationship between the size of the signal and the concentration or amount of analyte. However, other detectors have a non-linear response and this can either be shown as such or, with some modern instruments, displayed as a linear output by means of an appropriate calibration system within the electronics of the instrument. Where quantitation is based on the assumption of a linear response, serious errors can occur when this assumption does not apply. Analysts should always be aware of the performance characteristics of the detectors in the instruments they are using.

The detectors on almost all laboratory instruments produce an analogue signal. However, computers and almost all data processing systems are digital devices, and so the signal from the detector has to be converted into a digital form. A typical conversion system is shown in Figure 9.1.1. The electrical signals from the sensors or transducers (S1–S3) are amplified by preamplifiers (A1–A3). The amplified signals are connected to a multiplexer, which can switch between several analogue signals. The output from the multiplexer is then fed to the analogue-to-digital converter system, which may consist of several units and the resultant analogue-to-digital conversion is presented to the computer *via* a digital interface.

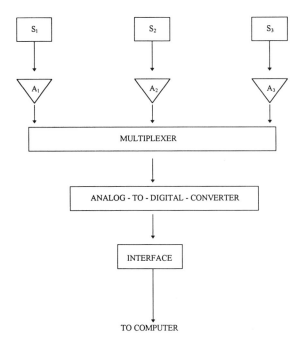

Figure 9.1.1 *Typical analogue-to-digital conversion system*

Most analysts generally leave the design of such systems to the instrument manufacturer. However, it is important that the main terms, the underlying laws, and the basic concepts are understood or wrong results may be obtained.[7]

Resolution of Analogue-to-Digital Converters (ADC)

The resolution of an ADC is measured in bits. It is the logarithm to base 2 of the number of discrete steps an analogue-to-digital converter can produce.

Dynamic Range of the Signal

This is defined as the ratio of the largest possible value of the signal to its smallest possible value, generally the noise limit. The dynamic range of the signal can exceed the resolution of the ADC, in which case an amplifier with a digital gain control must be inserted before the analogue-to-digital conversion takes place to adjust the signal level to the operating range of the ADC. If this is not done then the effective resolution of the ADC decreases for low level signals.

Sampling rate

Nyquist's sampling theory shows that to analyse a signal correctly, it is necessary to sample at a frequency of twice the value of the highest frequency component of the signal. However, practical problems with filter cut-offs and the shape of the signal often limit its applicability. It may be possible to determine the optimum sampling rate without knowing the frequency spectrum of the signal by considering the desired time resolution of the shortest event in a signal. For example, the signal produced by one single mass in a scanning mass spectrometer will last for about 1 ms. To get a good representation of the peak it needs to be 'sampled' 20 times, *i.e.* a sampling frequency of 20 kHz is needed. Sampling frequencies substantially less than this can give a distorted reproduction of the peak.

Noise

Any electronic circuitry can develop noise from a wide variety of sources. Intrinsic noise is developed from the signal-generating process itself whilst interference noise is picked up from outside. A data acquisition system has little effect on intrinsic noise. It is reduced in the first instance by selecting low noise components in the construction of the measuring system. Where it remains as a problem then, when the noise is random, it can be reduced by making repeated measurements and statistically averaging the resultant sets of signals to produce a usable signal. This approach is obviously facilitated by techniques which employ rapid scanning techniques, *e.g.* use of Fourier Transform in spectroscopy. In such cases it is possible to produce a very low noise level in the measurement process enabling the measurement of very low signals.

  Interference noise is minimised by:

(i)   reducing any impedance that is common to more than one electrical device in the system;
(ii)   reducing capacitative coupling between devices by introducing a conducting shield connected to a reference potential (generally earth) between them;
(iii) reducing inductive coupling caused by the magnetic fields in current carrying devices by avoiding loops, especially ground loops, in the system.

Where the electronics in the measuring system have been optimised, the need for signal manipulation to optimise the quality of the information contained in the signal arises from problems associated with the measuring method itself.

9.1.4 Spectroscopic Signals

In most cases the processing of spectroscopic signals is easier than chromato-graphic signals. In the majority of cases there is little background noise or interference and quantification is a matter of measuring one peak. The area under a particular peak in the spectrum is the true measure of the amount of substance present, but in most cases the peak height is an acceptable approximation. Where interferents are a problem, then pre-programmed software is often required to deal with it, but obviously this has its limitations when dealing with situations not taken into account when the programmes were developed.

Choosing a wavelength that is unique to the analyte of interest is rarely possible. The best way of avoiding interferences is to remove them from the sample solution by chemical methods. When the spectral peak of interest has been isolated the conventional form of background correction is to subtract the baseline, which is estimated by the two spectral intensities on either side of the peak. However, this has obvious limitations when there is overlap of peaks.

Even if there are no interfering peaks at the points where the baseline is measured, there are still two common cases that can cause error. The first is when the peak of interest falls on the slope of a large absorbance band, giving a sloping baseline (Figure 9.1.2). The second is when the peak is skew (Figure 9.1.3) so that

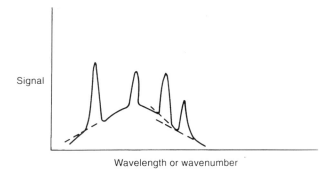

Figure 9.1.2 *Effect of overlapping absorbance bands*

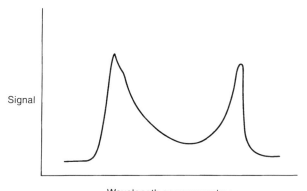

Figure 9.1.3 *Skewed absorbance bands*

the baseline correction measurement is taken from the sloping side of the peak. Both of these cases can go undetected when automated data processing equipment is used that does not give graphical representation of the spectrum. It is therefore essential that where an analyst is making spectroscopic measurements of new or unfamiliar samples, graphical representation of the spectral peaks of interest, and of adjacent portions of the spectrum, are obtained, so that these problems can be identified by a trained eye.

9.1.5 Chromatographic Signals

The quantity of an analyte passing through a chromatographic detector is proportional to the area of the peak produced as the detector output is plotted against time. Provided that the shape of the peak remains constant, this is in turn proportional to the peak height. The purpose of signal processing in chromatography is the accurate measurement of the area or height of the peak corresponding to the eluted component.

Although the peak area is the true measure of the amount of substance passing through the detector, there are severe problems with its measurement if peaks are poorly resolved. Depending on the integration parameters used, the area of the peak can be greatly diminished by partially co-eluting peaks (Figure 9.1.4) or greatly enhanced by unstable baselines or tailing solvent fronts (Figure 9.1.5). Peak height measurements are less affected by these phenomena. Since most trace analytical samples produce noisy chromatograms, height measurements are usually

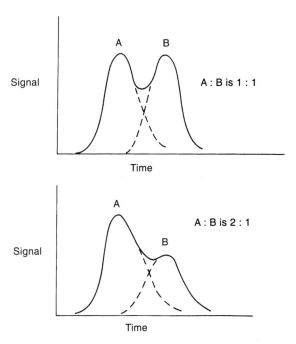

Figure 9.1.4 *Effect on co-eluting peaks*

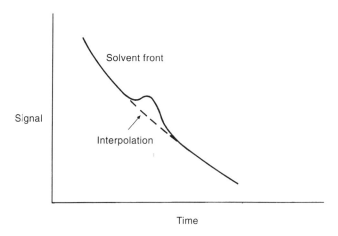

Time

Figure 9.1.5 *Effect of tailing solvent front*

used in preference to areas, although it is important that the retention time of the peak of interest does not change from that of the pure substance on the same column under the same operating conditions.

In chromatographic data processing most errors occur in integration.[8] Automatic integrators are now regarded as a necessity, for quick and reproducible quantification of peaks, but they should be used with caution. Integrators are by no means infallible and even the best can be configured wrongly. They are an excellent tool in the right hands. They measure peaks that are suitable for measuring, rapidly and without tedium. However, only analysts can improve the chromatography.

The most fundamental property of an integrator is the mathematical algorithm which integrates the peaks. The earliest integrators approximated each peak to a triangle which caused significant error in the estimation of the area of tailing peaks. Even modern integrators cannot perfectly model all shapes and forms. Although the algorithms used are commercially confidential, it is known that most treat the peak as a Gaussian curve, with an exponentially modified tail. Algorithm calibrators are available, which electronically generate standard peaks with known areas, and these can be very useful. Severely tailing peaks, partially co-eluting peaks, and peaks on sloping baselines can all be tested.

In order for an integrator to recognise peaks, it must be programmed with certain user-defined parameters. The manner in which these are set determines the validity of the integration. The three basic variables are peak width, minimum area (or height), and peak threshold (sometimes called slope). The peak width is the minimum baseline width (usually in minutes) which the integrator will consider as a peak. Any peaks with baselines widths below this value are not considered. Thus, the lower the width value, the more sensitive the integration, although more baseline noise is also integrated. The lower the peak threshold (a value relating to the rate of baseline gradient change at which the integrator considers that a peak is beginning), and the lower the minimum area, the more sensitive the integration.

These settings should be empirically tested on a real chromatogram, to ensure that a peak corresponding to the reporting limit will be recognised and integrated. It is not usually sufficient to set them using a standard chromatogram.

When the integration algorithm has been validated and the integration parameters set, the next critical factor is the positioning of the baseline. It is essential that this is both correct and reproducible. For this reason, the use of integrators which are not configured to draw a clear visible baseline on the chart or display is strongly discouraged. The two most common cases to guard against are the failure to recognise a small peak on the tail or side of a large peak, and the erroneous inclusion of negative spikes. These may not be recognised unless there is a visual baseline. If the integrator used does not include this facility, peaks in such situations should always be visually checked against well-resolved peaks of similar height on a flat baseline to ensure that the area counts roughly agree.

The height of the analyte and of the calibration peaks should be within the full scale deflection of the plotter. This way, if there is any doubt over the position of the baseline and the integrated areas it is always possible to manually measure the heights of the peaks. This situation does not arise with data stations that mostly have the option of manual setting of baselines. A further reason for keeping peaks on scale is to enable visual confirmation of retention times. With some integration parameters it is possible for integrators to fail to distinguish between two partially co-eluting compounds and thus assign them a single retention time. The only indication of partial co-elution can be a small cleft at the top of the peak which will not be noticed if the peak is off-scale.

Integrators may also be equipped with a combination of the following features: tangent skimming; negative peak corrections; time programmes and locks, usually employed to avoid the solvent front and associated noise; peak identification tables; a facility to calibrate against reference materials and calculate solution concentrations; manual baseline setting; and other add-on functions of varying usefulness. None of these should be used without careful reference to the relevant instruction manual and a full understanding of the way in which the function being used operates.

When a chromatographic separation step is coupled with spectroscopic detection there are additional considerations. Different conditions can apply as the separated peaks enter the spectroscopic detector and this may cause problems when using spectral library matching. When attempts are being made to distinguish a compound from others of the same class with similar spectra it is essential to run all the library spectra under the same conditions as the analysis. The problem is enhanced at low concentrations as spectra become less well defined as the signal-to-noise ratio decreases.

A problem peculiar to mass-selective detectors is caused by stepping baselines. As the detector is programmed to acquire different ions during a multi-analyte assay, the baseline changes to a new background level. The integration must be carefully checked to ensure that baseline steps have been taken into account.

9.1.6 References

1. Analytical Methods Committee, 'Internal Quality Control of Analytical Data', *Analyst*, 1995, **120**, 29.
2. Environmental Protection Agency, 'Good Automated Laboratory Practices. EPA Recommendations for ensuring data integrity in automated laboratory operations with implementation guidance', EPA, North Carolina, 1990.
3. 'The Application of GLP Principles to Computer Systems', GLP Advisory Leaflet No 1., Department of Health, London, 1989.
4. 'A Guide to Managing the configuration of Computer Systems', NIS 37, NAMAS Executive, National Physical Laboratory, Teddington, UK, 1993.
5. 'Use of word processing systems in NAMAS Accredited Laboratories', NIS 41, NAMAS Executive, National Physical Laboratory, Teddington, UK, 1991.
6. 'General Criteria of Competence for Calibration and Testing Laboratories', NAMAS Accreditation Standard M10, NAMAS Executive, National Physical Laboratory, Teddington, UK, 1989.
7. H. Lohninger and K. Varmuza, 'Data Acquisition in Chemistry', in 'PCs for Chemists', ed. J. Zupan, Elsevier, Amsterdam, 1990.
8. N. Dyson, 'Chromatographic Integration Methods', The Royal Society of Chemistry, Cambridge, UK, 1990.

9.2 Data Handling

9.2.1 Overview

After the chemical treatment, measurement process, and signal processing the data usually needs further manipulation. It goes without saying that data produced during an analysis, including all instrumental output, must be recorded in laboratory notebooks or other suitable media. Electronic data handling should minimise misreadings or transcription errors but, as described in Section 9.1, these systems need to be tested regularly to ensure they are performing correctly and that no errors occur.

When data are processed to produce the final result, the recommended procedure is to defer rounding off of numbers until all the calculations have been completed. The number of significant figures is determined once the statistical significance has been evaluated. When reporting a result, the use of exponential notation, *e.g.* 2.5×10^{-6} mg ml^{-1}, is recommended. This form of presentation expresses both the number of significant figures and the magnitude of a result. However, the number of significant figures should not be used to convey the uncertainty in the measurement.

When possible, a sufficient number of measurements should be made so that a statistical treatment of the results can be carried out.

9.2.2 Statistical Methods

The treatment and expression of the variability of results, uncertainty, and the comparison of results necessarily involves the use of statistics. Detailed treatment of statistics is outside the scope of this book and is readily available in specialist texts. However, a few relatively simple approaches cover the majority of situations encountered in trace analysis.

If a number of repeated measurements are made of the same parameter, then the number of results obtained which fall within a given range of values, when each range is of the same size, *e.g.* 0.0–0.1, 0.1–0.2, 0.2–0.3, *etc.,* will fall on a distribution curve. If there is an equal probability of a particular value being obtained at each measurement, then the results are said to be random. They will have a distribution that can be used to calculate parameters that describes the distribution, such as the mean value $\bar{x}$, and the sample standard deviation *s*. If the results obtained are not random, *i.e.* with equal probability, then bias may be introduced and the calculated parameters may show systematic deviations from their true values.

Various terms are required to represent the data in a quantitative manner. The most commonly used are:

Measures of *location*, giving the location of the central or typical value.
Measures of *dispersion*, showing the degree of spread of the data around the central value.
Measures of *skewness* or lack of symmetry of the distribution.
Measures of *kurtosis* or 'peakedness' of the data.

The most widely used measures of location, in chemical measurements, are the arithmetic mean of the results and the median.
The arithmetic mean, $\bar{x}$, is given by:

$$\bar{x} = \frac{\sum_{i=1}^{n} x_i}{n}$$

where *n* is the number of values of *x*. If the sample is random then $\bar{x}$ is the best estimate of μ (the population mean).

If the data are arranged in order of magnitude then the *median,* $\overleftrightarrow{x}$, is the central number of the series, *i.e.* there are equal numbers of observations smaller and greater than the median. For a symmetrical distribution, the median and the mean have the same value. However, the median is more robust in that it is unaffected by extreme values.

$$\overleftrightarrow{x} = x_m \quad \text{[when } m \text{ is odd } (1,3,5...)]$$

and

$$\overleftrightarrow{x} = \frac{x_m + x_{m+1}}{2} \quad \text{(when } m \text{ is even)}$$

where *m*= *n*/2 rounded up.
The mode of a set of values is the value that occurs most frequently. It is most often used with discrete distributions, *i.e.* small numbers of results. For example, in

$$1,2,2,2,3,3,3,3,4,5,5,6,7,8$$

The mode is 3. It is possible to have more than one mode, which usually indicates a non-homogeneous data set.

The measures of dispersion are variance, standard deviation, standard deviation of the mean, and range. Data can be classified in one of two forms, population and sample, according to the context in which it is used. Thus in a laboratory of 100 staff all the staff are the population of the laboratory but any sub-set of 99 or fewer staff is a sample. However, all the staff of that laboratory are only a sample of the laboratory staff in the UK.

The *variance* of the population is the mean squared deviation of the individual values from the population mean and is denoted or σ_n^2 or σ^2. Either s^2 or σ_{n-1}^2 is used to denote the variance of sample data. The variance is a measure of the extent to which the data differs in relation to the mean. The larger the variance, the greater the difference.

$$\sigma^2 = \frac{\sum_{i=1}^{n}(x_i-\overline{x})^2}{n} \qquad s^2 = \frac{\sum_{i=1}^{n}(x_i-\overline{x})^2}{n-1}$$

The *standard deviation* is the positive square root of the variance.

$$\sigma = \left(\frac{\sum_{i=1}^{n}(x_i-\overline{x})^2}{n}\right)^{1/2} \qquad s = \left(\frac{\sum_{i=1}^{n}(x_i-\overline{x})^2}{n-1}\right)^{1/2}$$

The standard deviation of the mean, also known as standard error (SE), is given by:

$$SE = s\, n^{-1/2}$$

It is an estimate of the standard deviations of the means of (sets of) samples from the population. It is less than the standard deviation of a sample because it is an estimation of the variation that would arise if repeated sets of samples were taken from the population.

The *range* is the difference between the highest and lowest values in a sample. It is only useful when dealing with very small sample sizes and, for $n=2$, it is the most effective measure of variance. However, it tends to show distorted information about large samples since the appearance of an outlier is then possible.

Probability distributions are used as models for the pattern of distribution of populations. In general, knowledge of the population is incomplete and it can only be assumed, on the basis of evidence or experience that a particular distribution will accurately model the distribution of a particular variable. Since there are two types of variable, discrete and continuous, there are two types of probability distribution. The most commonly used continuous distribution is the normal distribution; two discrete distributions are the binomial and Poisson.

Many of the distributions found in measurement and experimental work are of the normal type. The integral of the curve of a normal distribution is given by:

$$\int_{-\infty}^{+\infty} y \, dy = \int_{-\infty}^{+\infty} \frac{1}{\sigma\sqrt{2\pi}} \exp -\frac{(x-\mu)^2}{2\sigma^2} = 1$$

The curve is almost bell-shaped and is completely described by the values of μ and σ. In such a distribution, 68.27% of the values lie in the interval $(\mu \pm \sigma)$; 95.45% in the interval $(\mu \pm 2\sigma)$ and 99.73% in the interval $(\mu \pm 3\sigma)$.

Confidence limits are the values within which one is willing to assert with a given confidence that the true value lies. It may be necessary to use confidence limits in the following situations.

(i)   Population standard deviation known but population mean unknown. The 95% confidence limits (CL) for a mean value from a set of n observations will be

$$CL = \bar{x} \pm 1.960 \, (\sigma \, n^{-\frac{1}{2}})$$

(ii)   When both the mean and the standard deviation are unknown. In such cases, both these parameters can be estima ted from the data. However, since an estimate is being used, there is more uncertainty resulting in a wider confidence interval. This uncertainty is reflected in the use of Student's t distribution. The confidence interval can be calculated by

$$CL = \bar{x} \pm t \, (s \, n^{-\frac{1}{2}})$$

where s is the sample standard deviation and t is the value obtainable from the Tables of Critical Values of Student's t Tests. The appropriate value of t is that for $v = (n-1)$ degrees of freedom at various percentages of the confidence level.

The relationships outlined above can be illustrated with some simple, real examples.

If an analytical test, which is known to be highly variable, has been shown, from a large number of observations, to have a population standard deviation of 0.55 units, then if five tests on a particular sample give values of:

16.9, 16.6, 17.3, 17.7 and 16.4

the mean value is given by

$$\bar{x} = \sum_{i=1}^{n} \frac{x_i}{n} = \frac{84.9}{5} = 16.98$$

The 95% confidence interval of the mean is given by

$$\bar{x} \pm 1.96 \, \sigma n^{-1/2} = 16.98 \pm (1.96 \times 0.55) \, 5^{-1/2} = 16.98 \pm 0.48$$

i.e. 16.50 to 17.46, generally written as [16.50,17.46].

If the population standard deviation is not known, then the sample standard deviation (*s*) can be obtained from

$$s = \left(\frac{\sum_{i=1}^{n} (x_i - \bar{x})^2}{n-1} \right)^{1/2} = (1.108/4)^{1/2} = 0.5263$$

from Student's *t* tables the critical value for $\nu = 4$ is 2.776. The 95% confidence limits are

$$\bar{x} \pm (t \times s) \, n^{-1/2} = 16.98 \pm (2.776 \times 0.526) \, 5^{-1/2} = 16.98 \pm 0.65$$

i.e. 16.33 to 17.64, which is written as [16.33,17.64].

If one requirement of a measurement using this method was to obtain a value of the true mean to within ± 0.15 at 95% confidence level, then a number of repeated tests would have to be made. This sample size has to be calculated. For 95% CL, the confidence interval (CI) = $(1.96 \times \sigma n^{-1/2})$

$$n = \frac{(1.96 \times \sigma)^2}{CI} = \frac{(1.96 \times 0.55)^2}{0.15} = 51.65$$

Therefore 52 (always round up in order to obtain at least 95%) results are required to give a CI of ± 0.15.

9.2.3 Measurement Uncertainty

Spread of Results

Whenever a measurement of a particular parameter is made more than once, it will be found that even when the same method is used to make each measurement, there will be differences between the results obtained. If a large number of measurements of the parameter are made using the same method then there will be a spread of results in the form of a statistical distribution curve. Any individual measured value of the parameter therefore has an uncertainty associated with it since a repeat of the measurement will give a different result.

The uncertainty associated with any measurement will be increased if the method used to make the measurement contains a systematic error caused by some artifact within the measurement system. Two levels of systematic error are possible, the 'long-term' or 'persistent' error and the 'short-term' or 'run' effect. The persistent error is only identifiable after the analytical system has been in operation for a long time. The short-term error, if sufficiently large, will be identifiable by quality control procedures and indicate an out-of-control condition. A systematic error may arise from the extraction method; this results in additional consistent uncertainty associated with each result as it is obtained. This can mean that even when the spread of results obtained from repeated measurements of the parameter is very small, the mean of the measurements can differ from the 'true' value.

Therefore when the 'true' value of the parameter is not known and is derived from the set of measurements, then a single value for the parameter cannot be given. All that can be said is that there is a particular probability that the 'true' value lies within a range of values. This range of values and the associated probability is the uncertainty associated with the measurement.

To be able to quote either the accuracy or the error of a measurement, the 'true' value of the parameter being measured must be known. The purpose of most measurements, however, is to determine the value of a particular parameter and hence it is rare in trace analysis for the 'true value' of the analyte to be known. Indeed, in principle, the 'true' value can never be known. The best that can be achieved is a 'conventional true value' or 'assigned value' which are often values agreed by consensus from a large number of measurements. Therefore the terms 'accuracy' or 'error' in most measurements have little real meaning.[1,2] Uncertainty, however, is an estimate of the range of values, which is expected to include the true value, and obviously can be used when the true value is not known.[3]

Since all measurements have a degree of uncertainty associated with them, a major consideration is how the uncertainty can be assessed and where in the measurement process the main contributions to the uncertainty of the measurement arise. Information on the latter enables changes to be made to the process to reduce the uncertainty and also concentration of control procedures on the steps in the process where the greatest contributions to the final uncertainty take place.

However, it must always be borne in mind that the main consideration (in deciding what is an acceptable degree of uncertainty in a result) is the *purpose* of the analysis, *i.e.* the use to which the result will be put. There may be little point in using a method that has a very low degree of uncertainty if all that is required from the measurement is an indication of whether or not a particular substance is present in a sample. If, however, the amount of a particular component of a mixture has to be controlled to within tight limits for production or legislative reasons, then the method used to measure it must be capable of producing results with a low uncertainty.

In trace analysis measurements the uncertainty can be estimated by 'calibrating' the measurement system with a 'traceable' measurement standard, such as a certified reference material (CRM), whose value for the composition is traceable back to national standards. 'Traceability' means traceable to national or international measurement standards through an unbroken chain of comparisons with the uncertainty being stated at each stage.

When a reference material is used to 'calibrate' the system the sources of uncertainty that need to be taken into account are the following.[4]

(i)   The relative standard deviation of the complete measurement system, s_R. This is the spread of results obtained by passing a reference material through the system.

(ii) The relative standard deviation associated with the value of the calibration standard, s_{CS}.

(iii) The relative standard deviation associated with the calibration measurements, reproducibility of the measurement, s_{CM}.

(iv) The relative standard deviation associated with the suitability of the chosen reference material for calibrating the system, *e.g.* difficulty in matching the composition of the reference material to the sample which is being measured, s_G.

Then the overall uncertainty is given by Equation (1).

$$k \, [(s_R)^2 + (s_{CS})^2 + (s_{CM})^2 + (s_G)^2]^{1/2} \tag{1}$$

where k is a factor agreed with the user of the results. The size of the factor depends on the uncertainty that is acceptable to the user of the results, *i.e.* the degree of confidence required, and in most instances it has a value of two.

Obviously a number of components are contained in each of these individual uncertainties, especially s_R and s_G. Also there may be difficulties if s_{CM} is already included in s_R.

9.2.4 Sources of Uncertainty

Overall uncertainty is generally estimated from the spread of repeated or similar measurements. However, where possible the 'budget approach' should be used during method validation, since the results indicate (or confirm opinion on) where the largest contributions to the overall uncertainty arise in the chemical measurement process.

A detailed evaluation of the uncertainty in the measurement of the reference material used to calibrate breathalysers has been carried out.[4] This is a solution of ethanol in water. The ethanol concentration is measured by quantitative oxidation of the ethanol with potassium dichromate(VI), using a CRM for the preparation of the dichromate solutions. A known amount of dichromate(VI) solution is added to the ethanol solution and the excess is determined by titration with ammonium iron(II) sulfate, which is itself calibrated against the standard potassium dichromate(VI). The uncertainties associated with the different stages and the materials used in the procedure are shown in Table 9.2.1.

Table 9.2.1 *Sources of uncertainty*

Stage	Uncertainty (%)
Weighing dichromate by difference	0.005
Calibration titre	0.02
Sample titre	0.13
Ethanol density	0.10
Dichromate purity	0.06
Extent of oxidation	0.17
End-point accuracy	0.17

Using a value of $k = 2$, the overall uncertainty of the measurement is 0.5%.

Examination of Table 9.2.1 indicates where in a chemical measurement process the greatest contribution to the overall uncertainty of the measurement arises. Thus the chemical oxidation of the ethanol, which is assumed to go to completion rather than being measured in any way, makes a significant contribution. This can be compared with the sample preparation stage of other methods. This has been shown in interlaboratory comparisons when methods are being developed or compared. Table 9.2.2 compares the precision of results obtained by Atomic Absorption Spectroscopy (AAS) and X-ray Fluorescence Spectrometry (XRF) of lead in air samples taken on glass fibre filters. Analysis by AAS involves a dissolution step for the lead on the filter whilst filters can be measured directly by XRF with no sample preparation. The added stage leads to a much larger coefficient of variation.

Table 9.2.2 *Analysis of lead in air*

Method	No. of results	Coefficient of variation (%)
XRF	72	2.2
AAS	616	11.0

The data in Table 9.2.1 also indicate that the stages involving subjective estimations by the analyst such as the measurement of the titre and the estimation of the end-point also contribute to the uncertainty. This is supported by the data in Table 9.2.3, which were obtained by Whitehead as part of a proficiency testing scheme in clinical chemistry: the precision of an instrumental method of determining urea was compared with a manual method, using urease.[5]

The observations are not in themselves conclusive, but they do support the general observation that when a chemical measurement method is being developed, the overall uncertainty associated with it will be minimised if:

(i)   the number of stages in the method is minimised;
(ii) sample preparation is removed or kept to a minimum;
(iii) stages involving subjective estimation of any parameter by the analyst are removed or minimised.

Some of the sources and values of uncertainty in chemical measurements and methods of calculating overall uncertainty are summarised in the document 'Quantifying uncertainty'.[6] A standard spreadsheet can be used to simplify the calculations.[7]

Table 9.2.3 *Analysis of urea*

Method	No. of results	Coefficient of variation (%)
Flame photometry	138	3.3
Urease (manual)	54	9.2

9.2.5 References

1. 'Terms and Definitions used in Connection with Reference Materials', ISO Guide 30: 1992, Geneva.
2. 'Accuracy (trueness and precision) of measurement methods and results. Part 1, General principles and definitions', ISO 5725-1: 1994(E), Geneva.
3. 'Draft ISO/IUPAC Harmonized Guidelines for Internal Quality Control in Analytical Chemistry Laboratories', ed. M. Thompson and R. Wood, *Pure Appl. Chem.*, 1995, **67**, 649.
4. A. Williams, 'Measurement Uncertainty in Chemical Analysis', *Anal. Proc.*, 1993, **30**, 248.
5. T. P. Whitehead, 'Quality Control in Clinical Chemistry', John Wiley, Chichester, UK, 1977.
6. 'Quantifying Uncertainty in Analytical Measurement', Eurachem, 1995, ISBN 0-948926-08-2.
7. J. Kragten, 'Calculating Standard Deviations and Confidence Intervals with a Universally Applicable Spreadsheet Technique', *Analyst*, 1994, **119**, 2161.

9.3 Reporting of Results

9.3.1 Overview

It is clear from what has already been written in this chapter and elsewhere in the book that a great deal of processing is necessary to translate the raw data collected into a result that will be understood by the customer. There is scope for error and it is appropriate that someone independent of the analysis and data processing checks the result. It is unfortunate that mistakes made during the reporting stage can render the work useless. Good laboratory practice and proper quality assurance procedures help reduce such errors.

The final stage of any analysis is reporting the results to the customer. Where a laboratory is working to a particular quality standard, there may be specific requirements governing the level of information to be included in the report.

It is seldom possible to foresee all the potential uses of analytical results. In order to prevent inappropriate interpretation, it is important that every result has associated with it a statement of the intended accuracy and the procedures used to demonstrate that it has been achieved. The analyst should make it clear to customers the importance of an adequate statement of uncertainty. If the result obtained for an analyte is 1.90 $\mu g\ kg^{-1}$ and the limit is 2.0 $\mu g\ kg^{-1}$, has the limit been exceeded? If the uncertainty is ± 0.15, the answer is yes, if the uncertainty is lower than this value the answer may be no! Problems may arise when the analysis is part of litigation; lawyers usually wish to dispense with the uncertainty associated with the results. It is a case of public education and awareness.

9.3.2 Customer Requirements

One of the problems relating to uncertainty or confidence of measurements is that many customers require measurements because they have to comply with some numerical standard set by a regulatory authority. Such a standard could include exposure limits for toxic substances in the workplace or levels of contaminants such as pesticide residues in foodstuffs. Such limits have often been set without any indication of the degree of confidence required before it is possible to say that a standard has been met. Thus for an upper limit, the amounts of analyte found, the number of measurements and their associated uncertainties, would be different if it was required that the confidence that the level had not been exceeded was 90%, 95%, or 99%. A similar situation would apply if an authority had to be 90%, 95%, or 99% certain that the limit had been exceeded in order to obtain a conviction.

One measure of guidance in this area is the requirement for a successful prosecution for exceeding the alcohol in blood limit for driving a vehicle. The value measured must be above the value (limit + 3s). This means there is a 99.7% confidence that the limit has been exceeded. If this precedent were applied elsewhere in the regulatory field it would have a considerable impact on the field of chemical measurement.

Measurement results should be reported so that their meaning is not distorted by the reporting process. Printout from a computer may show the original data or it could report the results following a mathematical calculation on that data. It is important to define what is meant by raw data.

There are two schools of thought on how to represent analyte recovery data. Analytical results can be reported as measured (uncorrected for recovery) with full and complete supporting data involving recovery experiments. Alternatively, if measurements are reported as 'recovery corrected', in this case all calculations and experimental data should be documented so that the original uncorrected values can be derived if desired. There is usually some uncertainty in determining the percentage recovery and this must be included when reporting the results.

Laboratory reports must contain sufficient data and information so that users of the conclusions (even years later) can understand the interpretation of raw data, without having to make their own.

Both customers and laboratories should constantly be aware of the consequences if the data are wrong. These consequences are an indicator of the 'gearing effects' of measurements. Wrong results can mean the loss of thousands of pounds of production together with the damage to the good name of both customer and laboratory. Similarly, wrong results can have social consequences such as ill-health of members of the population from the consumption of contaminated foodstuff.

Trace analysis is difficult but results fit for purpose can be achieved consistently if the analyst and customer clearly define the problem and the analyst works in a quality framework and follows the protocol for trace analysis.[1,2]

The value of a measurement lies in the information it contains and the confidence that can be placed in that information rather than in the cost of producing that information. In this sense, chemical measurement is a very valuable activity.

9.3.3 References

1. M. Sargent, 'Development and Application of a Protocol for Quality Assurance. Annex 1 Protocol for Quality Assurance of trace analysis', *Anal. Proc.*, 1995, **32**, 71.
2. 'Guidelines for Achieving Quality in Trace Analysis', ed. M. Sargent and G. McKay, The Royal Society of Chemistry, Cambridge, 1995.

Appendix 1: Some Frequently Used Acronyms and Technical Abbreviations

AAS	atomic absorption spectroscopy
ADC	analogue-to-digital converter
AES	atomic emission spectroscopy
AFS	atomic fluorescence spectroscopy
AOAC	Association of Official Analytical Chemists
A_r	relative atomic mass
ASTM	American Society for Testing Materials
ASV	anodic stripping voltammetry
CCD	charge coupled device
CI	chemical ionisation
CID	charge injection device
CRM	certified reference material
CSV	cathodic stripping voltammetry
CV	coefficient of variation
DCP	direct current plasma
DMSO	dimethyl sulfoxide
DPP	differential pulse polarography
DPV	differential pulse voltammetry
EC	exclusion chromatography
ECD	electron capture detector
EDTA	ethylenediaminetetraacetic acid
EI	electron impact (ionisation)
EIA	enzyme immunoassay
EI-MS	electron impact/ionization mass spectrometry
ELISA	enzyme linked immunosorbent assay
emf	electromotive force
ESA	electrostatic analyser
ETAAS	electrothermal AAS
FAAS	flame AAS
FAB	fast atom bombardment
FABMS	fast atom bombardment mass spectrometry
FAES	flame AES
FAFS	flame AFS
FAO-WHO	Food and Agriculture Organization World Health Organization
FI	flow injection
FIA	fluorescence immunoassay
FIAS	flow injection analysis system
FID	flame ionisation detector
FPD	flame photometric detector
FT	Fourier transform
FTIR	Fourier transform infrared
FTMS	Fourier transform mass spectroscopy
GC	gas chromatography

GC-MIP	gas chromatography microwave-induced plasma
GDL	glow discharge lamp
GDMS	glow discharge mass spectroscopy
HCL	hollow cathode lamp
HG	hydride generation
HGAAS	hydride generation atomic absorption spectroscopy
HMDE	hanging mercury drop electrode
HPLC	high performance liquid chromatography
IC	ion chromatography
ICP	inductively coupled plasma
ICP-MS	inductively coupled plasma mass spectrometry
ICP-OES	inductively coupled plasma optical emission spectroscopy
IR	infrared
ISE	ion-selective electrode
ISFET	ion selective field effect transistor
LA ICP-MS	laser ablation ICP-MS
LC	liquid chromatography
LIMS	laboratory information management system
LOD	limit of determination
LOQ	limit of quantification
LSV	linear sweep voltammetry
M_r	relative molecular mass
MIP	microwave induced plasma
MS	mass spectrometry
MSD	mass selective detector
MS–MS	tandem mass spectrometry
NIST	National Institute of Standards and Technology
NMR	nuclear magnetic resonance
NPD	nitrogen phosphorus detector
OES	optical emission spectroscopy
PAH	polycyclic aromatic hydrocarbons
PC	paper chromatography
PCB	polychlorinated biphenyl
PDA	photodiode array
PDVB	poly(divinyl benzene)
PFK	perfluorokerosene
PMT	photomultiplier tube
ppb	parts per billion (10^9)
ppm	parts per million (10^6)
ppt	parts per trillion (10^{12})
PTFE	poly(tetrafluoroethylene)
PVC	poly(vinyl chloride)
QA	quality assurance
QC	quality control
RIA	radioimmunoassay
RSD	relative standard deviation

s	sample standard deviation
SCE	saturated calomel (reference) electrode
S/B	signal-to-background ratio
SE	standard error
SFC	supercritical fluid chromatography
SFE	supercritical fluid extraction
SIM	selected ion monitoring
SIMS	secondary ion mass spectrometry
S/N	signal-to-noise ratio
SPE	solid phase extraction
SRM	standard reference material
TGA	thermogravimetric analysis
TIMS	thermal ionisation mass spectrometry
TLC	thin layer chromatography
TOF	time-of-flight
u	atomic mass unit
UV	ultra violet
UV/VIS	ultraviolet/visible (spectroscopy)
WHO	World Health Organization
XRF	X-ray fluorescence
ν	degrees of freedom

Appendix 2: Some Sources of Reference Materials

Amersham International plc
Amersham Laboratories
White Lion Road
Amersham
Bucks
HP7 9LL, UK

Materials available: Radioactivity isotopes

Bureau of Analysed Samples Ltd
Newham Hall
Newby
Middlesbrough
Cleveland
TS8 9EA, UK

Materials available: Metal alloys, ores, slags, ceramics, minerals, and cement

BCR
Community Bureau of Reference
Commission of the European Communities
Rue De La Loi 200
Bl-1049
Brussels
Belgium

Materials available: General coverage

Johnson Matthey Chemicals Ltd
Orchard Road
Royston
Herts
SG8 5HE, UK

Materials available: Metals and alloys

Laboratory of the Government Chemist (LGC)
Queens Road
Teddington
Middlesex
TW11 0LY, UK

Materials available: General coverage

MBH Analytical Ltd
Holland House
Queens Road
Barnet
Herts
EN5 4DJ, UK

Materials available: Metals, alloys

McCrone Research Associates Ltd
2 McCrone Mews
Belsize Lane
London NW3 5BG

Materials available: UV/VIS calibration standards

Merck Ltd
Merck House
Poole
Dorset
BH15 1TD, UK

Materials available: Thermometric standards

NIST
National Institute of Standards & Technology
Building 202
Room 205
Gaithersburg
Maryland 20899
USA

Materials available: General coverage

Polymer Laboratories Ltd
Essex Road
Church Stretton
Shropshire
SY6 6AX, UK

Materials available: Polymers and resins

RAPRA Technology Ltd
Shawbury
Shrewsbury
Shropshire
SY4 4NR, UK

Materials available: Polymers and resins

Starna Ltd
33 Station Road
Chadwell Heath
Romford
Essex RM6 4BL

Materials available: UV/VIS calibration standards including luminescence

Subject Index